Springer Proceedings in Business and Economics

D1740424

Springer Proceedings in Business and Economics brings the most current research presented at conferences and workshops to a global readership. The series features volumes (in electronic and print formats) of selected contributions from conferences in all areas of economics, business, management, and finance. In addition to an overall evaluation by the publisher of the topical interest, scientific quality, and timeliness of each volume, each contribution is refereed to standards comparable to those of leading journals, resulting in authoritative contributions to the respective fields. Springer's production and distribution infrastructure ensures rapid publication and wide circulation of the latest developments in the most compelling and promising areas of research today.

The editorial development of volumes may be managed using Springer's innovative Online Conference Service (OCS), a proven online manuscript management and review system. This system is designed to ensure an efficient timeline for your publication, making Springer Proceedings in Business and Economics the premier series to publish your workshop or conference volume.

More information about this series at http://www.springer.com/series/11960

Hui Yang · Robin Qiu · Weiwei Chen
Editors

Smart Service Systems, Operations Management, and Analytics

Proceedings of the 2019 INFORMS
International Conference on Service Science

 Springer

Editors
Hui Yang
Harold and Inge Marcus Department of
Industrial and Manufacturing Engineering
Pennsylvania State University
University Park, PA, USA

Robin Qiu
Division of Engineering
and Information Science
Pennsylvania State University
Malvern, PA, USA

Weiwei Chen
Department of Supply Chain Management
Rutgers, The State University of New Jersey
Piscataway, NJ, USA

ISSN 2198-7246   ISSN 2198-7254   (electronic)
Springer Proceedings in Business and Economics
ISBN 978-3-030-30969-5   ISBN 978-3-030-30967-1   (eBook)
https://doi.org/10.1007/978-3-030-30967-1

This Springer imprint is published by the registered company Springer Nature Switzerland AG
The registered company address is: Gewerbestrasse 11, 6330 Cham, Switzerland

Contents

Cleaning and Processing on the Electric Vehicle Telematics Data

Shuai Sun, Jun Bi and Cong Ding

Abstract The development of the Internet of Vehicles (IoV) enables companies to collect an increasing amount of telematics data, which creates plenty of new business opportunities. How to improve the integrity and precision of electric vehicle telematics data to effectively support the operation and management of vehicles is one of the thorniest problems in the electric vehicle industry. With the purpose of accurately collecting and calculating the driving mileage of electric vehicles, a series of data cleaning and processing methodologies were conducted on the real-world electric vehicle telematics data. More specifically, descriptive statistics was conducted on the data, and the statistical results showed the quality of the data in general. Above all, the driving mileage data were segmented according to the rotate speed of the electric motor, and the anomaly threshold of the driving mileage data was obtained by the box-plot method. Then, the typical anomalies in the data were screened out by the threshold and analysed, respectively. Ultimately, the real-time and offline abnormal processing algorithms are designed to process real-time and offline data, respectively. After debugging and improvement, these two sets of abnormal processing algorithms we designed have been able to run on a company's big data cloud platform. According to the feedback of the operation results of real-world massive data, the two sets of algorithms can effectively improve the statistical accuracy of driving mileage data of electric vehicle.

Keywords Internet of Vehicles · Telematics data · Data cleaning and processing · Box-plot method

1 Introduction

The Internet of Vehicles (IoV) is a vast interactive network of vehicles around information such as location, speed and routes [1]. It can realize information sharing through vehicle-to-vehicle, vehicle-to-person, vehicle-to-road interconnection and

S. Sun (✉) · J. Bi · C. Ding
School of Traffic and Transportation, Beijing Jiaotong University, Beijing 100044, China
e-mail: 17114261@bjtu.edu.cn

© Springer Nature Switzerland AG 2020
H. Yang et al. (eds.), *Smart Service Systems, Operations Management,*
and Analytics, Springer Proceedings in Business and Economics,
https://doi.org/10.1007/978-3-030-30967-1_1

intercommunication, processing, sharing and releasing information collected from multiple sources on the information network platform. The Internet has brought society into the era of big data, and the Internet of vehicles has also brought vehicles into the era of big data [2]. This paper investigates the value of IoT data with particular attention to telematics data in the electric vehicle industry.

The commercial vehicle industry is one of the first to put the Internet of Vehicles technology into use due to its own industry background and industry demand [3]. The application of the Internet of Vehicles in the field of commercial vehicles, to a certain extent, solves the problem of lacking telematics data. However, in the actual process, incomplete, inaccurate and unreliable telematics data often fail to achieve the expected effect of supporting operation management. Therefore, how to improve the integrity and precision of electric vehicle telematics data to effectively support the operation and management of vehicles is one of the thorniest problems in the electric vehicle industry.

At the business level, the fleet management generally have the corresponding index requirements. Mileage is an important index of transportation cost accounting. Most of the driver's salary is linked to the mileage. Inaccurate mileage data is not conducive to the cost accounting and operation management of transportation enterprises. Accurate mileage and fuel consumption data is conducive to the promotion and application of commercial vehicle network in transportation enterprises. Therefore, our purpose is to find the existing problems in the driving mileage data of electric vehicles and propose solutions to these problems through cleaning, mining and processing.

2 Data Description

First of all, descriptive statistics have been carried out on the data to show the quality of data from the statistical results. There are 62 properties in each valid data file, and the key focus of this data analysis is some basic parameters, e.g. four mileage indicators and one standard mileage indicator, which is shown in Table 1.

Given that there is supplementary in data transmission, the missing distribution of mileage data was calculated. The statistical results of mileage data missing are shown in Table 2.

The statistical results show that some of the mileage data are missing. After preliminary observation, most of the missing cases were found in the supplementary data. Coincidentally, most of the supplementary data are out of driving trip. Therefore, it is very necessary to segment trips, which can effectively improve data quality.

Table 1 Basic information of key data items

Type	Attributes	Use for
Basic parameters	ID	Identify the terminal
	GPS time	Identify the time
	ACC	Segment the trip
	Motor speed	Segment the trip
	Supplementary	Identify the missing
Driving mileages	Mileage 1	Vehicle mileage
	Mileage 2	ECU mileage
	Mileage 3	GPS mileage
	Mileage 4	Integral mileage
	Standard mileage	Tagged mileage

Table 2 Percentage of mileage data missing

Mileage attributes	Missing percentage (%)
Mileage 1	6.18
Mileage 2	18.74
Mileage 3	0
Mileage 4	2.57
Standard mileage	10.51

3 Data Processing

The statistical analysis results of the data enable us to have a preliminary understanding of the data, but in order to find the problems in the data and propose solutions to the problems, the data needs to be processed.

3.1 Trip Segmentation

According to practical experience, ACC cannot be used as the basis of trip segmentation because it has been always ON in some cases. Therefore, the segmentation of trips is mainly based on motor speed. That is, the first point whose motor speed is not 0 counts as the start point of a trip, and the point whose motor speed is reduced to 0 and lasts for 5 min counts as the end point of that. According to the above rules, the vehicle's driving trip is segmented, and a typical example of trip segmentation is shown in Fig. 1.

As shown in the figure below, we can clearly see seven complete trips. Good trip segmentation results will help us to process and clean the data.

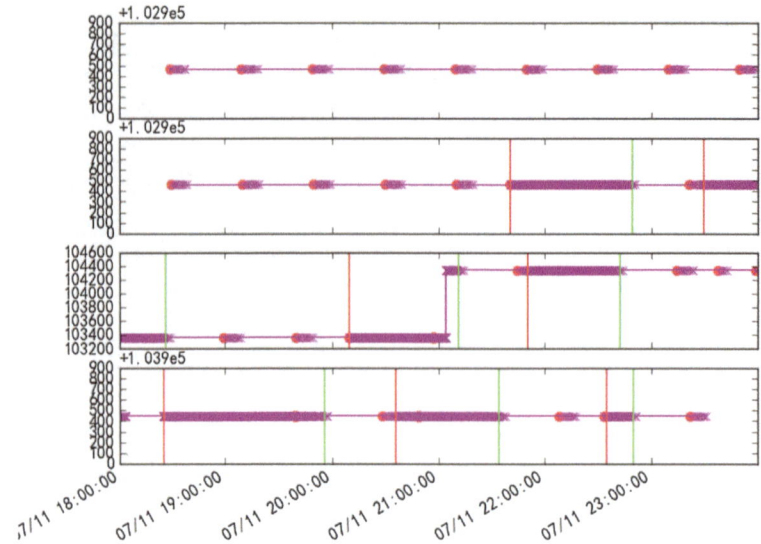

Fig. 1 Result of segmentation of trips

3.2 Threshold Selection

After the trip is segmented, outliers in the data can be filtered. Here, we define a new speed indicator, that is, the ratio of the distance difference between two adjacent points to the time difference between two adjacent points (unit: km/s), as a criterion to screen data outliers. For the speed indicator we defined, we used box-plot method and three-delta method to conduct statistical screening of all the data, and some results are shown in Table 3.

According to the comparison results in Table 3, the result of box-plot method is more stable than that of three-delta method, and the speed threshold obtained by the former method is of more practical reference value. Therefore, box-plot method is used in the threshold selection and subsequent data processing.

Table 3 Comparison of box diagram method and three-delta method

	Box diagram method	Three-delta method
Count	88	78
Mean	0.038069	0.20065
Std	0.031169	0.118598
Min	0.004833	0.049026
25%	0.0275	0.115383
50%	0.0325	0.176352
75%	0.037125	0.238261
Max	0.2	0.685202

4 Data Cleaning

4.1 Real-Time Algorithm

The basic idea of real-time algorithm is calculating the average speed of the latest two points of high-priority data, comparing to the predetermined speed threshold. If it is within the threshold range, the mileage difference between these two points is selected for accumulation; if it is beyond the threshold range, the two-point mileage difference with low priority in the simultaneous segment that meets the threshold requirements is selected for accumulation [4].

4.2 Offline Algorithm

Taking the data with the trip as the research object, the basic idea of offline algorithm is segmenting the data according to the threshold selected in advance. In the normal segment, the value of the maximum moment minus the value of the minimum moment is used to obtain the mileage difference. In the segment with abnormal conditions, the real-time algorithm is used to traverse the data to obtain the mileage accumulation value. Finally, the processing results of each segment are summed to obtain the total mileage difference value.

The results of the two algorithms are shown in Fig. 2. After debugging and improvement, these two sets of abnormal processing algorithms we designed have been able to run on a company's big data cloud platform. According to the feedback of the operation results of real-world massive data, the two sets of algorithms can effectively improve the statistical accuracy of driving mileage data of electric vehicle.

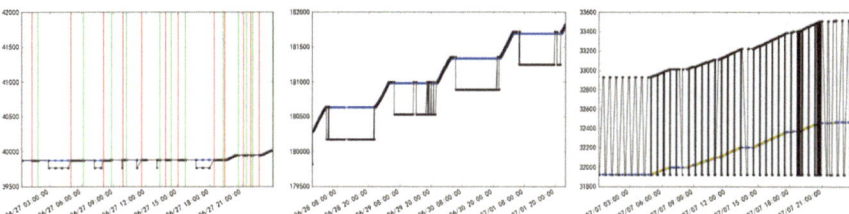

Fig. 2 Algorithm processing results

5 Summary

With the fundamental purpose of accurately counting the mileage of Internet of Vehicles data, a series of data cleaning and processing have been carried out for telematics data. In the following work, we will constantly adjust the existing algorithms according to the actual business rules, so that the real-time algorithm and the offline algorithm can meet the market demand.

The electric vehicle telematics data have not been made available because we signed a confidential agreement with an IoV company, and all telematics data related to commercial secrets is not suitable for disclosure.

Acknowledgements This research is supported by the National Key R&D Program of China under grant No. 2018YFC0706005 and No. 2018YFC0706000.

References

1. S. Duri, J. Elliott, M. Gruteser, X. Liu, P. Moskowitz, R. Perez et al., Data protection and data sharing in telematics. Mob. Netw. Appl. **9**(6), 693–701 (2004)
2. J.F. Ehmke, Data chain management for planning in city logistics. Int. J. Data Min. Model. Manag. **1**(4), 335–356 (2009)
3. I. Reimers, B. Shiller, *Welfare Implications of Proprietary Data Collection: An Application to Telematics in Auto Insurance* (Social Science Electronic Publishing, 2018)
4. J. Lauer, L. Richter, T. Ellersiek, A. Zipf, TeleAgro+: analysis framework for agricultural telematics data, in *ACM SIGSPATIAL International Workshop on Computational Transportation Science* (ACM, 2014)

Performance Analysis of a Security-Check System with Four Types of Inspection Channels for High-Speed Rail Stations in China

Chia-Hung Wang and Xiaojing Wu

Abstract In recent years, the High-Speed Rail (HSR) in China has continued to thrive rapidly with the development of China's booming economy, and it has become the preferred mode of transportation for many travelers. This paper investigates the stochastic process of security inspection for passengers in the high-speed rail station. A queuing model is developed for studying the proposed security-check system via computer simulation. In the numerical experiments, we illustrate the influence of varying model parameters on the average waiting time and safety level of the queuing system. The sensitivity analysis of our simulation model could contribute to improve the service level and efficiency on security check before implementing operation plans for high-speed rail stations.

Keywords Security check · Queuing system · Rail station · Homeland security management · System performance

1 Introduction

High-Speed Rail (HSR) in China, the country's network of passenger-dedicated railways, has been the world's longest high-speed rail network [1]. In 2018, HSR has extended to 30 of China's 33 provincial-level administrative divisions and accounted for almost two-thirds of the world's high-speed rails in commercial service. The total length of HSR network in China is going to reach 30,000 km in 2020 and 38,000 km in 2025 according to The Mid-to-Long-Term Railway Network Plan [2], issued by National Development and Reform Commission of the People's Republic

C.-H. Wang (✉) · X. Wu
College of Information Science and Engineering, Fujian University of Technology, Fuzhou City 350118, Fujian Province, China
e-mail: jhwang728@hotmail.com

C.-H. Wang
Fujian Provincial Key Laboratory of Big Data Mining and Applications, Fuzhou City 350118, Fujian Province, China

© Springer Nature Switzerland AG 2020
H. Yang et al. (eds.), *Smart Service Systems, Operations Management, and Analytics*, Springer Proceedings in Business and Economics, https://doi.org/10.1007/978-3-030-30967-1_2

of China. The demand for HSR in China steadily increases over time, whereas the public concerns about its efficiency and safety.

The security screening on passengers is a key step for entering the quarantine area of rail stations from the public areas. According to the Anti-Terrorism Act of the People's Republic of China [3], the management units of high-speed rail stations should conduct security checks on the passengers, personal belongings, and vehicles that enter the stations. The growing terrorist threats make the research on security-check systems become a hot issue [4–6]. Interested readers may refer to [7–12] and references therein. From the related works in [13], it shows that the security checkpoint is a bottleneck in the passenger traveling process at the high-speed rail station. With the increase of passenger flow, passengers often need to wait in a long line for security check due to the complexity of the security inspection process, which would significantly reduce passenger satisfaction [14]. Therefore, it's important and necessary to set up a research on improving the security-check efficiency while ensuring safety.

In this paper, we formulate the passenger security-check process as a queuing model with four types of inspection channels. Based on the Arena simulation platform, we observe the effect on the average waiting time of passengers and safety level of the security-check system through varying each interested model parameter. The main contribution of this work is supposed to perform a sensitivity analysis for managing the studied security-check system at the high-speed rail stations.

2 Security-Check Mode of High-Speed Rail Station

The purpose of the security check is to prevent passengers from carrying items that may endanger safety into the high-speed rail stations or trains [15]. The security screening on incoming passengers includes the inspection for personal identity, body check, and carry-on baggage. When passengers are going into the security check-points of rail stations, their identity cards, and train tickets will be inspected by the ticket inspectors. After the ticket is inspected, the passenger puts his/her belongings into a tray which is going to be inspected via the X-ray machine, while he/she passes the security door for body check. After the security-check procedure is completed, the passenger picks up and organizes his/her belongings or baggage, then leaves the security checkpoint. There are two possibilities in the inspection process of passengers' baggage: containing or without contraband. Some suspicious passengers who need to open their baggage would go through additional complete inspection.

There are two kinds of signal cases in the output of inspection process: alarm and clear (no alarm). It may result in the following four possibilities: (1) Correct Alarm: Raise the alarm when passenger carries a threat or contraband; (2) Correct Clear: Pass the inspection when there is neither a threat nor contraband; (3) False Alarm: Raise the alarm when there is neither a threat nor contraband; (4) False Clear: Pass the inspection when passenger carries a threat or contraband.

3   A Queuing Model with Four Types of Inspection Channels

As shown in Fig. 1, there are four types of inspection channels in the security-check system of high-speed rail station, including Green channel, Strict inspection channel, Normal inspection channel, and Fast-Pass inspection channel. Based on passengers' identities and attributes, they are differentiated into four different risk classes and then go through the corresponding type of inspection channels. It is assumed that incoming passengers are inspected based on a First-Come-First-Served order. All model parameters of the studied queuing model are summarized in Table 1.

The description of the proposed queuing model is introduced as follows. The overall arrival rate of passengers to the high-speed rail station is denoted as λ, and it holds that $\lambda = \lambda_g + \lambda_s + \lambda_n + \lambda_f$. The Green channel is designed to help special passengers quickly complete security checks, such as VIPs, government officials, and soldiers on duty. We denote μ_g as the average service rate of Green channel, and the symbol λ_g represents the average arrival rate of passengers to the Green channel. While in Normal inspection channel, the security personnel conduct a routine inspection of ordinary passengers. The average arrival rate of passengers to the Normal inspection channel is given as λ_n, and the average service rate is μ_n. Note that the queue length of Normal inspection channel is limited and denoted as the finite integer number M.

In Fast-pass inspection channel, those passengers without baggage are carried out a quick and simple inspection procedure. According to the practical experience, the inspection time for personal baggage is much longer than personal inspection time. The average arrival rate of passengers to the Fast-pass inspection channel is set as λ_f, and the average service rate is given as μ_f. On the other hand, there is a complicated and strict inspection procedure for suspicious passengers with baggage in Strict inspection channel, which results in longest security-check time, whereas the highest recognition rate for "threats". We denote λ_s as the average arrival rate of passengers to the Strict inspection channel, and the average service rate of Strict inspection channel is given as μ_s.

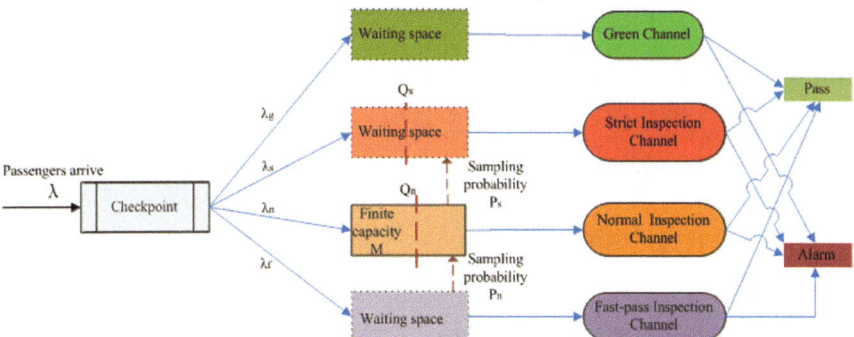

Fig. 1   An illustration of queuing modeling of a security-check system with four types of inspection channels

Table 1 Notations for the model parameters

Notation	Definition
λ_g	Average arrival rate of passengers to the Green channel
λ_s	Average arrival rate of passengers to the Strict inspection channel
λ_f	Average arrival rate of passengers to the Fast-pass inspection channel
λ_n	Average arrival rate of passengers to the Normal inspection channel
λ	The overall arrival rate of passengers to the high-speed rail station
μ_g	Average service rate of Green channel
μ_s	Average service rate of Strict inspection channel
μ_f	Average service rate of Fast-pass inspection channel
μ_n	Average service rate of Normal inspection channel
M	Finite queue length of Normal inspection channel
γ_n	Risk threshold used to differentiate the normal and low-risk passengers
γ_s	Risk threshold used to differentiate the normal and high-risk passengers
Q_n	Queue threshold used to draw out a proportion of passengers from Fast-pass inspection channel
Q_s	Queue threshold used to draw out a proportion of passengers from Normal inspection channel
P_n	Sampling probability of passengers from Fast-pass inspection channel
P_s	Sampling probability of passengers from Normal inspection channel

The security-check system will assign passengers into the corresponding inspection channels in accordance to the risk threshold value. The risk threshold γ_n and γ_s are used to differentiate the risk value of high-risk and low-risk passengers in the security-check system of the high-speed rail station. When a passenger's risk value is smaller than a given risk threshold γ_n, he/she could go to the Fast-pass inspection channel for security check. If his/her risk value is between two risk thresholds γ_n and γ_s, he/she would go to the Normal inspection channel. For those suspicious passengers whose risk values are larger than a given risk threshold γ_s, he/she ought to go through complete inspection in Strict inspection channel. It holds that $0 \leq \gamma_n \leq \gamma_s \leq 1$.

The queue thresholds Q_n and Q_s in the higher level inspection channels are used to draw out a proportion of passengers in the lower level inspection channels. The sampling probability P_n indicates the proportion of passengers entering Normal inspection channel from Fast-pass inspection channel when the current queue length in Normal inspection channel is less than Q_n. The sampling probability P_s represents the proportion of passengers entering Strict inspection channel from Normal inspection channel when the current queue length in Strict inspection channel is less than Q_s.

In this paper, the system performance of the studied security-check system includes the following two performance measures: the average waiting time of passengers and safety level o f the system. Here, the safety level of the studied high-speed rail station is defined as 100% minus the percentage of "simulation threats" passed the mistaken inspection channels.

4 Numerical Results

In this section, the relationship between the model parameters and the system performance will be illustrated by performing multiple sets of experiments on the change of each decision variable. As shown in Table 2, our experimental settings are taken from data collected at the Changsha South High-Speed Station. The settings of our model parameters are given as follows: $\gamma_n \in [0, 0.145]$, $\gamma_s \in [0, 0.99]$, $Q_n \in [1, 20]$, $Q_s \in [1, 20]$, $P_n \in [0, 1]$, $P_s \in [0, 1]$, and $M \in [1, 25]$. Here, the average service rate at Green channel is $\mu_g = 16.5$ people per minute. The average service rate of Fast-pass inspection channel is $\mu_f = 13.4$ people per minute. The average service rate of Normal inspection channel is $\mu_n = 11$ people per minute, and the average service rate at Strict inspection channel is $\mu_s = 3.083$ people per minute. Our numerical experiments are run by using Arena simulation software version 14.00 installed on the PC platform with Intel Core i5-2520M (2.5 GHz) and 8 GB RAM. In each simulation experiment, the simulation will be carried out with sufficient running time and repeated times.

In the following experiments, we analyze the effect on the average waiting time and the safety level of the proposed queuing system by varying a certain decision variable. We conduct a sensitivity analysis to study the managerial effect of model parameters on the interested system performance. Our numerical results are illustrated in Figs. 2, 3, 4, 5, 6, 7, and 8.

From Fig. 2, when we increase the value of risk threshold γ_n, it can be observed the average waiting time decreases first and then increases up, and there is a minimum value at the data point $\gamma_n = 0.025$. Nevertheless, there is no obvious changing trend in the safety level of the system with the increase of the risk threshold γ_n.

As shown in Fig. 3, with the increase of risk threshold γ_s, the average waiting time keeps decreasing, and it tends to be gentle after the point $\gamma_s = 0.24$. With the increase of risk threshold γ_s, the safety level of the system is unchanged before the point $\gamma_s = 0.60$. After that data point, it gradually becomes smaller, and finally stabilizes.

In Fig. 4, we find that there is no obvious changing trend in the average waiting time with the increase of the queue threshold Q_n. The minimum waiting time is

Table 2 The number of trains and passenger flow at various time periods in Changsha South High-Speed Station	Time period	Number of HSR trains	Passenger flow
	07:00–09:00	9	6720
	09:00–11:00	16	9228
	11:00–13:00	25	13,836
	13:00–15:00	21	12,276
	15:00–17:00	24	14,760
	17:00–19:00	22	12,996
	19:00–21:00	17	9120
	21:00–23:00	4	840

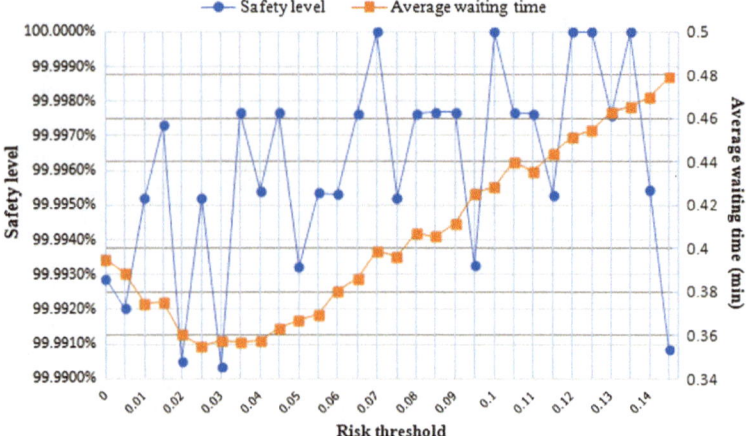

Fig. 2 Risk threshold γ_n versus two performance measures

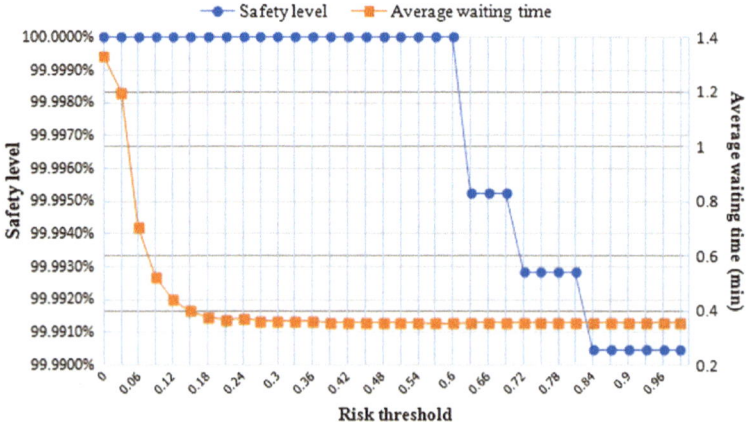

Fig. 3 Risk threshold γ_s versus two performance measures

obtained when we take $Q_n = 3$. The safety level of the studied system is not notably affected by the change of the queue threshold Q_n. The safety level has no significant changes while we take $1 \le Q_n \le 12$, and it tends to be flat and stable around a value after we take $Q_n \ge 12$.

In Fig. 5, we observe that the average waiting time first decreases slightly and then continues to increase when increasing the queue threshold Q_s. Besides, we obtain the minimal value of average waiting time as the queue threshold $Q_s = 3$. However, as the queue threshold Q_s increases, there is no obvious changes in the safety level.

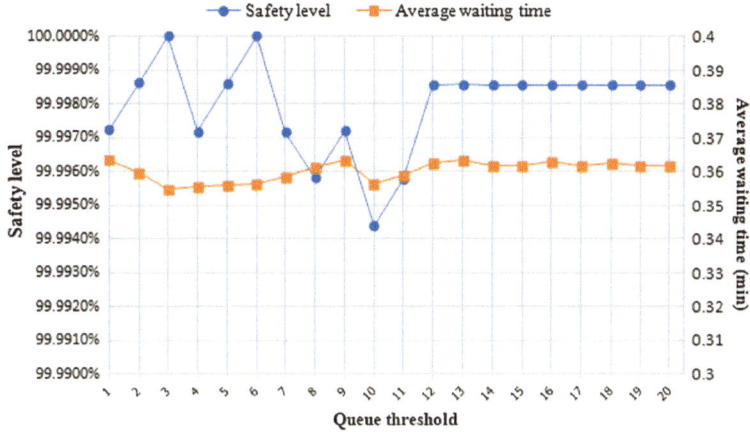

Fig. 4 Queue threshold Q_n versus two performance measures

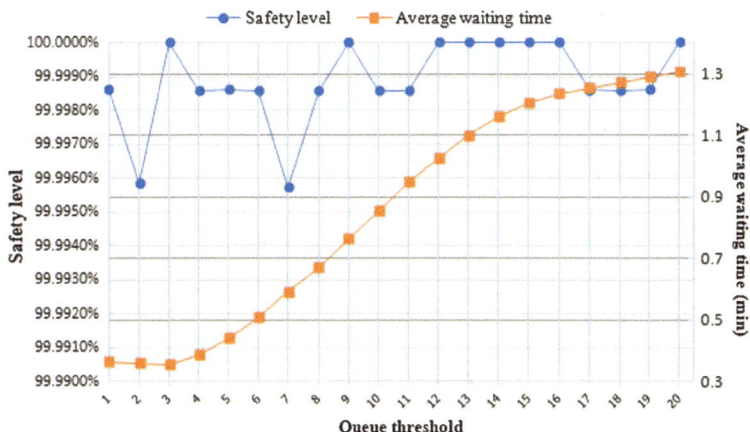

Fig. 5 Queue threshold Q_s versus two performance measures

We can observe from Fig. 6 that as the sampling probability P_n increases, there is no obvious changing rule for the average waiting time of passengers. When we set $P_n = 20\%$, it gives a minimum of the average waiting time. In addition, there is no significant change in the safety level as the sampling probability increases.

As illustrated in Fig. 7, it shows a downward trend in the average waiting time of passengers, and finally tends to be flat as the sampling probability P_s increases. At a sampling probability of 45%, there is a minimum waiting time for passengers. Nevertheless, when we vary the value of sampling probability P_s, there is no significant change in the safety level.

It can be observed in Fig. 8 that with the increases of the finite capacity M of Normal inspection channel, it illustrates a downward trend in the average waiting

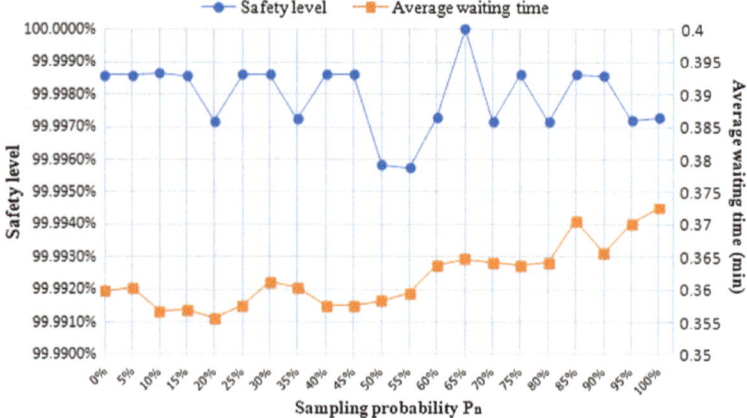

Fig. 6 Sampling probability P_n versus two performance measures

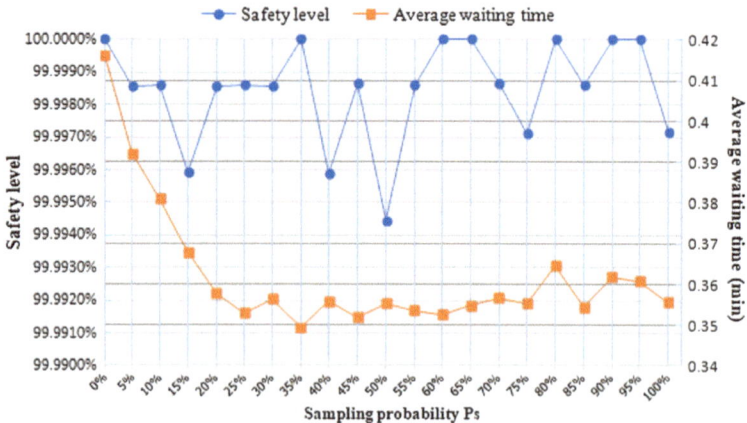

Fig. 7 Sampling probability P_s versus two performance measures

time of passengers and tends to be gentle. Besides, when the finite capacity is set as $M = 9$, we have the smallest value of these average waiting time of passengers. It can also be observed in Fig. 8 that there is no obvious change in the safety level when we vary the finite capacity of Normal inspection channel.

From Fig. 2, 3, 4, 5, 6, 7, and 8, we have studied the service performance of a security-check queuing system with data from a high-speed rail station in China. In the numerical experiments, we have discussed the influences of seven decision variables on the interested performance indicators for the presented security-check system with four types of inspection channels. Note that not all seven decision variables could have a significant impact on the average waiting time. There is a clear trend in the influence of several decision variables, such as risk threshold γ_n, risk threshold γ_s, queue threshold Q_s, sampling probability P_s, and finite capacity M.

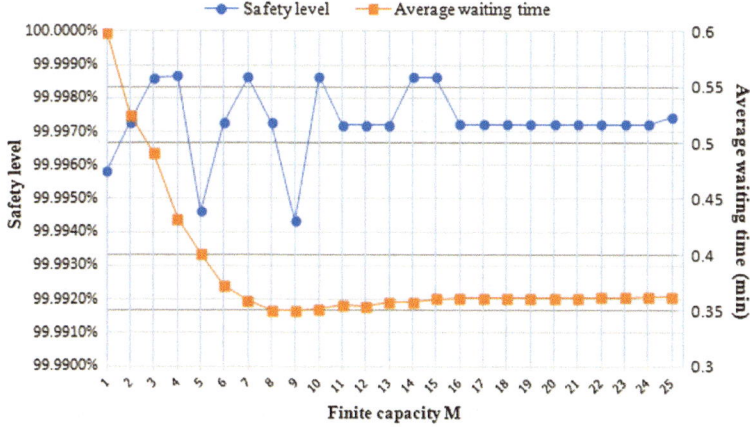

Fig. 8 Finite capacity M versus two performance measures

On the other hand, only one decision variable can have a significant impact on the safety level, i.e., risk threshold γ_s. The other six decision variables have no significant impact on the safety level.

5 Conclusions

In this paper, we studied the service performance of security-check queuing system for high-speed rail stations in China. A simulation model was developed for studying the security-check process via using Arena simulation platform. With data of a train station, we investigated the stochastic behavior of the presented queuing model through a series of computer simulations, where the numerical results are discussed and managerial implications are obtained. A sensitivity analysis was also conducted to understand the relationship between the decision variables and system performance. From the presented data analysis, it illustrates that there are several significant trends in the change of system performance. The data analysis in this work could provide effective and practical guidance for not only service management but also security issues of high-speed rail stations in China.

Acknowledgements The authors are grateful to the anonymous referees for their constructive comments, which led to the improvement of this paper in numerous ways. This study was supported in part by Fujian Provincial Department of Science and Technology under Grant No. 2016J01330, and the Research Fund from the Fujian University of Technology under Grant No. GY-Z18148.

References

1. C.-H. Wang, A queueing analysis of a security-check system with two types of inspection channels, in *Proceedings of 2018 International Conference on Mathematics, Modelling, Simulation
   and Algorithms (MMSA2018), Advances in Intelligent Systems Research*, vol. 159, Chengdu,
   China (2018), pp. 102–106
2. The National Development and Reform Commission (NDRC) of the People's Republic of
   China, *Mid-to-Long-Term Railway Network Plan* (2016 Revision No. 1536), issued on July
   13, 2016, http://www.ndrc.gov.cn/zcfb/zcfbtz/201607/t20160720_811696.html
3. The National People's Congress of the People's Republic of China, *The Anti-Terrorism Act of
   People's Republic of China*, issued on December 27, 2015, http://www.npc.gov.cn/npc/xinwen/
   2015-12/28/content_1957401.htm
4. Z.G. Zhang, H. Luh, C.-H. Wang, Modeling security-check queue. Manag. Sci. **57**, 1979–1995
   (2011)
5. X. Nie, G. Parab, R. Batta, L. Lin, Simulation-based selectee lane queuing design for passenger
   checkpoint screening. Eur. J. Oper. Res. **219**, 146–155 (2012)
6. C.-H. Wang, J. Lan, Data analysis and optimization strategy for a risk-based three-level inspection channel. Mod. Manag. **9**(1), 9–23 (2019)
7. F. Kaakai, S. Hayat, A.E. Moudni, A hybrid petri nets-based simulation model for evaluating
   the design of railway transit stations. Simul. Model. Pract. Theory **15**, 935–969 (2007)
8. C.-H. Wang, M.-E. Wu, C.-M. Chen, Inspection risk and delay for screening cargo containers
   at security checkpoints, in *Proceedings of the 11th International Conference on Intelligent
   Information Hiding and Multimedia Signal Processing (IIH-MSP-2015)*, Adelaide, Australia
   (2015), pp. 211–214
9. C.-H. Wang, A modelling framework for managing risk-based checkpoint screening systems
   with two-type inspection queues, in *Proceedings of the 3rd International Conference on Robot,
   Vision and Signal Processing (RVSP 2015)*, Kaohsiung, Taiwan (2015), pp. 220–223
10. P. Huang, H. Luh, Z.G. Zhang, A queueing model for tiered inspection lines in airports. Int. J.
    Inf. Manag. Sci. **27**, 147–177 (2016)
11. C.-H. Wang, Arena simulation for aviation passenger security-check systems, in *Proceed-
    ings of 10th International Conference on Genetic and Evolutionary Computing (ICGEC2016),
    Advances in Intelligent Systems and Computing*, vol. 536, Fuzhou, China (2016), pp. 95–102
12. C.-H. Wang, R. Chen, A simulation analysis of airport security-check queues based on passen-
    gers' risk classification. Manag. Sci. Eng. **7**(2), 110–124 (2018)
13. C.-H. Wang, A review of operational mechanisms and simulations for security screening sys-
    tems. Comput. Sci. Appl. **7**(11), 1067–1078 (2017)
14. C.-H. Wang, J. Lan, Performance evaluation of a risk-based three-tier inspection system, in *Pro-
    ceedings of 2nd International Conference on Computational Modeling, Simulation and Applied
    Mathematics (CMSAM2017)*, DEStech Transactions on Computer Science and Engineering,
    Beijing, China (2017), pp. 464–468
15. E. Matsika, C. O'Neill, U. Battista, M. Khosravi, A. de Santiago Laporte, E. Munoz, Devel-
    opment of risk assessment specifications for analysing terrorist attacks vulnerability on metro
    and light rail systems. Transp. Res. Procedia **14**, 1345–1354 (2016)

LSTM-Based Neural Network Model for Semantic Search

Xiaoyu Guo, Jing Ma and Xiaofeng Li

Abstract To improve web search quality and serve a better search experience for users, it is important to capture semantic information from user query which contains user's intention in web search. Long Short-Term Memory (LSTM), a significant network in deep learning has made tremendous achievements in capturing semantic information and predicting the semantic relatedness of two sentences. In this study, considering the similarity between predicting the relatedness of sentence pair task and semantic search, we provide a novel channel to process semantic search task: see semantic search as an atypical predicting the relatedness of sentence pair task. Furthermore, we propose an LSTM-Based Neural Network Model which is suitable for predicting the semantic relatedness between user query and potential documents. The proposed LSTM-Based Neural Network Model is trained by Home Depot dataset. Results show that our model outperforms than other models.

Keywords LSTM · Deep learning · Semantic search · RNN

1 Introduction

Having a better understanding of user's intention is very important to improve search engine's quality and optimize the user experience. Traditional search engine is mainly based on matching keywords in documents with search queries. However, this technology cannot distinguish synonymous words from different sentences. Moreover, users have to consider a lot about how to organize query words in order to get the right information they want. This brings too much inconvenience to users. As a result, semantic search engine emerges in order to better serve users.

X. Guo · J. Ma (✉) · X. Li
College of Economics and Management, Nanjing University of Aeronautics and Astronautics, Nanjing, Jiangning District 211106, China
e-mail: majing5525@126.com

© Springer Nature Switzerland AG 2020
H. Yang et al. (eds.), *Smart Service Systems, Operations Management, and Analytics*, Springer Proceedings in Business and Economics, https://doi.org/10.1007/978-3-030-30967-1_3

Recently, Long short-term memory (LSTM) networks have shown remarkable performance in several kinds of Natural Language Processing (NLP) tasks, including image captioning [1], machine translation [2] and semantic search [3]. LSTM networks have also been used to predict semantic relatedness score [4] in order to find the relevant questions from the existing questions and find the relevant answers to a new question for Community Question Answering forums. Inspired by this task, we propose an LSTM-Based Neural Network Model reformed from predicting semantic relatedness score model based on similarities between predicting the relatedness of sentence pair and semantic search. We perform experiments to validate the utility of our model in Home Depot dataset, and compare the method to other models.

The contributions of this paper are as follows:

(1) Explore a novel method to process semantic search task. In this field, user query contains less information than search result does, so this paper proposed to split search result document and splice user query and related documents many times to balance the information quantity.

(2) Based on the method mentioned in (1), this paper built a novel neural network model for semantic search field.

(3) Perform experiments to validate the utility of our model, and compare our model to other model using different variants of LSTM architecture.

2   Related Work

The traditional language model, without taking sequence factor into account, cannot capture semantic information thoroughly. For example, "look after" and "after looking" have different meaning and traditional language model cannot tell any differences from each other. Mikolov et al. [5] proposed Recurrent Neural Network Language Model (RNNLM) in order to process sequence data. Recurrent neural network (RNN), which is different from normal neural network, introduces constant circulation into its model so that it could process sequence information. RNN shows remarkable performance in processing many tasks concerning sequence, but RNN has a Long-Term Dependencies problem [6] when it processes longer passages that contain too much information. The LSTM architecture, proposed by Hochreiter and Schmidhuber [7], addresses this problem of Long-Term Dependencies by introducing a memory cell that is able to store state information over long period of time into RNN structure. LSTM has recently been used in information retrieval field for extracting sentence-level semantic vectors [8] and context-aware query suggestion [9].

Commonly used variants of LSTM architecture are the Bidirectional Long Short-Term Memory (BLSTM) Network and the Multilayer Long Short-Term Memory (MLSTM). Several scholars pay much attention to exploring novel LSTM variants, including Tree-Structured Long Short-Term Memory Networks proposed by

Tai et al. [10] and multiplicative LSTM proposed by Stephen Merity et al. [11] in order to obtain a better performance on NLP tasks. In this paper, we conducted our experiments using different variants of LSTM.

The setup of our work is closely related to what was proposed already by Nassif H et al. [4]. Nassif et al. aim to obtain the semantic relatedness score between two questions a s shown in Fig. 1. The model was built on two bidirectional LSTM whose output can be augmented with extra features and fed into the multilayer perceptron. Other similar methods include the combination of recurrent and convolutional models [12]. In this paper, we regard semantic search as predicting semantic relatedness between user query and potential document aiming to find the closest result in semantic meaning. The main difference of our method compared to these models is that we balance the quantity of information between inputs by exploiting user query several times. We also compare our method to these methods.

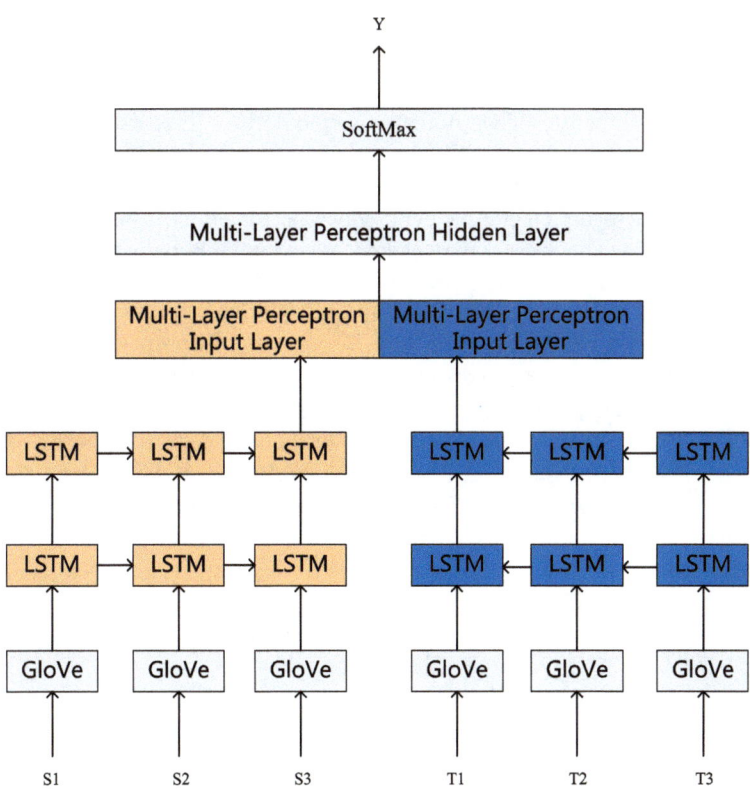

Fig. 1 The general architecture of predicting semantic similarity score model

3   Method

The essence of search engine is to calculate the relatedness score between user query and possible documents. Hence, predicting the semantic similarity bears strong resemblances to semantic search. We could see a semantic search task as described below:

$$f(Query, Doc) = SemanticScore\{a|a \in R, a \in [1, 3]\} \tag{1}$$

where Query is user input query words, Doc is a document and SemanticScore is a real number between 1 and 3.

That is to say, given user query and documents, processed by functional transformation and then output a real number between 1 and 3, in which 1 denotes "not relevant" and 3 denotes "extremely relevant".

Based on the model mentioned in 2, we proposed a LSTM-Based Neural Network Model which is suitable for semantic search, as is shown in Fig. 3. If we directly apply the model mentioned in 2 into semantic search, we would see a model shown in Fig. 2. Apparently, there exists an imbalance in information quantity between user queries and documents. That is to say, documents contain more information than user queries do. Therefore, if we directly use model in Fig. 2, we could not gain remarkable performance theoretically.

To balance the information quantity, we reform the model in Fig. 2 and the model in Fig. 3 is the general architecture of our model. There are three differences between Figs. 2 and 3. (1) We splice user query and related documents many times. The input layer of our model consists of two parts. One is LSTM representation of user query and another is LSTM representation of sentences in related documents. In this way, we make the best use of user query so that we balance the information quantity. (2) For every pair of user query and a document, we gain several semantic scores. (3) We use Ridge Regression as the last step to output the final semantic relatedness score.

The input of our model consists of two parts: one is user query words (QW) and another is Related Document. For QW, the model directly transforms them to vectors using GloVe and then encodes vectors using LSTM. For Related Document, the model splits the documents into sentences first and then processes them using the same methods as QW. After LSTM encoding, the model splice user query to every sentence twice. Then we get several semantic scores and our goal is to gain a final semantic score. Therefore, we see this task as a regression problem:

$$\hat{S} = Xw$$
$$= X(X^T X + \lambda E)^{-1} X^T S$$

where S denotes actual semantic scores, $\hat{S}$ denotes the final forecasting semantic score, and X denotes a matrix consisted of sub-semantic scores.

In the last step, this model uses Ridge Regression to predict the final score. The output of our model is a final semantic score representing the semantic relatedness

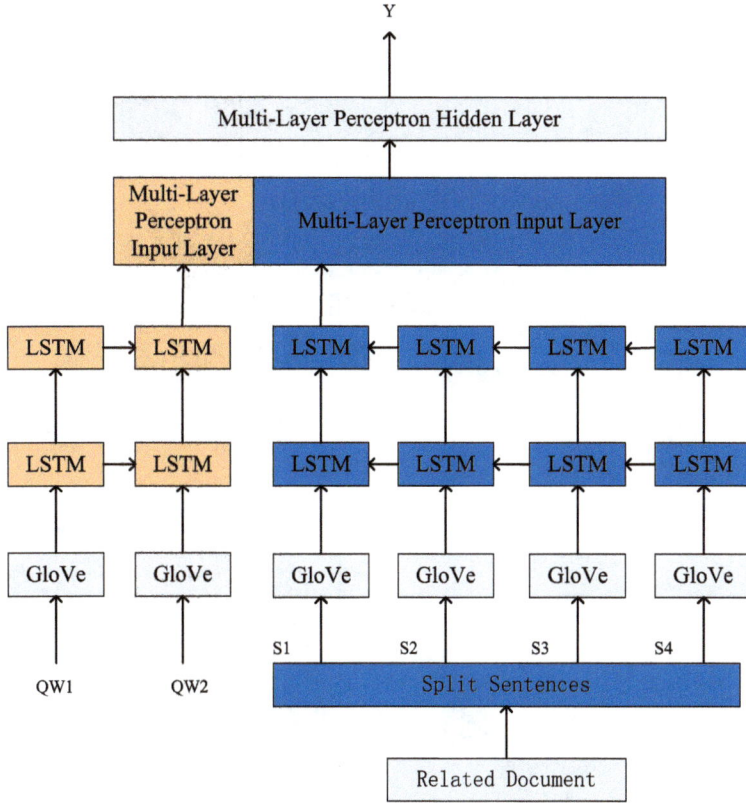

Fig. 2 Directly apply predicting semantic score model into semantic search

between user query and document. This model is designed to process English semantic search task. If you want to apply this model to other languages, all you have to do is to change the input part of the model. Take Chinese into example, we can split words first and then uses this model.

4 Experimental Setup and Results

In this section, we describe our experimental setup. The dataset and Evaluation methodology is described in Sects. 4.1 and 4.2. The experiments that we conduct to test the performance of our model is given in Sect.4.3.

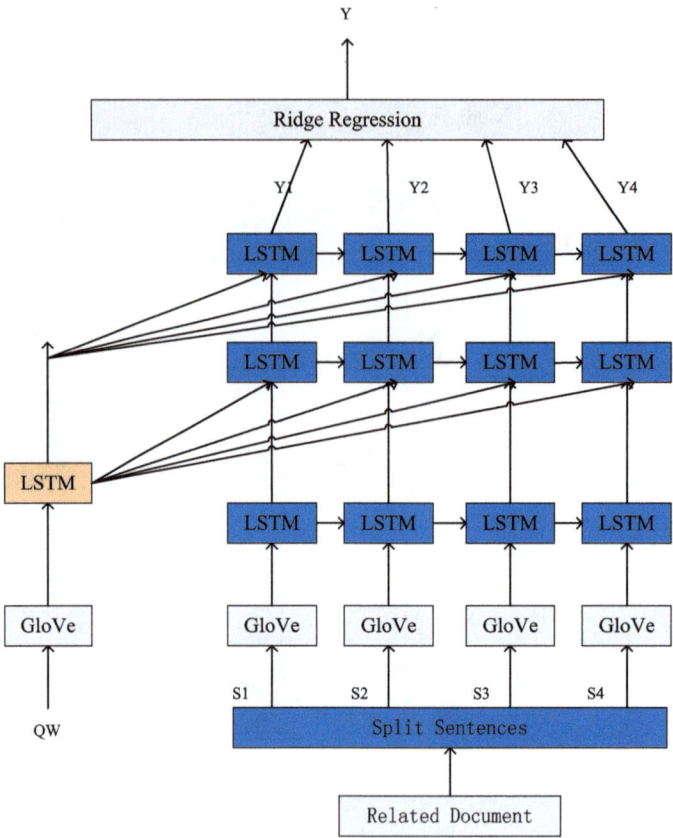

Fig. 3 LSTM-based neural network for semantic search

4.1 Dataset

In this paper, the dataset [13] we used to test the performance is from The Home Depot, an e-commerce website in America. This platform sells building materials. As a user who wants to buy products from this website, all you have to do is to type your objective and then click "search" button. For example, type "I want to lay the foundation" in input field and then click "search". The results will show tools for laying the foundation. Therefore, this dataset is suitable for semantic tasks.

This dataset contains 124429 query–document pairs and 37034 unique product description documents. Each query–document pair contains a human-generated relevance label, which is the average of 5 ratings assigned by different human annotators. We split the query sessions into three parts, and use 60% to train LSTM-Based Neural Network Models, 10% to trial models (aim to choose a best one then apply it to the test dataset) and 30% as test data to evaluate the prediction performance.

4.2 Evaluation Methodology

We use root mean squared error (RMSE) as evaluation metrics. Actually, RMSE score indicates the gap between prediction and real value. Therefore, we aim to gain a smaller RMSE value.

$$\text{RMSE} = \sqrt{\frac{1}{m} \sum_{i=1}^{m} \left(y_i - \hat{y}_i \right)^2} \tag{3}$$

4.3 Experiments

For a given pair of user query and document, our task is to predict a human-generated rating of the similarity of user query and document in meaning.

Here, we use the similarity model described in Fig. 3. We initialized our word representations using publicly available 300-dimensional Glove vectors [14]. We trained our models with a learning rate of 0.05 and a minibatch size of 25, mentioned by AdaGrad [15].

The last step is Ridge Regression [16], in which we predict the final relatedness score using sub-relatedness scores. In this step, we choose a proper Alpha value to get the best result. Figure 4 shows the relationship between Alpha and Cross-validation (CV) value of different LSTM variants.

We compare our model with predicting semantic similarity model using the same dataset and the results are summarized in Table 1. The first group is the performances of predicting semantic similarity model and the second group is the performances

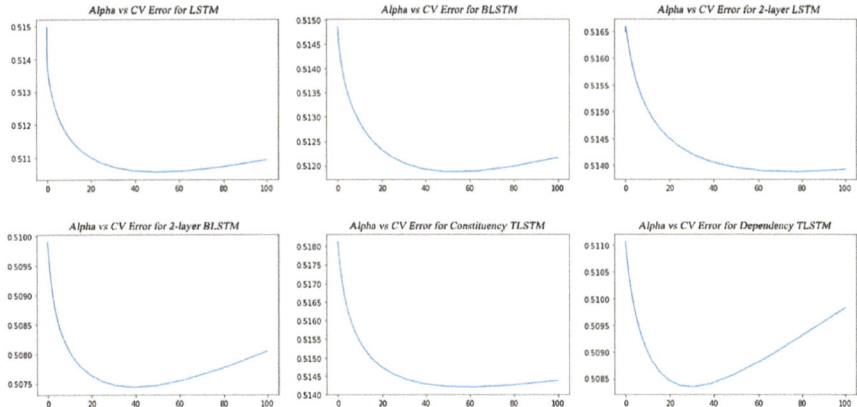

Fig. 4 The relationship between alpha and Cross-Validation (CV) value of different LSTM variants

Table 1 Test set results on the home depot dataset

Method		RMSE
Predicting semantic similarity model (Henry Nassif et al.,2016)	LSTM	0.6664
	BLSTM	**0.6747**
	2-layer LSTM	0.6830
	2-layer BLSTM	0.6608
LSTM-based neural network for semantic search	LSTM	0.4682
	BLSTM	**0.4611**
	2-layer LSTM	0.4643
	2-layer BLSTM	0.4668
	Constituency TLSTM	0.4646
	Dependency TLSTM	0.4645

of LSTM-Based Neural Network for Semantic Search. Overall, the performances of LSTM-Based Neural Network is better than predicting semantic similarity model's, with approximately 0.2 improved. More specifically, BLSTM shows best performance among all the LSTM variants on both two models, while there are slight differences within each group.

5 Conclusion and Future Work

In this paper, we propose an LSTM-Based Neural Network model, which is reformed from an existent predicting relatedness score task model. We see the semantic search problem as an atypical predicting relatedness score task. The results show that our model has a remarkable performance in predicting semantic relatedness score between user query and potential documents without adding any additional man-made feature engineering. Therefore, our model saves time to make feature engineering and reduce the influence of human factors for sorting search results.

We may foresee that it will serve a better search experience for users if we use our model in semantic search field because there is no need to think too much about search words. For example, people who have little knowledge of search engine principle may consider too much about which word should be included and which word should not be included in query because the extra word may have a bad influence on search results. Therefore, our model may save too much time that users used to consider constructing query and provide a better search service for users.

In the future, we would pay much attention to the output period. In this paper, we just use Ridge Regression to output our semantic relatedness score. But Ensemble learning is famous for its performance in machine learning field. Consequently, we would try to combine Ridge Regression with other machine learning methods to output the final result in future work. In addition, we just use one English dataset to test model because it is hard to obtain semantic search-related dataset. Therefore, we will dedicate to obtaining different datasets and testing our model.

References

1. O. Vinyals, A. Toshev, S. Bengio et al., *Show and Tell: A Neural Image Caption Generator [J]* (2014)
2. I. Sutskever, O. Vinyals, Q.V. Le, *Sequence to Sequence Learning with Neural Networks [J]* (2014)
3. P.S. Huang, X. He, J. Gao et al., Learning deep structured semantic models for web search using clickthrough data [C], in *Proceedings of the 22nd ACM International Conference on Conference on Information and Knowledge Management* (ACM, 2013)
4. H. Nassif, M. Mohtarami, J. Glass, Learning semantic relatedness in community question answering using neural models [C], in *Proceedings of the 1st Workshop on Representation Learning for NLP* (2016), pp. 137–147
5. T. Mikolov, *Statistical Language Models Based on Neural Networks [J]* (Presentation at Google, Mountain View, 2012), p. 80
6. R. Pascanu, T Mikolov, Y. Bengio, On the difficulty of training recurrent neural networks [C], in *International Conference on Machine Learning* (2013), pp. 1310–1318
7. S. Hochreiter, J. Schmidhuber, Long short-term memory[J]. Neural Comput. **9**(8), 1735–1780 (1997)
8. H. Palangi, L. Deng, Y. Shen et al., Deep sentence embedding using long short-term memory networks: Analysis and application to information retrieval [J]. IEEE/ACM Trans. Audio Speech Lang. Process (TASLP) **24**(4), 694–707 (2016)
9. A. Sordoni, Y. Bengio, H. Vahabi et al., A hierarchical recurrent encoder-decoder for generative context-aware query suggestion[C], in *Proceedings of the 24th ACM International on Conference on Information and Knowledge Management* (ACM, 2015), pp. 553–562
10. K.S. Tai, R. Socher, C.D. Manning, *Improved Semantic Representations from Tree-Structured Long Short-Term Memory Networks [J]* (2015). arXiv:1503.00075, 2015
11. B. Krause, L. Lu, I. Murray et al., *Multiplicative LSTM for Sequence Modelling [J]* (2016) arXiv:1609.07959
12. H.S. Joshi, *Finding Similar Questions in Large-scale Community QA Forums [D]*. Massachusetts Institute of Technology (2016)
13. https://www.kaggle.com/c/home-depot-product-search-relevance/data
14. J. Pennington, R. Socher, C. Manning, Glove: global vectors for word representation [C], in *Proceedings of the 2014 Conference on Empirical Methods in Natural Language Processing (EMNLP)* (2014), pp. 1532–1543
15. J. Duchi, E. Hazan, Y. Singer, Adaptive subgradient methods for online learning and stochastic optimization[J]. J. Mach. Learn. Res. 2121–2159 (2011)
16. A.E. Hoerl, R.W. Kennard, Ridge regression: biased estimation for nonorthogonal problems [J]. Technometrics **12**(1), 55–67 (1970)

Research on the Evaluation of Electric Power Companies' Safety Capabilities Based on Grey Fixed Weight Clustering

Yijing Wang, Jie Xu, Chuanmin Mi and Zhipeng Zhou

Abstract With the development of power grid and the widespread use of power, the impact of power safety is becoming more and more important. Meanwhile, Internet and Internet of Things have both provided new technical and management ideas for power safety. Combining literature research and field investigation on safety management of Jiangsu Electric Power Company, this paper focuses on solving the problem of electric power companies' safety capabilities. First, based on the idea of data-driven management and decision-making, power safety risks are identified according to electric power companies' actual business situation and therefore, safety capability evaluation index system is created. Second, the weight of each indicator is determined according to the proportion of safety risks in historical accidents. Finally, considering the problem of information lack in electric power safety management in China, this paper uses Grey Fixed Weight Clustering Theory including determining thresholds of each indicator and constructing whitening weight functions to make an evaluation of electric power companies' safety capabilities which could help managers learn about their own existing safety levels and provide them a reference for future decisions.

Keywords Safety management · Electric power companies' safety capabilities · Evaluation index system · Safety risks · Grey fixed weight clustering

1 Introduction

Nowadays, since China's economy is in a period of rapid development, the demand for electricity is becoming larger and much more urgent than before. With the expansion of the scale of power grid, the issue of electric power grid safety management

Y. Wang · C. Mi (✉) · Z. Zhou
College of Economics and Management, Nanjing University of Aeronautics and Astronautics, Nanjing 210016, Jiangsu, China
e-mail: michuanmin@163.com

J. Xu
State Grid Wuxi Power Supply Company, Wuxi 214061, Jiangsu, China

© Springer Nature Switzerland AG 2020
H. Yang et al. (eds.), *Smart Service Systems, Operations Management, and Analytics*, Springer Proceedings in Business and Economics, https://doi.org/10.1007/978-3-030-30967-1_4

is increasingly related to the overall situation of national economic development and social stability.

Because of the much more important role which electric power plays in the social production, the government and related electric power companies have paid more attention to the management of power grid. In order to strengthen the safety management of power grid, they have formulated some safety production standard documents, actual control measures and mechanisms in recent years. Through the establishment of electric power safety management system, the pre-control measures have been improved and the standardization of power safety production has been gradually realized to some degree.

Apart from practice, there are also many studies and researches in the field of electric power safety and safety management home and abroad which lead to multi-theory safety risk assessment indicators. The continuous advancement of science and technology and productivity has made humans' lives more and more colourful, but it has also created safety problems that threaten human safety and health. In the twelfth century, the United Kingdom enacted the "Fire Protection Act", and in the seventeenth century, the "Personal Protection Law" was promulgated, and then safety management started to have its own content. In the middle of the eighteenth century, with the progress of the industrial revolution, the health and safety of workers gradually received attention. Some scholars began to study the issue of labour safety and health, and thus the content and scope of safety production management have been greatly developed. In the 1930s, many countries set up government agencies for safe production management and issued laws and regulations on labour safety and health, thus gradually establishing a relatively complete safety education, management, and technical system at which time modern safety production management has taken shape. In the 1950s, some risk management theories such as Safety Production Management Principles, Accident Cause Theory and Accident Prevention Principles were produced. Since the beginning of the twenty-first century, some scholars in China have proposed a systematic prototype of enterprise safety production risk management. They believe that enterprise safety production management is indeed risk management and it includes hazard identification, risk assessment, hazard warning, monitoring management, accident prevention, risk control management emergency management, etc. As for the field of electric power, safety management has also aroused widespread attention in society and academia due to the high risks of work contents related to electricity production. At present, the research on electric power safety management is mainly divided into the following three aspects: (1) Identify sources of critical risks. Since electric accidents are an important cause of occupational deaths in the world [1], seeking out the sources of critical risks that lead to these accidents is certainly a good way to decrease deaths. Scholars believe that climate factors [2], safety equipment [3], management factors [1] and human factors [4] are mainly sources of risks, which could bring about safety problems in the electric power industry. (2) Evaluate safety capabilities. As to the evaluation of electric power companies' safety capabilities, an index system is a common tool. In general, the index system often has multiple levels and is interconnected to reflect the current status of safety capabilities. For instance, based on the hidden dangers of human factors, the safety level of electric power enterprises can be comprehensively evaluated through constructing an index system including the basic quality of personnel, the characteristics of operations and the management of personnel [5]. (3) Develop accident prevention measures. Even if electric accidents do not occur as frequently as other types of occupational injuries, they can have serious consequences once they occur [6]. Therefore, the issue about how to avoid similar accidents should arouse our attention. Apart from improving workers' safety awareness [7], strengthening safety management [7], improving safety and stability control systems [8] and so on are all good methods for preventing dangers and accidents in electric power industry.

What's more, with the rapid development of Internet and Things of Internet, it has become quite common that electric power companies apply Information and Technology (IT) to their daily safety management in which the wide use of information management platforms is the typical representative. Thanks to exclusive information management platforms, the data acquisition for research becomes simple and all of these have provided a solid theoretical foundation and technical support for the safety management within the power sector.

Although the implementation of such standards and measures has well guided the electric power companies, risks still pose a great threat to its safety management and sometimes even influence the convenience of people's lives in reality. Risks refer to hidden dangers which arise from the unsafe state of the workplace, equipment and facilities, people's unsafe behaviour and management defects and they often lead to safety production accidents directly. Therefore, apart from related safety production guidelines and standardized information management platforms, it is a necessity to put more emphasis on pre-control measures, thus for an electric power company, seeking out risks hidden in its normal production and then making an assessment of its current safety capability could make it clearly know about its weaknesses and take targeted measures to improve its level of safety management.

The paper is organized as follows: Sect. 1 offers a brief introduction of background and literature review to explore researches related to this paper. In Sect. 2, Grey Fixed Weight Clustering Theory is proposed. Based on the field investigation on safety management of Jiangsu Electric Power Company, an evaluation index system of electric power companies' safety capabilities is proposed in Sect. 3. According to the evaluation index system and Grey Fixed Weight Clustering Theory, taking an electric power company as a practical example, Sect. 4 evaluates its safety levels from 2016 to 2018 by way of Grey Fixed Weight Clustering Theory. In the end, Sect. 5 presents the major conclusions and managerial implications of this paper.

2 Grey Fixed Weight Clustering Theory

Grey System Theory is first proposed by a Chinese professor Julong Deng in 1982. In reality, a varying environment always results in inappropriate human judgments and imprecise information, which tend to increase vagueness and biasness of final results. Grey System Theory takes the uncertain system which has partial information and small quantity of data as the research object. The main purpose is to realize the correct description of the operational behaviour of the system by generating and developing some known information, and then realizing the quantitative prediction of its future changes. What's more, a main advantage of this theory is that it could generate satisfactory results using a relatively small quantity of data which to some extent could solve the problem of lack of data sometimes [9].

Grey Fixed Weight Clustering Theory, which belongs to Grey System Theory, is mainly applied to check whether an observation object belongs to predefined grey classes [10]. Compared with other evaluation methods like Analytic Hierarchy Process (AHP), Fuzzy Theory (FT) and Rough Set Theory (RST), Grey Fixed Weight Clustering Theory does not require a large set of classification criteria and classification rules base. Therefore, it has been successfully applied in a wide range of research fields such as economy, technology and environment these years [11–13].

If there are n objects, m clustering indicators and p grey classes, d_{ij} is the whitening value of the ith clustering object ($i = 1, 2,...,n$) for the jth clustering indicator ($j = 1, 2,...,m$) and f_{jk} is the whitening weight function of the jth clustering indicator for the kth ($k = 1, 2,...,p$) grey class. The detailed process is as follows:

Step 1: Construct the sample matrix D according to the given d_{ij}.

$$D = \begin{bmatrix} d_{11} & d_{12} & \cdots & d_{1m} \\ d_{21} & d_{22} & \cdots & d_{2m} \\ \vdots & \vdots & \ddots & \vdots \\ d_{n1} & d_{n2} & \cdots & d_{nm} \end{bmatrix} \tag{1}$$

Step 2: Determine the whitening weight function f_{jk}. There are mainly three whitening weight functions including the upper limit whitening weight function, the lower limit whitening weight function and the moderate measure whitening weight function.

For the jth indicator ($(j = 1, 2,...,m)$, p thresholds are set as $\lambda_{j1}, \lambda_{j2}, \ldots, \lambda_{jp}(p \geq 3)$ which divide objects into different grey classes.

$$\lambda_{jk} = \begin{bmatrix} \lambda_{11} & \lambda_{12} & \cdots & \lambda_{1p} \\ \lambda_{21} & \lambda_{22} & \cdots & \lambda_{2p} \\ \vdots & \vdots & \ddots & \vdots \\ \lambda_{m1} & \lambda_{m2} & \cdots & \lambda_{mp} \end{bmatrix} \tag{2}$$

I. The Upper Limit Whitening Weight Function

As shown in Fig. 1a, λ_{p-1} is the starting point and the value of the turning point is λ_p. The formula is

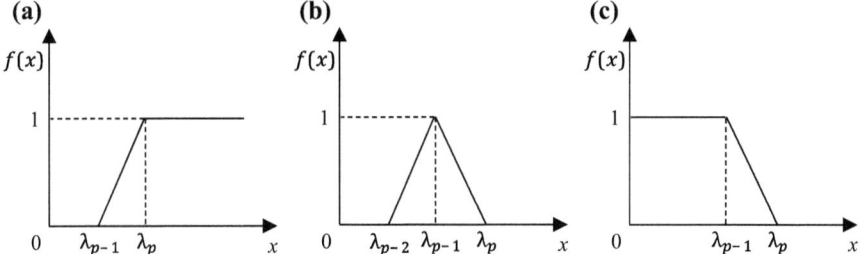

Fig. 1 Three whitening weight functions

$$
f(x) = \begin{cases} 0 & x < \lambda_{p-1} \\ \frac{x-\lambda_{p-1}}{\lambda_p-\lambda_{p-1}} & \lambda_{p-1} \le x < \lambda_p \\ 1 & x \ge \lambda_p \end{cases} \tag{3}
$$

II. The Moderate Measure Whitening Weight Function

As shown in Fig. 1b, λ_{p-2} is the starting point, λ_{p-1} is the turning point with the end of λ_p. The formula is

$$
f(x) = \begin{cases} 0 & x \notin [\lambda_{p-2}, \lambda_p] \\ \frac{x-\lambda_{p-2}}{\lambda_{p-1}-\lambda_{p-2}} & \lambda_{p-2} \le x < \lambda_{p-1} \\ \frac{x-\lambda_p}{\lambda_{p-1}-\lambda_p} & x \in [\lambda_{p-1}, \lambda_p] \end{cases} \tag{4}
$$

III. The Lower Limit Whitening Weight Function

As shown in Fig. 1c, 0 is the starting point, λ_{p-1} is the turning point with the end of λ_p. The formula is

$$
f(x) = \begin{cases} 0 & x \notin [0, \lambda_p] \\ 1 & 0 \le x < \lambda_{p-1} \\ \frac{x-\lambda_p}{\lambda_{p-1}-\lambda_p} & x \in [\lambda_{p-1}, \lambda_p] \end{cases} \tag{5}
$$

Step 3: Calculate clustering coefficients according to the clustering weight η_{jk} and the whitening weight function f_{jk}. The clustering coefficient of the ith clustering object for the kth grey class is

$$
\sigma_{ik} = \sum_{j=1}^{m} \left[f_{jk}(x_{ij}) \cdot \eta_{jk} \right] \tag{6}
$$

Step 4: Construct clustering vectors.

$$
\sigma_i = (\sigma_{i1}, \sigma_{i2}, \ldots, \sigma_{ik}) \tag{7}
$$

Step 5: Clustering. The largest one among σ_i represents the class to which the ith clustering object belongs. For example, for the ith clustering object, if σ_{i1} is larger than any other element in this row, the ith clustering object is classified into the first grey class.

In case of the evaluation of electric power companies' safety capabilities, there are two main issues needed to be thought about. On one hand, the required data of applying an evaluation index system is always uncertain and incomplete and since electric power is a thriving industry, the quantity of available data is now quite small. On the other hand, it is a necessity to determine classification criteria with respect to some evaluation indicators. Consequently, considering the advantages of Grey Fixed Weight Clustering Theory, it is thought to be a suitable way to evaluate safety capabilities of electric power companies.

3 The Proposed Evaluation Index System

As we all know, when it comes to evaluating a person or an organization, an evaluation index system is always indispensable since everything in the world has its versatility and we could not understand anything from a single angle. Evaluation index systems are developed and used in different areas such as medical care [14], environmental science [15], economic development [16] and personnel performance [17]. In a word, the selection of index system is required in almost all evaluation programs [18].

As to making an evaluation of an electric power company's safety capability, we also propose an index system which is constructed according to related papers and electric power companies' actual business situation. After listening to experts of State Grid Wuxi Power Supply Company and implementing field investigation on safety management, a safety capability evaluation index system of electric power companies is determined as Table 1.

As we could see from Table 1, there are four primary indicators including staff factors, management factors, tool factors and environmental factors with eleven secondary indicators including work skill level, regulatory violations, unplanned work, drawing accuracy, manager arrival rate, work load level, safety appliance rate, complete rate of signs, working hours in bad weather, working hours at high and low temperatures and working hours at night in this safety capability evaluation index system of electric power companies among which $x_1, x_4, x_5, x_6, x_7, x_8$ are benefit indicators while $x_2, x_3, x_9, x_{10}, x_{11}$ are cost indicators. That is to say, when the values of $x_1, x_4, x_5, x_6, x_7, x_8$ are larger and the values of $x_2, x_3, x_9, x_{10}, x_{11}$ are smaller, an electric power company always has a higher safety capability.

Table 1 A safety capability evaluation index system of electric power companies

Primary indicators	Secondary indicators
Staff factors	Work skill level x_1
Staff factors	Regulatory violations x_2
Management factors	Unplanned work x_3
Management factors	Drawing accuracy x_4 (%)
Management Factors	Manager arrival rate x_5 (%)
Management factors	Work load level x_6 (%)
Tool factors	Safety appliance rate x_7 (%)
Tool factors	Complete rate of signs x_8 (%)
Environmental factors	Working hours in bad weather x_9 (h)
Environmental factors	Working hours at high and low temperatures x_{10} (h)
Environmental factors	Working hours at night x_{11} (h)

4 Evaluation of Electric Power Companies' Safety Capability Based on GFWC Model

4.1 Data Collection and Weight Determination

When it comes to the evaluation of electric power companies' safety capabilities based on Grey Fixed Weight Clustering Model, the result is divided into three levels: excellent, medium and low, that is, three grey classes.

Thanks to Jiangsu Electric Power Company, all the data needed for modelling are available and as below Table 2 displayed.

Table 2 Safety capability indicator data of an electric power company from 2016 to 2018

Primary indicators	Secondary indicators	2016	2017	2018
Staff factors	Work skill level x_1	77	82	90
Staff factors	Regulatory violations x_2	80	76	60
Management factors	Unplanned work x_3	13	13	8
Management factors	Drawing accuracy x_4 (%)	80	88	95
Management factors	Manager arrival rate x_5 (%)	90	92	98
Management factors	Work load level x_6 (%)	88	90	94
Tool factors	Safety appliance rate x_7 (%)	95	97	99
Tool factors	Complete rate of signs x_8 (%)	96	97	99
Environmental factors	Working hours in bad weather x_9 (h)	20	15	10
Environmental factors	Working hours at high and low temperatures x_{10} (h)	50	47	45
Environmental factors	Working hours at night x_{11} (h)	19	14	8

As can be seen from Table 2, according to Grey Fixed Weight Clustering Model, this evaluation has three objects ($i = 1, 2, 3$) and eleven clustering indicators ($j = 1, 2,...,11$). The observation value x_{ij} of the ith clustering object with respect to the jth clustering indicator has been given in Table 2. Therefore, the sample matrix is

$$X = \begin{bmatrix} 77 & 80 & 13 & 80 & 90 & 88 & 95 & 96 & 20 & 50 & 19 \\ 82 & 76 & 13 & 88 & 92 & 90 & 97 & 97 & 15 & 47 & 14 \\ 90 & 60 & 8 & 95 & 98 & 94 & 99 & 99 & 10 & 45 & 8 \end{bmatrix}$$

In addition, it is also a necessity to determine the weight in evaluation since each indicator always has a different impact on electric power companies' safety capabilities in reality. In order to determine the weight of each indicator, about 180 historical accidents in electric power industry offered by State Grid Wuxi Power Supply Company are analysed. The weight of each indicator is finally determined according to the proportion of safety risks which, respectively, reflect the frequency of each indicator occurring in these historical accidents. Details are as follows (Table 3).

Table 3 The weight of each indicator in the index system

Primary indicators	Secondary indicators	Weight
Staff factors	Work skill level x_1	0.1587
Staff factors	Regulatory violations x_2	0.4762
Management factors	Unplanned work x_3	0.0071
Management factors	Drawing accuracy x_4 (%)	0.0141
Management factors	Manager arrival rate x_5 (%)	0.2893
Management factors	Work load level x_6 (%)	0.0071
Tool factors	Safety appliance rate x_7 (%)	0.0190
Tool factors	Complete rate of signs x_8 (%)	0.0127
Environmental factors	Working hours in bad weather x_9 (h)	0.0053
Environmental factors	Working hours at high and low temperatures x_{10} (h)	0.0053
Environmental factors	Working hours at night x_{11} (h)	0.0053

4.2 Evaluation Process

As there are three grey classes ($k = 1, 2, 3$), the whitening weight function of each grey class should be selected according to the actual situation.

In this paper, the indicators $x_1, x_4, x_5, x_6, x_7, x_8$ use the upper limit whitening weight function for the grey class "excellent" and the thresholds are, respectively, 95, 95, 95, 95, 95, 95 while the indicators $x_2, x_3, x_9, x_{10}, x_{11}$ select the lower limit whitening weight function for the grey class "excellent" whose thresholds are 60, 10, 10, 40 and 10, respectively. Similarly, for the grey class "low", the indicators $x_2, x_3, x_9, x_{10}, x_{11}$ use the upper limit whitening weight function and the thresholds are, respectively, 80, 20, 20, 80, 20 while the indicators $x_1, x_4, x_5, x_6, x_7, x_8$ select the lower limit whitening weight function whose thresholds are 80, 80, 80, 80, 80, 80. At last, all the indicators use the moderate measure whitening weight function for the grey "middle", and the thresholds are 88, 70, 15, 88, 88, 88, 88, 88, 15, 60, 15, respectively. The matrix of all the thresholds of these eleven indicators is

$$
\lambda_{jk} = \begin{bmatrix}
\lambda_{11} & \lambda_{12} & \lambda_{13} \\
\lambda_{21} & \lambda_{22} & \lambda_{23} \\
\lambda_{31} & \lambda_{32} & \lambda_{33} \\
\lambda_{41} & \lambda_{42} & \lambda_{43} \\
\lambda_{51} & \lambda_{52} & \lambda_{53} \\
\lambda_{61} & \lambda_{62} & \lambda_{63} \\
\lambda_{71} & \lambda_{72} & \lambda_{73} \\
\lambda_{81} & \lambda_{82} & \lambda_{83} \\
\lambda_{91} & \lambda_{92} & \lambda_{93} \\
\lambda_{101} & \lambda_{102} & \lambda_{103} \\
\lambda_{111} & \lambda_{112} & \lambda_{113}
\end{bmatrix} = \begin{bmatrix}
95 & 88 & 80 \\
60 & 70 & 80 \\
10 & 15 & 20 \\
95 & 88 & 80 \\
95 & 88 & 80 \\
95 & 88 & 80 \\
95 & 88 & 80 \\
95 & 88 & 80 \\
10 & 15 & 20 \\
40 & 60 & 80 \\
10 & 15 & 20
\end{bmatrix}
$$

In addition, each whitening weight function $f_j^k(x)$ is as follows.

I. The Whitening Weight Functions of Each Clustering Indicator on Grey Class "Excellent"

$$
f_1^1(x) = \begin{cases} 0 & x < 88 \\ \frac{x-88}{7} & 88 \leq x < 95 \\ 1 & x \geq 95 \end{cases}, \quad f_2^1(x) = \begin{cases} 0 & x \notin [0, 70] \\ 1 & 0 \leq x < 60 \\ \frac{x-70}{-10} & x \in [60, 70] \end{cases},
$$

$$
f_3^1(x) = \begin{cases} 0 & x \notin [0, 15] \\ 1 & 0 \leq x < 10 \\ \frac{x-15}{-5} & x \in [10, 15] \end{cases}, \quad f_4^1(x) = \begin{cases} 0 & x < 88 \\ \frac{x-88}{7} & 88 \leq x < 95 \\ 1 & x \geq 95 \end{cases},
$$

$$
f_5^1(x) = \begin{cases} 0 & x < 88 \\ \frac{x-88}{7} & 88 \leq x < 95 \\ 1 & x \geq 95 \end{cases}, \quad f_6^1(x) = \begin{cases} 0 & x < 88 \\ \frac{x-88}{7} & 88 \leq x < 95 \\ 1 & x \geq 95 \end{cases},
$$

$$
f_7^1(x) = \begin{cases} 0 & x < 88 \\ \frac{x-88}{7} & 88 \leq x < 95 \\ 1 & x \geq 95 \end{cases}, \quad f_8^1(x) = \begin{cases} 0 & x < 88 \\ \frac{x-88}{7} & 88 \leq x < 95 \\ 1 & x \geq 95 \end{cases},
$$

$$f_9^1(x) = \begin{cases} 0 & x \notin [0, 15] \\ 1 & 0 \le x < 10 \\ \frac{x-15}{-5} & x \in [10, 15] \end{cases}, f_{10}^1(x) = \begin{cases} 0 & x \notin [0, 60] \\ 1 & 0 \le x < 40 \\ \frac{x-60}{-20} & x \in [40, 60] \end{cases},$$

$$f_{11}^1(x) = \begin{cases} 0 & x \notin [0, 15] \\ 1 & 0 \le x < 10 \\ \frac{x-15}{-5} & x \in [10, 15] \end{cases}$$

II. The Whitening Weight Functions of Each Clustering Indicator on Grey Class "Medium"

$$f_1^2(x) = \begin{cases} 0 & x \notin [80, 95] \\ \frac{x-80}{8} & 80 \le x < 88 \\ \frac{x-95}{-7} & x \in [88, 95] \end{cases}, f_2^2(x) = \begin{cases} 0 & x \notin [60, 80] \\ \frac{x-60}{10} & 60 \le x < 70 \\ \frac{x-80}{-10} & x \in [70, 80] \end{cases},$$

$$f_3^2(x) = \begin{cases} 0 & x \notin [10, 20] \\ \frac{x-10}{5} & 10 \le x < 15 \\ \frac{x-20}{-5} & x \in [15, 20] \end{cases}, f_4^2(x) = \begin{cases} 0 & x \notin [80, 95] \\ \frac{x-80}{8} & 80 \le x < 88 \\ \frac{x-95}{-7} & x \in [88, 95] \end{cases},$$

$$f_5^2(x) = \begin{cases} 0 & x \notin [80, 95] \\ \frac{x-80}{8} & 80 \le x < 88 \\ \frac{x-95}{-7} & x \in [88, 95] \end{cases}, f_6^2(x) = \begin{cases} 0 & x \notin [80, 95] \\ \frac{x-80}{8} & 80 \le x < 88 \\ \frac{x-95}{-7} & x \in [88, 95] \end{cases},$$

$$f_7^2(x) = \begin{cases} 0 & x \notin [80, 95] \\ \frac{x-80}{8} & 80 \le x < 88 \\ \frac{x-95}{-7} & x \in [88, 95] \end{cases}, f_8^2(x) = \begin{cases} 0 & x \notin [80, 95] \\ \frac{x-80}{8} & 80 \le x < 88 \\ \frac{x-95}{-7} & x \in [88, 95] \end{cases},$$

$$f_9^2(x) = \begin{cases} 0 & x \notin [10, 20] \\ \frac{x-10}{5} & 10 \le x < 15 \\ \frac{x-20}{-5} & x \in [15, 20] \end{cases}, f_{10}^2(x) = \begin{cases} 0 & x \notin [40, 80] \\ \frac{x-40}{20} & 40 \le x < 60 \\ \frac{x-80}{-20} & x \in [60, 80] \end{cases}, f_{11}^2(x) = \begin{cases} 0 & x \notin [10, 20] \\ \frac{x-10}{5} & 10 \le x < 15 \\ \frac{x-20}{-5} & x \in [15, 20] \end{cases}$$

III. The Whitening Weight Functions of Each Clustering Indicator on Grey Class "Low"

$$f_1^3(x) = \begin{cases} 0 & x \notin [0, 88] \\ 1 & 0 \le x < 80 \\ \frac{x-88}{-8} & x \in [80, 88] \end{cases}, f_2^3(x) = \begin{cases} 0 & x < 70 \\ \frac{x-70}{10} & 70 \le x < 80 \\ 1 & x \ge 80 \end{cases},$$

$$f_3^3(x) = \begin{cases} 0 & x < 15 \\ \frac{x-15}{5} & 15 \le x < 20 \\ 1 & x \ge 20 \end{cases}, f_4^3(x) = \begin{cases} 0 & x \notin [0, 88] \\ 1 & 0 \le x < 80 \\ \frac{x-88}{-8} & x \in [80, 88] \end{cases},$$

$$f_5^3(x) = \begin{cases} 0 & x \notin [0, 88] \\ 1 & 0 \le x < 80 \\ \frac{x-88}{-8} & x \in [80, 88] \end{cases}, f_6^3(x) = \begin{cases} 0 & x \notin [0, 88] \\ 1 & 0 \le x < 80 \\ \frac{x-88}{-8} & x \in [80, 88] \end{cases},$$

$$f_7^3(x) = \begin{cases} 0 & x \notin [0, 88] \\ 1 & 0 \le x < 80 \\ \frac{x-88}{-8} & x \in [80, 88] \end{cases}, f_8^3(x) = \begin{cases} 0 & x \notin [0, 88] \\ 1 & 0 \le x < 80 \\ \frac{x-88}{-8} & x \in [80, 88] \end{cases},$$

$$f_9^3(x) = \begin{cases} 0 & x < 15 \\ \frac{x-15}{5} & 15 \le x < 20 \\ 1 & x \ge 20 \end{cases}, f_{10}^3(x) = \begin{cases} 0 & x < 60 \\ \frac{x-60}{20} & 60 \le x < 80 \\ 1 & x \ge 80 \end{cases}, f_{11}^3(x) = \begin{cases} 0 & x < 15 \\ \frac{x-15}{5} & 15 \le x < 20 \\ 1 & x \ge 20 \end{cases}$$

At last, the final step is to calculate the clustering coefficient of the ith clustering object for the kth grey class according to Formula (6) mentioned above. Therefore, the matrix of clustering coefficients is as below.

$$\sigma = \begin{bmatrix} \sigma_{11} & \sigma_{12} & \sigma_{13} \\ \sigma_{21} & \sigma_{22} & \sigma_{23} \\ \sigma_{31} & \sigma_{32} & \sigma_{33} \end{bmatrix} = \begin{bmatrix} 0.1198 & 0.2216 & 0.6586 \\ 0.2063 & 0.3889 & 0.4047 \\ 0.8843 & 0.1157 & 0.0000 \end{bmatrix}$$

As is apparently displayed in the result, for the first and the second row, respectively, their third elements are both the largest indicating the safety capabilities of this electric power company are "low" in 2016 and 2017 while its safety capability belongs to "excellent" in 2018 since the first element is the largest within the third row. As time passes by, the safety capability of the electric power company is ultimately improved.

5 Conclusions and Managerial Implications

In common sense, establishing an index system is always the first choice when faced with the evaluation. As for the electric power industry, it is also a necessity to make an evaluation of electric power companies' safety capabilities to understand its current safety level, along with to avoid serious safety accidents since electric power has been playing a much more important role in economic activities and national daily lives with the rapid development of power grid. Considering the dilemma of lack of data and incomplete information in the field of electric power safety, it is a more suitable way to evaluate electric power companies' safety capabilities based on Grey Fixed Weight Clustering Theory rather than conventional approaches such as Analytic Hierarchy Process (AHP) or Rough Set Theory. RST before which a specific evaluation index system is ought to be proposed.

The contributions of this paper are to propose an actual evaluation index system of electric power companies' safety capabilities after listening to experts of State

Grid Wuxi Power Supply Company. In this evaluation index system, four assessment dimensions—staff factors, management factors, tool factors and environmental factors are included. Besides, unlike many related researches which similarly apply conventional evaluation methods, such as Analytic Hierarchy Process (AHP) and Fuzzy Theory (FT), in electric power industry, Grey Fixed Weight Clustering Model fits the available data and evaluation requirements quite well via which electric power companies' present safety levels could be clearly categorized and correctly understood.

In view of the final evaluation results, the empirical electric power company's safety capabilities have been gradually improved from 2016 to 2018 according to which there are two main managerial implications left for reference. The first one is that based on different weights of different clustering indicators, staff factors and management factors are much more significant than any other factor which could affect the safety capability of an electric power company, thus guiding managers to focus much more on human factors, no matter subordinates or superiors. Last but not least, as for improving one's safety capabilities to the utmost extent with the least amount of time, efforts and funds, two issues about how to decrease regulatory violations and how to ensure manager arrival rate happen to be the most worth taking actions to dispose of.

In a word, this paper concentrates on the evaluation index system of electric power companies' safety capabilities and certainly, a number of works could be extended based on this research to make much more comprehensive assessments within the field of electric power. Future efforts of this research contain proposing a much more complete, mature evaluation index system with further development of some practicable dynamic evaluation models which could better describe the changes of electric power companies' safety capabilities by way of some safety prevention and control countermeasures.

References

1. J.D. Koustellis, S.D. Anagnostatos, Contact of heavy vehicles with overhead power lines. Saf. Sci. **49**(6), 951–955 (2011)
2. Y.B. Li, Y. Han, Research on impact model of meteorological factors on the power accidents. Power Syst. Technol. **37**(6), 1683–1687 (2013)
3. J.D. Koustellis, C.D. Halevidis, Analysis of a fatal electrical injury due to improper switch operation. Saf. Sci. **53**, 226–232 (2013)
4. A. Williamson, A.M. Feyer, The causes of electrical fatalities at work. J. Saf. Res. **29**(3), 187–196 (1998)
5. C.Y. Qu, J. Liu, Safety situation assessment model of power enterprise based on human factor. Adv. Technol. Electr. Eng. Energy **33**(10), 76–80 (2014)
6. J.C. Cawley, G.T. Homce, Occupational electrical injuries in the United States, 1992–1998, and recommendations for safety research. J. Saf. Res. **34**(3), 241–248 (2003)
7. R. Liu, Problems and solutions in electric power production safety management. Coal Technol. **32**(11), 88–89 (2013)
8. Y.L. Li, Power grid security management based on risk theory. Sci. Technol Progress Policy **27**(18), 96–99 (2010)

9. R. Rajesh, V. Ravi, Supplier selection in resilient supply chains: a grey relational analysis approach. J. Clean. Prod. **86**, 343–359 (2015)
10. Y.G. Dang, S.F. Liu, *Optimization of Regional Industrial Structures and Applications* (CRC Press, Boca Raton, US, 2010)
11. C. Hamzacebi, A.E. Huseyin, Forecasting the annual electricity consumption of turkey using an optimized grey model. Energy **70**, 165–171 (2014)
12. X.F. Tian, F. Chen, Y.L. Chen, Grey relational analysis on the influencing factors of China's comprehensive grain production capacity, in *1st International Conference on Industrial Economics and Industrial Safety* (Springer, Heidelberg, 2014), pp. 387–391
13. P.H. Hsin, C. Chen, The evolution of grey forecasting and its application on stock price prediction. Stock Forex Trading **3**(3), 1–2 (2014)
14. H.G. Xie, S.K. Wang, Qualified kidney biomarkers and their potential significance in drug safety evaluation and prediction. Pharmacol. Ther. **137**(1), 100–107 (2013)
15. Y.Y. Zhao, Y.S. Pei, Risk evaluation of groundwater pollution by pesticides in China: a short review. Procedia Environ. Sci. **13**, 1739–1747 (2012)
16. J.F. Fu, G.Y. Zhuang, Conceptual identification and evaluation index system for low carbon economy. China Popul. Resour. Environ. **20**(8), 38–43 (2010)
17. Y.D. Li, *The Construction and Empirical Study of the Performance Assessment and Evaluation System of the Leading Group in China's Insurance Industry: A Case Study of T Company* (Shandong University, Shandong, China, 2018)
18. L. Chong, K.J. Chen, An integrated framework for effective safety management evaluation: application of an improved grey clustering measurement. Exp. Syst. Appl. **42**(13), 5541–5553 (2015)

Analysis of Crude Oil Price Fluctuation and Transition Characteristics at Different Timescales Based on Complex Networks

Jiao Yan and Jing Ma

Abstract Based on the theory of complex network, the crude price data for 30 years is processed by coarse-graining method and the price fluctuation network of crude oil is constructed. According to the price of crude oil at different timescales, this paper establishes the different complex network models to compare the topological properties of crude oil price fluctuation network and explores the internal characteristics of the original oil price fluctuation system. Meanwhile, using the K-core, the obtained fluctuation loops at different timescales are analyzed to achieve certain prediction effect. As is shown in the study, small-world phenomenon and scale-free characteristics exist in the crude oil price fluctuation network at different timescales. However, with the increase of timescale, the small-world phenomenon of the crude oil price fluctuation network is enhanced, while the scale-free properties are weakened. Ultimately, this paper summarizes the differences and relations between the evolution of crude oil price fluctuation under different scales, and the next research direction is put forward.

Keywords Crude oil price · Complex network · Timescales · Topological property

1 Introduction

The energy issue has always been a hot topic of global concern, and crude oil, as a scarce, nonrenewable, but indispensable resource, its trend has naturally been taken seriously by people. However, the fluctuation of crude oil prices is not only related to its supply and demand, but also affected by the economic and noneconomic factors, such as the global economic situation, currency exchange rate, war, and geopolitics [1]. Thus, the fluctuation of crude oil price is not a simple linear change, but a complex and nonlinear one [2].

J. Yan · J. Ma (✉)
College of Economics and Management, Nanjing University of Aeronautics and Astronautics, Nanjing 211106, China
e-mail: majing5525@126.com

© Springer Nature Switzerland AG 2020   41
H. Yang et al. (eds.), *Smart Service Systems, Operations Management, and Analytics*, Springer Proceedings in Business and Economics, https://doi.org/10.1007/978-3-030-30967-1_5

In predicting the crude oil price, most scholars indicated that the uncertainty of forecasting was increased due to the complexity of the factors affecting crude oil prices. Therefore, Cheng et al. [3] combined vector error correction (VEC) and nonlinear autoregressive neural network (NAR) models to predict crude oil price and indicated that their prediction effects were better than traditional prediction models, but they did not take geopolitical factors into consideration. Safari and Davallou [4] use PHM model which was combined by exponential smoothing model (ESM), autoregressive integrated moving average (ARIMA), and NAR and the prediction error will be significantly reduced. Wang et al. [5] visualized the crude oil price data from the perspective of data fluctuation network to predict the price trend, but did not elaborate on the topological nature of crude oil prices.

For the characteristics of price fluctuations, the study found that the sequence of crude oil price return has the characteristics of "spike and thick tail" which showed a non-normal distribution [6] Meanwhile, considering the financial nature of crude oil prices, its futures prices will inevitably affect the spot price. With the development of network science, An et al. [7] researched on the relationship between crude oil futures and spot prices by introducing complex network theory and using the topological properties of complex network models to explore the complex relationships. However, the paper only described the direction of fluctuation between them and did not study the fluctuation characteristics of crude oil prices separately. Chen et al. [8] confirmed the complexity of crude oil price fluctuations from the topological properties of the network model, concluded that the crude oil price fluctuation network, which belonged to the small-world network and scale-free network had the large clustering coefficient and the short average path length. However, the fluctuating range of crude oil price and the relationship between it and external factors had not been discussed.

In summary, with the improvement of the methods, the research on crude oil price fluctuations is more in-depth and thorough. Comparing various methods, we find that the complex network can better reveal the essential characteristics of the system through its complicated structure and dynamic characteristics [9]. The existing literature has summarized the characteristics of crude oil price fluctuations such as sustainability and leverage. However, although the literature has built the complex network model to analyze crude oil price fluctuations, it hasn't analyzed the price fluctuations under different periods for comparative analysis in detail.

The structure of this paper is as follows. In Sect. 2, we introduce the network model of crude oil price fluctuation at different timescales and the methods in detail. In Sect. 3, we analyze and discuss the topological properties of the constructed crude oil price fluctuation network, then we predict the future crude oil price fluctuation to some extent. Finally, we summarize the research results of this paper in Sect. 4.

2 Network Models of Crude Oil Price Fluctuation at Different Timescales

2.1 Data Processing

Global crude oil trading is mainly based on West Texas Intermediate (WTI) and North Sea Brent (Brent) as the two benchmarks. Considering that Brent crude oil as the benchmark oil accounting for 65% of global crude oil trading volume [10], we choose Brent crude oil spot price from April 5, 1987 to January 1, 2019 as the sample data. The data comes from the US Energy Information Administration (EIA). In order to explore the characteristics of the fluctuation of crude oil prices from different angles, we select daily price, weekly price (average of five consecutive days), and monthly price (average of monthly price) as the object to compare and analyze the fluctuation of crude oil price at different timescales.

Since there is no direct relationship between the daily spot price of crude oil, the change of crude oil price can't be reflected. We define the change of price to represent the volatility of each period of the crude oil price. The crude oil price fluctuations series is denoted as

$$\Delta P(t) = P(t) - P(t - 1) \tag{1}$$

where $P(t)$ ($t = 1, 2, 3, ..., N$) is the current price and $P(t - 1)$ is the previous price.

Based on the crude oil price fluctuation sequence, we coarsen the price volatility sequence to highlight the essential characteristics of crude oil price volatility [11]. Then, we symbolize the data and transform the crude oil price fluctuation sequence into the symbol sequence S_t, and select four letters (G, g, d, D) for each fluctuation state. The sequence S_t is shown as follows:

$$S_t = \begin{cases} G, \ T_{pos} > T_{pos.int} \\ g, 0 < T_{pos} \leq T_{pos.int} \\ d, 0 \geq T_{neg} \geq T_{neg.int} \\ D, \ T_{neg} < T_{neg.int} \end{cases} \tag{2}$$

where the symbol sequence owns four kinds of symbol, including sharp rise (G), minor rise (g), minor decline (d), and sharp decline (D). Hence, the fluctuations of crude oil prices can be expressed by a continuous sequence of different symbols as the following $S_t = \{S_1, S_2, S_3, ...\}, (S_i \in (G, g, D, d))$.

2.2 Construction Method of Crude Oil Price Fluctuation Network at Different Timescales

According to the release time interval of Brent crude oil spot price, we select the five continual day's symbol as a modal unit. And we slide the data in steps of one day and use the conversion between strings as the edge. Since the former mode is the basis of the next mode, the complex network model is the directed network. Meanwhile, the conversion of the mode to the next same mode isn't single so that the edge of the network is given weight, which is, the network is a directed weighted network. If the former mode is the same as the next, we do nothing. In theory, $4^5 = 1024$ modes can be obtained from the symbol sequence, which is {GGGGG, GGGGg, GGGGd, GGGGD, ...}. In fact, we get 973 modes and it represents most of the theoretical modes and reflects the complexity of crude oil price fluctuations. Figure 1 shows the

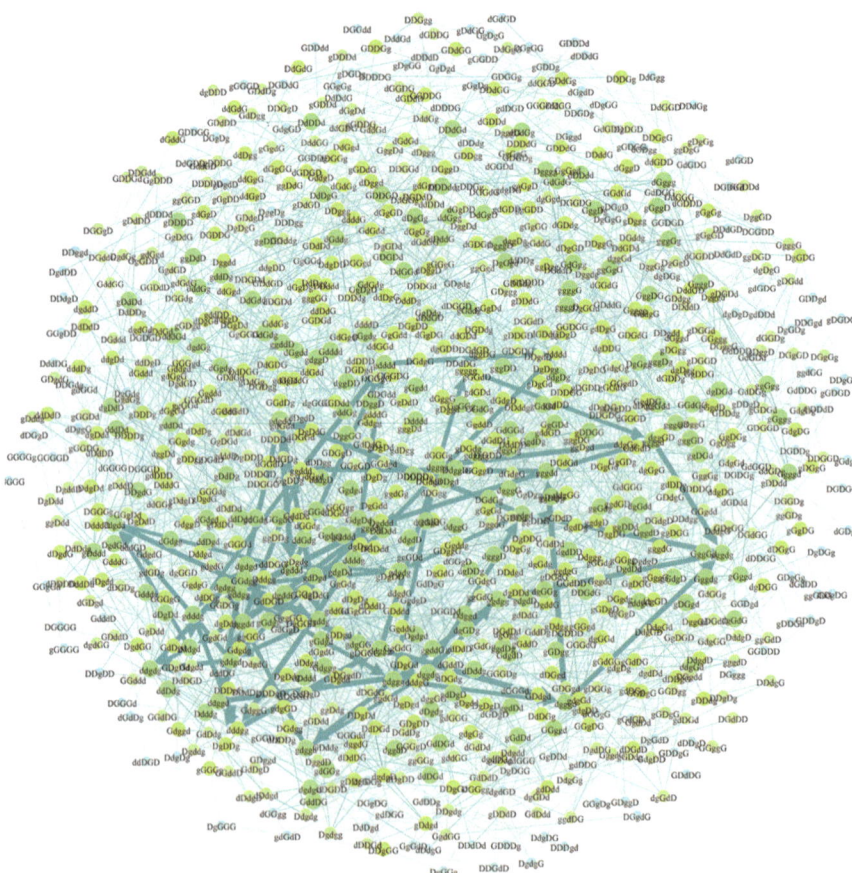

Fig. 1 The fluctuation network of crude oil daily price

established fluctuation network of daily price.

In addition, for the fluctuation network of crude oil weekly price, we select the four continual weekly price's symbol as a modal unit, which is, one node represents one month. Similarly, in theory, $4^4 = 256$ modes can be obtained from the symbol sequence and we acquire 229 modes. Meanwhile, for the fluctuation network of crude oil monthly price, we select the three continual monthly price's symbol as a modal unit and we obtain 62 modes.

3 The Dynamic Characteristics of the Crude Oil

This paper will analyze the crude oil price fluctuation network from four aspects: the node strength and the distribution of it, the shortest average path length, the clustering coefficient, and the betweenness centrality to explore the price fluctuation characteristics.

The strength distribution indicates the connection between the node and other nodes. In the actual network, the degree distribution can be described in the form of power law, which shows that the degree of most nodes is very low, while a small number of nodes own high degree. The average path length of the network means the average distance between any two nodes in the network. The shorter the distance is, the faster the crude oil price modes change. The clustering coefficient refers to the tightness of a node and its surrounding nodes. When the clustering coefficient of one node is large, it means the small cluster is formed around the node, and the nodes in it transform frequently. From the time period of the nodes with large clustering coefficient, we can judge whether these small clusters are periodic or not. This will be of great help to the prediction of crude oil price fluctuation. Finally, the betweenness centrality of the nodes will show us which nodes in the network are in the transition. The nodes with large betweenness centrality indicate the transition between different modes and play the great role in predicting the trend of crude oil price.

3.1 The Node Strength and the Distribution of the Node Strength

In the directed weighted network, the strength of the node is divided in-strength and out-strength. The out-strength (s_i^{out}) refers to the numbers of neighbor nodes pointed by node v_i, and the in-strength(s_i^{in}) is numbers of neighbor nodes that others point to node v_i. Then we get the strength calculation formula as follows [12]:

$$s_i^{in} = \sum_{j=1}^{N} a_{ji} w_{ji}, \ s_i^{out} = \sum_{j=1}^{N} a_{ij} w_{ij} \tag{3}$$

where a_{ij} is the element adjacency matrix, if node v_i has an edge pointing to v_j, then $a_{ij} = 1$, otherwise $a_{ij} = 0$. w_{ij} is the weight of the edge between node v_i and node v_j, and N is the total number of nodes in the network.

The connection rules we choose make the out-strength and in-strength of each node the same except for the first and last nodes. Thus, we select the out-strength of the node for analysis. In order to analyze the distribution of node strength quantitatively, we draw the double logarithmic image as shown in Fig. 2a–c. By using the least square method, we obtain the regression equations of the logarithmic distribution of strength in the daily crude oil price network, the weekly price network, and the monthly crude oil price network, which are $y = -1.084x - 1.679$, $y = -1.046x - 1.254$, and $y = -0.898x - 1.3$, where x is the logarithm of strength and y is the logarithm of strength distribution at different timescales. The trend line regression coefficients R^2 are 0.7879, 0.8983, and 0.7201, respectively. It indicates that they all

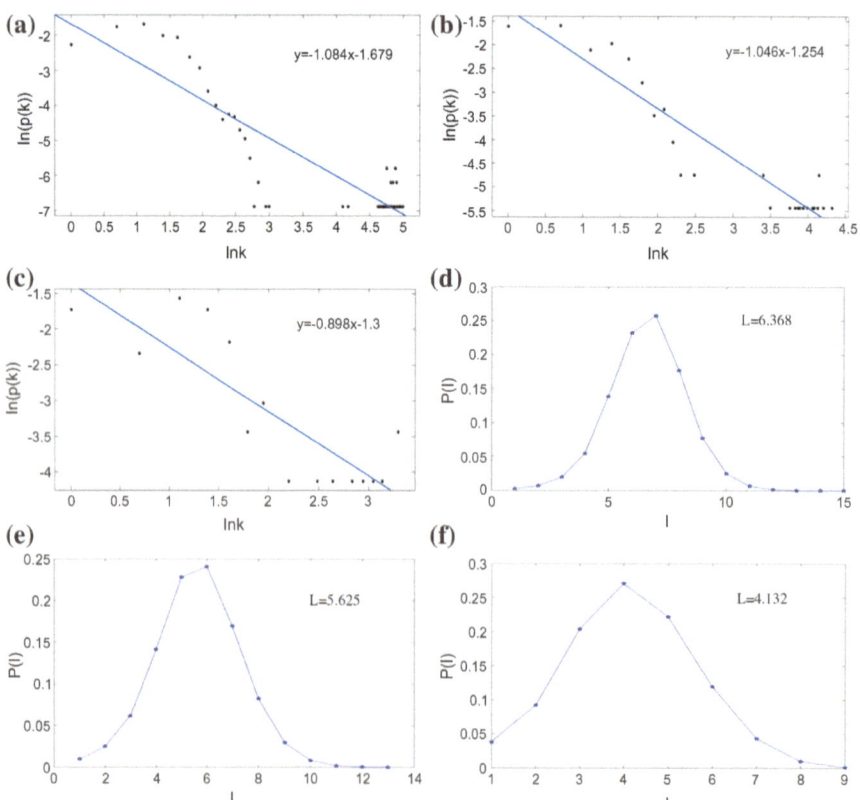

Fig. 2 The double logarithmic distribution of the strength of nodes in the daily price network (**a**); the double logarithmic distribution of the strength in the weekly price network (**b**); the double logarithmic distribution of the strength in the monthly price network (**c**); the distribution of path length in crude oil daily price network (**d**); the distribution of path length in crude oil weekly price network (**e**); the distribution of path length in crude oil monthly price network (**f**)

obey the power law distribution and have scale-free characteristics [13]. However, as the timescale increases, the power law index decreases gradually. It means that the scale-freeness features that most of nodes strength are small while a few nodes are higher strength are no longer obvious.

3.2 The Shortest Average Path Length

We define d_{ij} as the distance which is the shortest path between node v_i and v_j. And let the average path length L be the average of the distance between any two nodes in the network. The formula is as follows:

$$L = \frac{1}{\frac{1}{2}N(N-1)} \sum_{1 \leq i \leq j \leq N} d_{ij} \tag{4}$$

where N is the number of nodes in the network, and the formula doesn't include the distance from itself to itself.

In the network, the weight of the edge is composed of the number of transitions from mode to mode. Based on that, we use the Floyd algorithm [14] to obtain the distance between any two nodes in the fluctuating network at different timescales, respectively. The calculation results of the path length distribution at different timescales are shown in Fig. 2d–f. It indicates that the daily price, the weekly price, and the monthly price fluctuation of crude oil can be completely converted from one mode to another by 6–7, 5–6, and 4–5 modes, respectively. At the same time, the distance between the nodes of the price fluctuation network at different timescales is mostly less than 10, which accounts for 99.66%, 99.7%, and 98%, respectively. It means that the larger the timescale is, the shorter the average path length of the network and the more frequent the conversion between the fluctuation modes will be.

3.3 The Weighted Clustering Coefficient

The clustering coefficient can measure the level of aggregation among nodes in the network and reflect the local characteristics of the network. For directed weighted networks, the weight of the edges has an effect on the clustering coefficients. Barrat, Onnela, and Holme redefined clustering coefficient, respectively. We synthesize the applicable scenarios of three definitions, and select the following calculation formula defined by Holme [15]:

$$C_i = \frac{\sum_{j,k} w_{ij} w_{jk} w_{ki}}{\max_j (w_{ij}) \sum_{j,k} w_{ij} w_{ki}} \tag{5}$$

where w_{ij}, w_{ki}, and w_{jk} is the weights of the edges between the node v_i and the node v_j, the node v_i and the node v_k, and the node v_j and the node v_k, respectively. The definition assumes that if there is no edge connection between the two nodes, then the weight of edge between the nodes is zero. In addition, the clustering coefficient of the entire network is equal to the average of the clustering coefficients of all nodes.

According to the definition given by Holme, we calculate the weighted clustering coefficients of the networks at different timescales respectively, which are 0.0116, 0.0410, and 0.0875. In the actual large-scale network, the clustering coefficient is much smaller than 0 and far greater than $O(N^{-1})$ [16], which indicates that the crude oil price fluctuation network has the clustering effect. From the clustering coefficient of the network, it can be seen that as the time span increases, the nodes in the network are more closely connected. For the proportion of nodes whose clustering coefficients aren't 0, the daily price fluctuation network is 3.51%, the weekly price fluctuation network is 17.18%, and the monthly price fluctuation network is 52.46%, respectively. It further confirms that the larger the timescale is, the closer the network nodes are. For the single node, the degree of closeness between each node and the neighboring node decreases with the increase of the timescale. In general, there are differences between the macro and micro networks of crude oil price fluctuation.

3.4 Analysis of the Transition Between the Fluctuation Modes

The strength of the nodes can measure the importance of the nodes in the whole network, however, some nodes with small strength may be the transit stations of the two communities. If they are removed, the transitions and connections between the two communities will be interrupted. Therefore, we introduce betweenness centrality to measure the hub level of nodes in the network. The formula of the betweenness centrality ($C_B(v_i)$) is as follows:

$$C_B(v_i) = \frac{B_i}{[(N-1)(N-2)]}, \quad B_i = \sum_{\substack{1 \le j \le l \le N \\ j \ne i \ne l}} \frac{n_{jl}(i)}{n_{jl}} \tag{6}$$

where n_{jl} is the number of shortest paths between node v_j and v_l, $n_{jl}(i)$ is the number of paths between node v_j and v_l passing node v_i. We normalize the betweenness to obtain the betweenness centrality and its formula is as follows:

We calculated that the nodes with the largest betweenness centrality in the crude oil price fluctuation network at different timescales are dgggG, gGGD and gGG, and their strengths are 12, 8, and 4. The betweenness centrality of them is high betweenness centrality and the strength is low . Therefore, although these nodes do not appear frequently in the price fluctuation sequence, they bear the important role of the transfer station. The relationship between betweenness centrality and strength of the nodes will be given in the following section.

3.5 The Connection and Difference of the Dynamic Characteristics

In order to explore the difference and connection of the dynamic characteristics of the crude oil price network at different timescales, we, respectively, selected the 20 nodes with the strongest strength, the 20 nodes with the largest clustering coefficient, and the 20 nodes with the highest betweenness centrality from the fluctuation networks and put them into the same three-dimensional coordinate system, as shown in Fig. 3a.

In Fig. 3a, the betweenness centrality of the crude oil monthly price fluctuation network is at the highest level, and the betweenness centralities of most other nodes are also much larger than that of the other two networks. The clustering coefficient and strength of the monthly price network is at a low level and more concentrated. It indicates that most of the modes are in transition period. The appearance of any node in the network may herald the arrival of fluctuating modes in the next period. At the same time, the clustering coefficient of the crude oil monthly price network is large, and the clustering coefficient of the single node modal is small. To some extent, it shows that the nodes of crude oil monthly price fluctuation network are of equal importance and each node is indispensable. Although the betweenness centrality of daily price fluctuation network is the smallest among the network at these timescales, the node strength and the clustering coefficient are both large. From the micro perspective, the nodes with nonzero clustering coefficient in the daily price network are closely related to their neighbors. For the crude oil weekly price network, its node strength, clustering coefficient and betweenness centrality are all at the medium level.

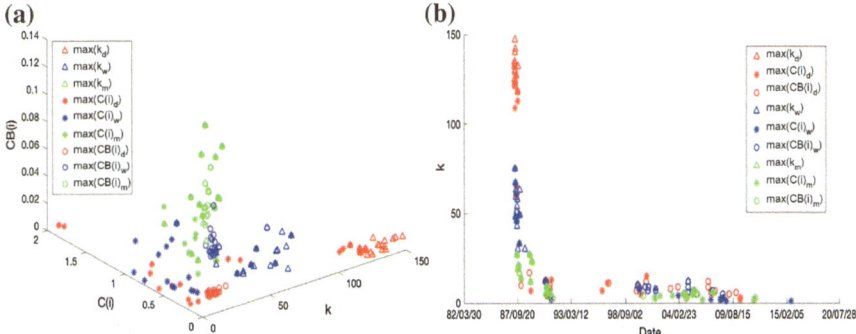

Fig. 3 The dynamic characteristics of core nodes of crude oil price network at different timescales (**a**) and the time distribution of the core nodes first appeared of crude oil price network (**b**)

In addition, in the crude oil price network at different timescales, the node strength, clustering coefficient, and betweenness centrality of some core nodes are all large. It demonstrates that these nodes occupy the quite important position in the network, and with the enlargement of timescale, the number of these core nodes also increases. Among them, there are 5 such core nodes in the daily price network, 12 such core nodes in the weekly price network, and 27 such core nodes in the monthly price network, which indicates that the larger the timescale of the mode, the more important each node in the network is. The time when the core nodes of the price fluctuation network at different timescales first appear in the price series is shown in Fig. 3b. As shown in the figure, regardless of the length of the time period represented by the modes, the nodes which appears in the same time period play similarly important roles in their respective networks.

3.6   Analysis of Fluctuation Modes Transition Characteristics at Different Timescales

According to the above analysis of topological properties of the core modes in the crude oil price fluctuation network at different timescales, we find that the importance of different modes in the propagation process is different. In order to predict the crude oil price to some extent, we introduce the concept of k-core to research the transition characteristics of modes in the transition process.

After filtering the crude oil daily price network, we keep edges whose weights are greater than 9. The final 32 nodes and 62 edges remaining are shown in Fig. 4. In the remaining network, we use the closed-loop generation algorithm of the directed graph to find that nodes and edges form a total of 15 fluctuation loops whose longest propagation path includes 20 modes, while the shortest one only contains 3 modes. Then we separately filter the weekly price network and the monthly price network, and we retain the edges whose weights are over the average strength. In the residual network of weekly prices, we obtain 16 nodes and 28 edges with 7 closed loops in the network, of which the longest path includes 9 modes while the shortest one is 3 modes. In the monthly price residual network, we finally get 8 nodes and 14 edges forming 3 fluctuation closed loops. The longest path includes 7 modes and the shortest includes 3 modes. They all exhibit an evident periodicity of properties. All the nodes in these remaining networks only contain g and d fluctuation symbols, which means that in the transition process, the fluctuation of crude oil price is mainly dominated by small fluctuations.

In the main transition path of crude oil price, from the perspective of the single mode, most will only experience the same fluctuation twice in a row at most and then be converted into opposite fluctuation, that is, the duration of the price fluctuation is short, and the larger the timescale is, the weaker the duration will be. In addition, with the scale-up of node time, the longest propagation path of crude oil price changes from 20 to 9 and then to 7, which indicates that the longer the period represented by

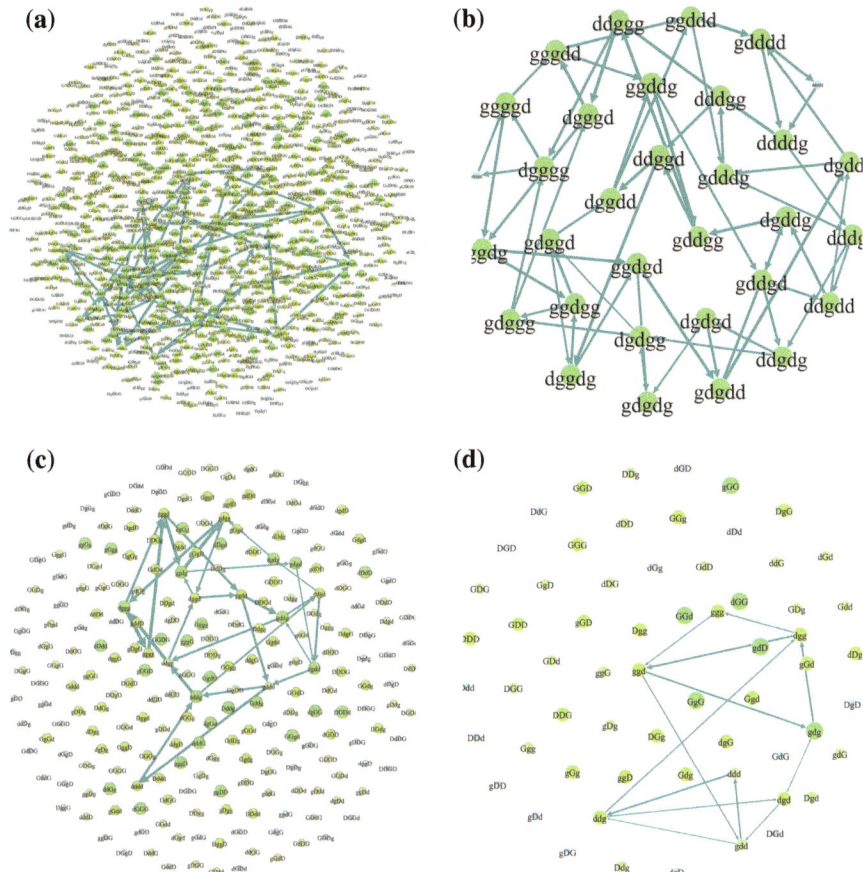

Fig. 4 The fluctuation transition network of crude oil price at different timescales

the mode is, the more concentrated the price transition is. In transition network, in most cases, if a mode has emerged, then there are only two possibilities for its next modal at most, such as the next modes of "gdggd" are only "dggdd" and "dggdg", and the next mode of "dgd" is "gdd". As the timescale of the mode is amplified, the next mode of the mode is more certain. It lays the foundation for us to predict the fluctuations of crude oil prices in the next week, next month, and even next quarter.

4 Conclusions

In this paper, we select the spot price of Brent crude oil from 1987 to 2019 issued by EIA as sample data, and we coarse grain the original price data into the crude oil price fluctuation symbol sequence composed of characters $\{G, G, D, D\}$. Using the symbol sequence, we construct the complex network model to research the characteristics of crude oil price fluctuation. Meanwhile, in order to explore the differences between crude oil price fluctuations in the long-term and short-term, we construct the crude oil price fluctuation networks at different timescales. By analyzing the topological properties of fluctuation networks, we have mainly come to some conclusions:

(1) The crude oil price networks at different timescales all belong to the scale-free network and the small-world network, but their characteristics are different. The larger the timescale is, the shorter the average path length of the crude oil price fluctuation network is, the larger the weighted clustering coefficient is and the stronger the small-world effect of the network is.

(2) There are differences between the macro and micro in the networks of crude oil price fluctuation. From the perspective of the whole network, with the expansion of the timescale, the fluctuation networks have the higher level of closeness. Nevertheless, for the single node, the larger the timescale is, the less the connection between the nodes will be. It indicates that the role of nodes is homogenized when the network is at large timescale.

(3) No matter how the timescale changes, the fluctuation modes appearing in the same time period occupy a similar position in different networks. The first occurrence time of core nodes in different networks has similar distribution characteristics. They are characterized by the fact that the core nodes with large strength have appeared in the early stage. Meanwhile, some core nodes with large clustering coefficient appear early, while others are scattered in time series. And the core nodes with large betweenness centrality appear randomly in the whole time series.

(4) The transition process of crude oil price has the characteristic of periodicity and the transition path of the core node can predict the crude oil price to some extent. For example, for nodes with prominent topological properties, we can give them greater weight in prediction. In the residual network, each closed loop contains at most nine modes which are mainly small fluctuations. In addition, as the timescale of the mode is amplified, the price transition is more concentrated and the next mode of the mode is more certain. In fact, when predicting crude oil prices, we can use Markov model to obtain possible modal groups in the future by combining the core modal and its weights.

Admittedly, there are still some deficiencies in the work of this paper. First, there are limitations in data selection. The price data, which is issued by EIA only covers three decades and is in lack of earlier data. Second, we take some subjective ideas in the process of coarse-grained data. In the future research, we consider combining theory with practice to examine into the essential causes of crude oil price fluctuation.

References

1. J. Chai, J.E. Guo, L. Meng et al., Exploring the core factors and its dynamic effects on oil price: An application on path analysis and BVAR-TVP model. Energy Policy **39**(12), 8022–8036 (2011)
2. J. Chai, L.M. Xing, X.Y. Zhou et al., Forecasting the WTI crude oil price by a hybrid-refined method. Energy Econ. **71**, 114–127 (2018)
3. F.Z. Cheng, T. Li, Y. M. Wei et al., The VEC-NAR model for short-term forecasting of oil prices. Energy Econ. (2018)
4. A. Safari, M. Davallou, Oil price forecasting using a hybrid model. Energy **148**, 49–58 (2018)
5. M.G. Wang, L.X. Tian, P. Zhou, A novel approach for oil price forecasting based on data fluctuation network. Energy Econ. **71**, 201–211 (2018)
6. K. Kanjilal, S. Ghosh, Dynamics of crude oil and gold price post 2008 global financial crisis— new evidence from threshold vector error-correction model. Resour. Policy **52**, 358–365 (2017)
7. H.Z. An, X.Y. Gao, W. Fang et al., Research on patterns in the fluctuation of the co-movement between crude oil futures and spot prices: a complex network approach. Appl. Energy **136**, 1067–1075 (2014)
8. W.D. Chen, H. Xu, Q. Guo, Dynamic analysis on the topological properties of the complex network of international oil prices. Acta Physica Sinica **59**(7), 4514–4523 (2010)
9. A. Chacoma, G. Mato, M.N. Kuperman, Dynamical and topological aspects of consensus formation in complex networks. Phys. A: Stat. Mech. Appl. **495**, 152–161 (2018)
10. L.T. Zhao, J.L. Yan, L. Cheng et al., Empirical study of the functional changes in price discovery in the Brent crude oil market. Energy Procedia **142**, 2917–2922 (2017)
11. X.L. Jia, H.Z. An, X.Q. Sun et al., Evolution of world crude oil market integration and diversi-fication: a wavelet-based complex network perspective. Appl. Energy **185**, 1788–1798 (2017)
12. M.G. Wang, Y. Chen, L.X. Tian et al., Fluctuation behavior analysis of international crude oil and gasoline price based on complex network perspective. Appl. Energy **175**, 109–127 (2016)
13. W. Yang, Y.Q. Hu, Z.R. Di et al., The effect of hub nodes on the community structure in scale-free networks. Phys. A: Stat. Mech. Appl. **390**(21–22), 4027–4033 (2011)
14. L. Ying, D. Li, Y. Yu, Multi-point shortest path in the complex road network based on floyd algorithm, in *International Conference on Information Computing and Applications* (Berlin, Germany, 2013), pp. 364–372
15. P. Holme, Scale-free networks with a large- to hypersmall-world transition. Phys. A: Stat. Mech. Appl. **377**(1), 315–322 (2006)
16. A.P. Kartun-Giles, G. Bianconi, Beyond the clustering coefficient: a topological analysis of node neighbourhoods in complex networks. Chaos, Solitons Fractals **X** (2019)

Understanding of Servicification Trends in China Through Analysis of Interindustry Network Structure

Yunhan Liu and Dohoon Kim

Abstract This study analyses the core changes in the entire industrial structure of China from 2002 to 2015. First, utilizing the input–output model (or interindustry relation model), we evaluate the production-inducing effects across the various manufacturing and service sectors. Employing standard centrality measures that are often used in network analysis, we identify the status of various industry sectors together with their structural roles on the input–output network and track the evolutionary paths as they change over time. Furthermore, the clustering analysis of industrial sectors also examines critical changes in the overall industrial structure. We will discuss the policy implications of these evolutionary paths and explore the trends and sources of Chinese economic development regarding manufacturing and service productions. As a result, it can be seen that the production-inducing effects in the manufacturing sectors—particularly, chemical, electric power, and primary metal products—play the core role in driving the economy. On the other hand, the distribution sector like the traditional wholesale and retail services still shows unstable connections with other industries. However, the sectors like finance and transportation, which constitute another axis of the service industry, have strengthened their connection with the manufacturing sectors since 2012. Moreover, the real estate and lease and the other business support services continue to absorb production resources, deepen their linkages with the key service industries (e.g., transportation), and the various manufacturing sectors. In conclusion, China still has manufacturing as a central driving force for economic development, but with the increasing integration with the service sectors, both manufacturing and service industries show an apparent convergence in their production activities.

Keywords Input–Output model · Network analysis · Interindustry network · Servicification · Chinese economy

Y. Liu · D. Kim (✉)
School of Management, Kyung Hee University, Seoul 02447, South Korea
e-mail: dyohaan@khu.ac.kr

© Springer Nature Switzerland AG 2020
H. Yang et al. (eds.), *Smart Service Systems, Operations Management, and Analytics*, Springer Proceedings in Business and Economics,
https://doi.org/10.1007/978-3-030-30967-1_6

1   Introduction

The economy of China is gradually slowing down after experiencing rapid growth. The manufacturing industry faces many challenges. For example, the original low-cost advantage has been weakened, resources and environmental constraints have been tightened, and labor costs also have risen. At the same time, the service industry has recently played an important role in gross domestic product (GDP) and employment. Regarding the link between manufacturing and service industries, Krugman [1] pointed out that non-single industry, related industries will continue to concentrate and cooperate to improve the economic productivity and competitive advantage of the country or region. In addition, manufacturers do not need to increase their investment to achieve economies of scale, as well as structural optimization and transformation, but to increase industrial efficiency by establishing interconnections between various production activities [2]. For example, with the deepening of the division of labor (service outsourcing), the proportion of producer service industries (Producer Services), which is an intermediate input factor in manufacturing, has increased. The dependence between the manufacturing industry and the service industry will further deepen.

The purpose of this paper is to clarify the basic characteristics of the relationships between China manufacturing and service industries, and to explore the overall industrial structure formed by these relationships. Based on the input–output model, this study builds the IO networks and uses social network indicators to analyze. We track the degree distribution between the various industry sectors, the direction of action, and the status of each sector across the network. This will give a more detailed understanding of the role of China industry sectors in economic production activities (input and output processes).

2   Theoretical Background and Methodology

2.1   Basic Input–Output Model

Interindustry analysis is a common method of exploring industrial relevance, especially in the interaction between manufacturing and service industries. Kwak et al. [3] used this method to study the correlation effects of national industries on the growth of Korean marine industry. Hauknes and Knell [4] analyzed the direct and indirect knowledge flows among different technology-intensive industries based on input–output data from France, Germany, Norway, Sweden, and the United States. Xing et al. [5] measured the integration of Chinese ICT sectors using the 2002 input–output table according to the classification in supply/demand side and supplement/substitution relationship. Through the input–output table, this research constructs the IO network by calculating the linear matrix. Then social network analysis

(SNA) method will be used to visualize the characteristics and structure of national industries [6].

In IO table, x_{ij} is the intermediate input of sector j from sector i, and X_j is total input of sector j. The input coefficient divides the intermediate input (raw materials required) of each sector into the total output. This means that sector j consumes the input from the sector i for each unit of product produced. The generated coefficient are called input coefficient (a_{ij}). And it is expressed by the following Eq. (1):

$$a_{ij} = \frac{x_{ij}}{X_j} \tag{1}$$

Since adding Final Demand to Intermediate Input is equal to Total Output, a relationship exist between the Input Coefficient Matrix A, the Final Demand Vector Y, and the Total Output Vector X. Expression (2) is as follows. Finally, the Leontief inverse matrix $(I - A)^{-1}$ can be obtained as

$$AX + Y = X \tag{2}$$

$$X = (I - A)^{-1}Y \tag{3}$$

Each element of the Leontief inverse matrix (L) shown in Eq. (3) is represented by the symbol l_{ij}. The implication is that the final demand of the sector j is increased by one unit, which leads directly or indirectly to the total output of l_{ij} in the sector i. It is defined as production-inducing coefficient (l_{ij}). According to the derivation of the Leontief inverse matrix, this study processes the data contained in the IO tables. That is, the direction and weight of the ties constituting the IO network are assigned numerical values based on the Leontief inverse matrix. In order to avoid the key features of connectivity that not sufficiently identify, meanwhile, it is necessary to further reveal the structural characteristics of the research object network. This paper only analyzed a certain degree of ties.

2.2 Defining the Input–Output Network Structure

As mentioned above, the purpose of this study is to build the IO network in both the adjacency matrix and the complete network using the basic data (the intermediate input). This research used the SNA method to analyze the national industrial structure by showing not only the input–output relation of each sector but also the connectivity of the overall industry. Recently, some studies have applied the SNA method to input–output analysis. Acemoglu et al. [7] employed the method of network analysis to conduct the economic shock analysis through the input–output structure. Based on this research, Cho and Kang [8] applied the IO model to the empirical study on Korean economic variability. In addition, Hidalgo et al. [9] and McNerney et al. [10]

adopted national input–output tables and applied network methods to describe the global production space and the characteristics of national interconnection.

In this paper, we interpret the related concepts as described in Table 1.

This study used the latest input–output table released by the National Bureau of Statistics in January 2017. The input–output tables for 2002, 2007, 2012, and 2015 were used as the analysis targets. By analyzing the relationships between 42

Table 1 Social network terminology and interpretation

Terms	Explanation	Examples of this study
Node	An actor in the network, such as individuals or organizations	An industry sector
Arc	A directed and weighted edge between a pair of nodes	Input–output relationship between sectors (Direction: the direction of the spillover effect; Weight: production-inducing coefficient)
Density	The ratio of the number of arcs actually connected to the number of possible arcs in the network	The ratio of the number of existing IO relationships to the total number of possible total relationships
Degree centrality	The number of edges each node has [11]	Number of neighboring industries directly connected to each industry – Indegree: $C_D^{IN}(i)$; Outdegree: $C_D^{OUT}(i)$ – Matrix form: Adjacency Matrix
Closeness centrality	Calculating the average geodesic distance that each node is from every other node in the network [12]	The ability of an industrial sector to receive production spillover effects or spread the effects to other sectors – Incloseness: $C_C^{IN}(i)$; Outcloseness: $C_C^{OUT}(i)$ – Matrix form: Adjacency Matrix
Betweenness centrality	The extent to each actor in a network lies on the shortest paths (i.e., geodesics) connecting all pairs of nodes in the network [12]	A sector that plays a medium or control role in the input–output flow – Marked as $C_B(i)$ – Matrix form: Adjacency Matrix
Minimum spanning tree	A network with three conditions that contain all nodes, no loops, and a minimum weight sum	The core structure of the industrial network with directionality and maximum production-inducing effects – Matrix form: Directed negative weight matrix

industrial sectors that constitute IO tables, the basic industrial structure (IO network structure) was established from 2002 to 2015. Using the intermediate input matrix of each annual table, their inverse matrix of Leontief was calculated. The ties are weighted using the production-inducing coefficient to implement a one-mode IO network matrix (42 × 42) with direction.

2.3 Input–Output Network Construction

First, this study tested the threshold of the weight (production-inducing effect) of the arc by observing the change in the number of arcs (input–output relationship), and finally determined the number of arcs to build the appropriate IO network size. As shown in Fig. 1, the arc number (the size of the network scale) of IO network in each year tends to be stable in the range of 0.3–0.5. When the threshold is equal to 0.1, a trend in which the number of arcs is decreasing the most. From this, it is judged that a tie having the production-inducing coefficient larger than 0.1 is determined as the meaningful arc. For example, in a total of 1,722 (42 × 41) arcs of the IO network in 2015, the number of arcs used for research was determined to be 228.

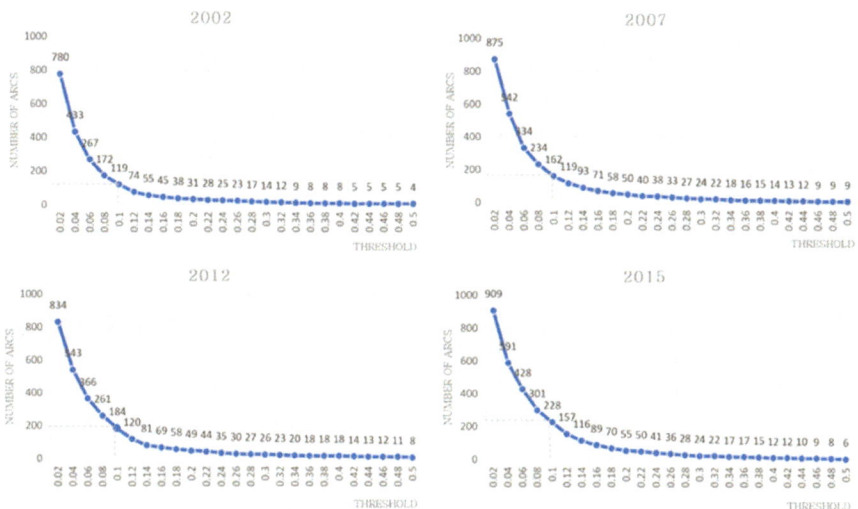

Fig. 1 Relationship between production-inducing coefficient and number of arcs

3 Measurement of Input–Output Network

3.1 Centrality Analysis

First, this study measures the number of other industries directly connected to each industry as the degree centrality. As the number of connected industries increases, the sector has a direct impact on economic production activities. Since the IO networks analyzed in this study is the directional graphs, the degree centrality of a node (e.g., sector i) is divided into indegree centrality $C_D^{IN}(i)$ and outdegree centrality $C_D^{OUT}(i)$. $C_D^{IN}(i)$ is the connection from the outside to the inside of sector i, which refers to the impact on sector i due to the fluctuation of final demand from other industries. $C_D^{OUT}(i)$ is the connection from the sector i to the outside, that is, the final demand of the sector i increases the production-inducing effect to other industries.

Similarly, the closeness centrality is measured in the directional network by distinguishing incloseness $C_C^{IN}(i)$ and outcloseness $C_C^{OUT}(i)$. According to SNA, if a sector connects all other industry sectors (nodes) in the network with minimal steps, it can be said that it occupies the most central position in the entire network structure. It means a sector in the IO network that enables production spillover effects to reach other sectors quickly, or (like transmitting information) having fast-spreading influence (or perception). This indirectly assesses the impact ability or perceived ability of a sector that can propagate production-inducing effects to other sectors with the shortest steps in an industrial network.

The third analytical indicator for measuring the central position of the industry structure is betweenness centrality and is expressed in $C_B(i)$. In order to reach other nodes, this node must be frequently passed. The more the number of times, the higher the degree of the center. From this point of view, if sector i can connect the two industry sectors well in the cooperative production process, then it can be regarded as the core industry playing the controlling role.

The Ucinet software package for analyzing SNA was used in this study to visualize the topology of IO networks and analyze three centrality metrics. In the process of analyzing network centrality, Ucinet does not measure the weighted arcs very well. That is to say, in order to measure the closeness centrality and the betweenness centrality, Ucinet does not test the weight of the arc as a parameter, when calculating the shortest path. Therefore, in the analysis of the three central indicators of the IO network, this study used the asymmetric adjacency matrix of 2002, 2007, 2012, and 2015, and summarized the analysis results in the next chapter.

3.2 Clustering Analysis

Next, the core structure of the network is determined by using the minimum spanning tree (MST). That is, extracting the industrial clusters with the largest production-inducing effects in the national industrial network.

The generated MST is an epitome of the analysis object network. It includes all the nodes of the existing network, but there is no loop and has the smallest weight sum. In this study, the ties of the network are directional and should have the maximized production-inducing effects. That is, the principle of preferentially confirming and connecting the arc of the maximum weight is applied. Therefore, this study also used the software package of Gephi. It takes into account both directionality and weight of the arcs, and can calculate the MST. First, we switch each cell value of the existing network matrix into the negative number (−) before entering the data into the software. In other words, directed negative weight graphs are created and put into Gephi for calculation (using the Kruskal's algorithm). Based on the algorithm model, the final composition of the industrial clusters must fully meet the following conditions. An industrial cluster consists of at least three nodes while includes one channel. Meanwhile, the two arcs at the point of intersection between the two clusters are blocked by each other. The relatively suitable structures are found to identify the clusters of IO networks through this method.

4 Research Results

4.1 The Characteristics of China Industrial Network

First, from the perspective of indegree centrality, our analysis results show that a large proportion of goods and services from other sectors are invested into the wearing apparel and leather sector, the wood, and products sector and the textiles sector in Chinese economic production activities. In addition, the general- and special-purpose machinery two sectors also maintained a relatively high degree of connectivity in 2015. The indegree centrality of nonmetallic mineral product manufacturing sector has been on the rise from 2002 to 2015. This shows that the input demand for intermediate products from other sectors has shown an increasing trend in the production process. Regarding the service sectors, the real estate service sector has become more connected with other sectors since 2012. The leasing and business service sector were highly variable in its ranking of indegree centrality. From the perspective of outdegree centrality, first, the chemical product manufacturing sector has the highest degree of connectivity in each year, which is considered to be the core industry of China. It shows that this sector has the largest number of sectors connected in the IO network, which produced the production-inducing effects on these industries. In addition, the electricity supply sector and the basic metal product manufacturing sector have relatively strong connectivity each year. The impact of the agriculture, forestry, and fishing sector and the ICT products manufacturing sector (Manufacture of computer, communications and other electronic equipment) on production activities is moderate. On the other hand, the wholesale and retail sector fluctuates greatly during the production process from 2002 to 2015. Since 2012, the degree of connectivity between the financial and insurance sector and other industries has shown a sharp upward trend. Moreover, the transportation service sector has also shown a similar development trend. In conclusion, the outdegree centrality of manufacturing industry is generally higher than the indegree centrality.

Second, in terms of incloseness centrality, the construction sector can perceive the spillover effects from other industries as quickly as possible. The nonmetallic mineral manufacturing sector, instrumentation manufacturing sector and other product manufacturing sector maintain relatively high values. This means that, if the final demand fluctuations in other sectors during the production process, they can react quickly. However, as the repair of machinery and equipment sector rose to the first (0.232) and third (0.211) in 2012 and 2015, it can be seen that the development of Chinese machinery manufacturing industry has turned to the trend of mechanical use and maintenance. The sewerage and waste management sector is second (0.211) in the analysis result of 2015, and it can be judged that the production spillover effects from other industries have a great impact on this industry. In terms of outcloseness centrality, the chemical product manufacturing sector still plays the most important role in the national economic system. That means, it have strong capabilities to spread the spillover effect to other industries, and the influence has continued to grow since 2002. A little different from the degree centrality results, the basic metal product manufacturing sector and the coke and refined petroleum product manufacturing sector have grown rapidly since 2007, thus having a stable and important impact on the transmission of production-inducing effects to other sectors.

Third, the results of betweenness centrality show that, the chemical product manufacturing sector are always positioned as the core industries by measuring all analytical indicators. In addition, the transport service sector and the basic metal product manufacturing sector were in the lead in 2002 (130.42) and 2007 (108.57) as the industries that control the consumption of production resources. In 2012 and 2015, leasing and business service sector soared to the second place. The similar situation has occurred in the coke and refined petroleum product manufacturing sector that first appeared in the third place (116.65) of the 2015 ranking. In comparison, it can be observed that the financial and insurance sector is not stable that supporting other industry sectors from 2012 to 2015. Moreover, the transportation service sector and the electricity supply sector play the role as a bridge to control the development of other industry sectors. The analysis index can better illustrate that China service industry has improved its position in economic development since 2012, and the types of service industries involved in manufacturing production activities have also increased.

4.2 The Clustering of China Industrial Network

The core structure with strong production-inducing effects was selected by MST method in the original 2002, 2007, 2012, and 2015 IO tables. A total of seven connected industry clusters are formed centering on leading industry sectors. The characteristics of structural change are summarized as follows.

First, the relationship between Cluster 1 (the industrial cluster centered on the chemical product manufacturing sector, called as "Representative node") and adjacent other industrial clusters. The two industrial clusters separately centered on the ICT product manufacturing sector and the coke and refined petroleum product manufacturing sector, were indirectly connected to Cluster 1 in 2002. However, the two industrial clusters converged as parts of Cluster 1 in 2015. Also, the leasing and business service sector which was previously owned by the industry cluster centered on ICT product manufacturing sector, has also joined Cluster 1 in 2012. The connection of the wholesale and retail sector with other industrial clusters has been unstable. With the rapid development of the chemical product manufacturing, Cluster 1 can be regarded as the core part, leading the entire industry clusters.

Second, the observation of the cluster centered on the agriculture, forestry and fishing sector. This industry cluster was ranked in Cluster 6 from 2002 to 2007. As the number of sectors that make up this cluster increased in 2012 and 2015, the ranking rose to Cluster 3. It shows that, with the development of manufacturing and service industry, agriculture still occupies an important position in Chinese economic production activities. From the perspective of the internal structure change, the culture, sports, and amusement sector entered this industrial cluster in 2012, and has a close relationship with the beverages and tobacco product manufacturing sector and the accommodation and food service sector.

Third, the observation of the changes in isolated points. The financial and insurance sector and the real estate sector remained at the isolated points from 2002 and 2007 IO networks. After taking part in the 2012 IO network, the financial and insurance sector connected the real estate sector with the transportation service sector. These three industries were centered on the financial and insurance sector and formed the Cluster 6. In 2015, the financial and insurance sector was still connecting with the real estate sector, but changed to link the gas supply sector at the other end. The waste management sector (abbreviation for the sector of "waste collection, treatment and disposal activities") as a new entrant joined the industrial network in 2015. With the transportation service sector as the center, they formed a new cluster (Cluster 5 i n 2015) together with the public administration and social organizations sector. It can be seen that the financial and insurance sector is closely related to the real estate sector, and their status in the industrial network has changed greatly in 2012 and 2015. Moreover, the industrial network has a tendency to eco-friendly development.

5 Conclusion

In general, the manufacturing industry of China still plays the central role in national economy. However, the number of service sectors participating in productive activities is increasing yearly. The links between the manufacturing and service sectors are gradually integrated and show synergies. On the other hand, industrial development was dominated by infrastructure construction and machinery and equipment manufacturing from 2002 to 2007, and by 2015, it was developing toward service industries closely related to technological innovation and environmental protection. It can be said that the above characteristics are fully reflected in our research on the changes in Chinese industrial structure.

References

1. P.R. Krugman, *Geography and Trade* (MIT Press, Cambridge, 1993)
2. D.I. Riddle, Service-led growth. Int. Exec. **28**(1), 27–28 (1986)
3. S. Kwak, S. Yoo, J. Chang, The role of the maritime industry in the Korean national economy: an input-output analysis. Mar. Policy **29**(4), 371–383 (2005)
4. J. Hauknes, M. Knell, Embodied knowledge and sectoral linkages: an input-output approach to the interaction of high- and low-tech industries. Res. Policy **38**(3), 459–469 (2009)
5. W. Xing, X. Ye, L. Kui, Measuring convergence of China's ICT industry: an input-output analysis. Telecommun. Policy **35**(4), 301–313 (2011)
6. F. Schweitzer, G. Fagiolo, D. Sornette, F. Vega-Redondo, A. Vespignani, D.R. White, Economic networks: the new challenges. Science **325**(5939), 422–425 (2009)
7. D. Acemoglue, V. Carvalho, A. Ozdaglar, A. Tahbaz-Salehi, The network origines of aggregate fluctuation. Econometrica **80**, 1977–2016 (2012)
8. S.S. Cho, S.W. Kang, An empirical study on network theory, economic structure and economic impact: technological and economic implications. J. Technol. Innov. **16**, 937–953 (2013)
9. C.A. Hidalgo, B. Klinger, A.L. Barabási, R. Hausmann, The product space conditions the development of nations. Science **317**(5837), 482–487 (2007)
10. J. McNerney, B.D. Fath, G. Silverberg, Network structure of inter-industry flows. Phys. A: Stat. Mech. Appl. **392**(24), 6427–6441 (2013)
11. L.C. Freeman, Centrality in social networks conceptual clarification. Soc. Netw. **1**(3), 215–239 (1978)
12. L.C. Freeman, D. Roeder, R.R. Mulholland, Centrality in social networks: II. Experimental results. Soc. Netw. **2**(2), 119–141 (1979)

Machine Learning Methods for Revenue Prediction in Google Merchandise Store

Vahid Azizi and Guiping Hu

Abstract Machine learning has gained increasing interests from various application domains for its ability to understand data and make predictions. In this paper, we apply machine learning techniques to predict revenue per customer for Google Merchandise Store. Exploratory Data Analysis (EDA) was conducted for the customer dataset and feature engineering was applied to the find best subset of features. Four machine learning methods, Gradient Boosting Machine (GBM), Extreme Gradient Boosting (XGBoost), Categorical Boosting (CatBoost), and Light Gradient Boosting Machine (LightGBM) have been applied to predict revenue per customer. Results show that LightGBM outperforms other methods in terms of RMSE and running time.

Keywords Feature engineering · GBM · XGBoost · CatBoost · LightGBM

1 Introduction

Google Merchandise Store (also known as GStore) is an online store which sells clothing, bags, drinkware, office supplies, and other accessories. Same as regular retailers, marketing teams are constantly challenged to design promotional strategies that are customized for individual customers, and online stores have advantages since data are often tracked at individual customer level. GStore is interested in analyzing customer dataset to predict revenue per customer since it has been well known that the 80/20 rule applies in retail business. The 80/20 rule refers to the phenomenon that 20% of the customers generate 80% of the revenue. Making accurate customer consumption predictions is significant for operational strategy and marketing investment.

This paper aims to predict revenue per customer for GStore. Four decision tree-based machine learning algorithms, Gradient Boosting Machine (GBM), Extreme Gradient Boosting (XGBoost), Categorical Boosting (CatBoost), and Light Gradient

V. Azizi · G. Hu (✉)

Industrial and Manufacturing Systems Engineering Department, Iowa State University, Ames, IA 50010, USA

e-mail: gphu@iastate.edu

© Springer Nature Switzerland AG 2020

H. Yang et al. (eds.), *Smart Service Systems, Operations Management, and Analytics*, Springer Proceedings in Business and Economics, https://doi.org/10.1007/978-3-030-30967-1_7

Boosting Machine (LightGBM), have been applied to the GStore dataset. All methods are developed based on boosting idea. Boosting is an ensemble technique that attempts to make a strong classifier from a number of weak classifiers. GBM and XGBoost have been two successful decision tree-based boosting techniques during recent years and LightGBM and CatBoost algorithms are the most advanced boosting methods for machine learning that have been developed lately. Compared to the previous machine learning algorithms, these algorithms improve prediction accuracy and reduce computational cost and the possibility of over-fitting. Both algorithms have been recently considered as prediction tools in high-dimensional large datasets in research projects and forecasting competitions. Therefore, these methods are suitable tools for large and high-dimensional dataset analyzed in this paper. Because the methods are relatively novel, the applications are not very extensive, and the articles related to them are very rare. Therefore, this paper can be considered as an extension of the application scope of the two algorithms.

There has been an expanding body of literature on making predictions for various applications. However, LightGBM and CatBoost have not been studied extensively. Li et al. developed a stacking model by combining LightGBM, XGBoost, and GBDT to detect phishing webpages using URL and HTML features [1]. Their proposed method outperforms existing machine learning models in terms of accuracy and missing alarm rate and false alarm rate. Ma et al. used "multi-observation" and "multi-dimensional" data cleaning method and applied LightGBM and XGBoost, which are based on real P2P transaction data from Lending club to predict default risk of loans in the platform [2]. The results of different methods were compared and LightGBM algorithm was shown to outperform. Sun et al. adopted LightGBM to forecast the price trend (falling or not falling) of cryptocurrency market [3]. Results showed that the robustness of the LightGBM model is better compared to the other methods. Wang et al. proposed transportation modes classification method based on LightGBM from GPS trajectory data [4]. The transportation modes include walking, cycling, taking a bus, taking a taxi, driving a car, taking the subway, and taking a train. The results show that LightGBM outperforms XGBoost in classifying three modes. Li et al. used LightGBM to predict remaining useful life of the turbofan engine to design best maintenance strategies [5]. Experiments show that LightGBM works well with the high-dimensional data of study and is insensitive to noise. Cao and Gui proposed a multivariate model using LightGBM to predict wind power generation in order to induce system operators to reduce risks [6]. The results prove that proposed method can effectively predict wind power for the next 6 h. Mei et al. developed a combined model prediction method based on XGBoost and LightGBM algorithms to improve the accuracy of short-term traffic flow prediction [7]. Experimental results indicate that combined model outperforms single model. Chen et al. proposed a weighted bagging LightGBM model to predict lncRNA-disease association [8].

The rest of the paper is organized as follows. The data structure is described in Sect. 2. Exploratory data analysis is provided in Sect. 3. Feature engineering steps are discussed in Sect. 4. Prediction methods are detailed in Sect. 5. Numerical results are discussed in Sect. 6 and conclusions are outlined in Sect. 7.

2 Data Description

In this section, data structure is described. The dataset contains user transactions from August 1, 2016 to April 30, 2018 and includes 1,708,337 instances and 10 features.

2.1 Features for Prediction

Features of train dataset are listed as follows:

- fullVisitorId—A unique identifier for each user of the Google Merchandise Store.
- channelGrouping—The channel via which the user came to the Store.
- date—The date on which the user visited the Store.
- device—The specifications for the device used to access the Store.
- geoNetwork—This section contains information about the geography of the user.
- trafficSource—This section contains information about the Traffic Source from which the session originated.
- visitId—An identifier for this session. This is part of the value usually stored as the utmb cookie. This is only unique to the user. For a completely unique ID, you should use a combination of fullVisitorId and visitId.
- visitNumber—The session number for this user. If this is the first session, then this is set to 1.
- visitStartTime—The timestamp (expressed as POSIX time).
- totals—This set of columns mostly includes high-level aggregate data.

2.2 Prediction Evaluation Criterion

Predictions are evaluated with the Root Mean Squared Error (RMSE). RMSE is defined as

$$\text{RMSE} = \sqrt{\frac{1}{n}\sum_{i=1}^{n}\left(y_i - \widehat{y}_i\right)^2} \tag{1}$$

where $\widehat{y}_i$ is the natural log of the predicted revenue for customer i and y_i is the natural log of the actual summed revenue value plus one for customer i.

3   Exploratory Data Analysis (EDA)

In this section, we analyze the information provided with each feature. Analyzing features can contribute to feature engineering. By having enough information about features, we can decide which features should be kept (feature selection) or what kind of new features can be extracted (feature extraction).

- Feature: channelGrouping
  This feature has information about the channels which customers have used to visit Google Store pages. As illustrated in Fig. 1, most of the users visited the Google Store pages by organic search.
- Feature: device.deviceCategory
  This feature provides useful information about the type of devices customers used for their visit. As shown in Fig. 2, more than 60% of page views come from desktop device.
- Feature: device.operatingSystem

Fig. 1 Distribution of visits based on channel

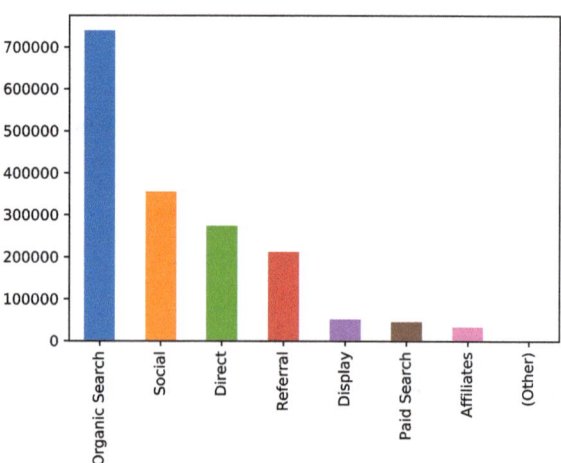

Fig. 2 Distribution of visits based on used device

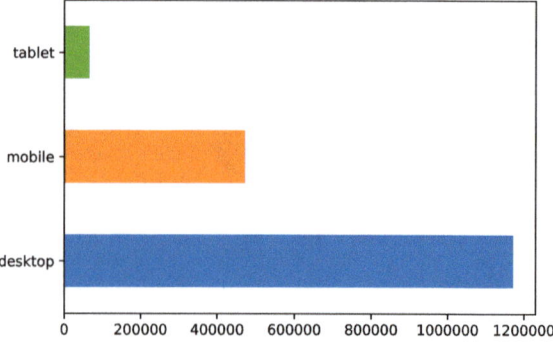

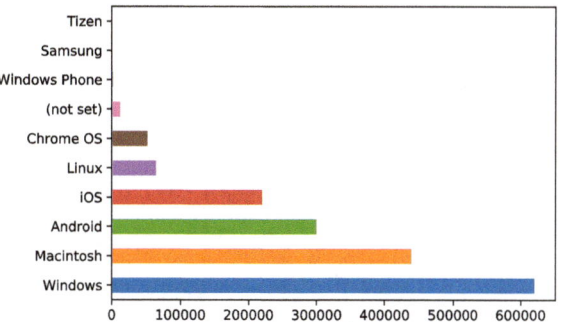

Fig. 3 Distribution of users based on operating systems

In addition to device, another feature is operating system for browsing. Figure 3 shows the number of customers for each operating system category. Windows, Macintosh, and Android are the top three.

- Feature: geoNetwork.continent
  Distribution of users based on the geographical location (continent) is provided by geoNetwork.continent. According to this feature, more than 40% of customers are from America as shown in Fig. 4.

- Feature: totals.transactionRevenue
  This feature is considered as a response feature and shows revenue obtained by each customer. The number of nonzero revenue is 18,514 and ratio is 0.0108. It shows that the dataset is highly unbalanced and carefully designed and feature engineering process is needed before making predictions. In addition, the prediction method choice has to take the imbalance of the data into consideration. Figure 5 illustrates the number of transactions including nonzero and zero transactions. Figure 6 shows the number of nonzero transactions in the same time interval.

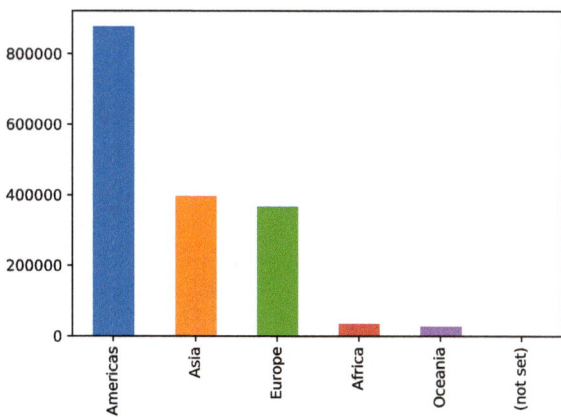

Fig. 4 Distribution of users based on continent

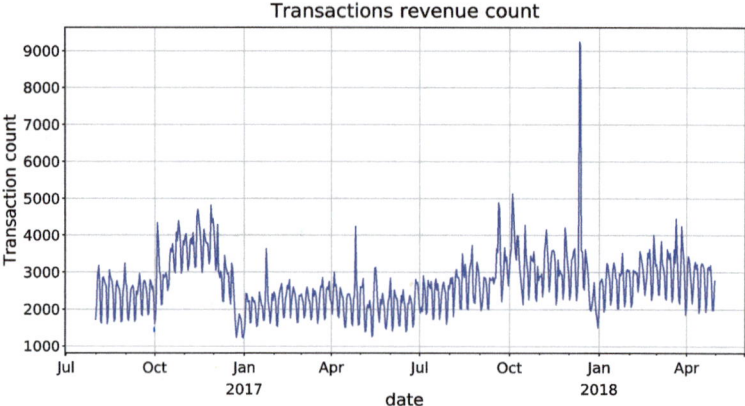

Fig. 5 Number of all transactions

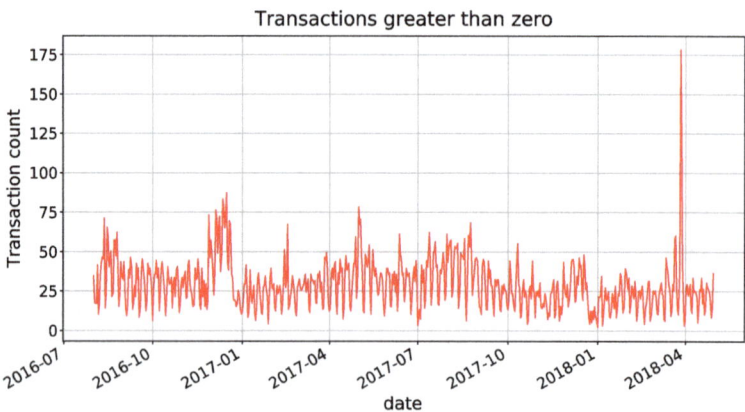

Fig. 6 Number of nonzero transactions

4 Feature Engineering

- JSON features
  In computing, JavaScript Object Notation (JSON) is an open-standard file format that uses human-readable text to transmit data objects consisting of attribute–value pairs and array data types. This dataset includes following JSON features: device, geoNetwork, totals, and trafficSource. The raw dataset has 10 features, but after converting JSON predictors, there are 29 features.
- New features
  By grouping existing features 33, new features were created which increased the number of features up to 62.
- Features with None values

Features with more than 50% of their value were "None" were deleted. At this step, trafficSource.isTrueDirect and totals.bounces were removed.

• Features with unique values
  Features with unique values were deleted.

After feature engineering process, the dataset includes 55 features ready to pass predictions models.

5 Prediction Methods

5.1 Gradient Boosting Machine

Gradient Boosting Machine (GBM) is introduced by Friedman [9]. It is also known as MART (Multiple Additive Regression Trees) and GBRT (Gradient Boosted Regression Trees). GBM is a machine learning method for regression and classification problems, which produces a prediction model in the form of an ensemble of weak prediction models. The ensemble of weak models leads to a strong model which reduces over-fitting and increases prediction accuracy [10]. GBM builds the model in a stage-wise fashion as other boosting techniques do, and it generalizes them by allowing optimization of an arbitrary differentiable loss function.

5.2 Extreme Gradient Boosting (XGBoost)

The Extreme Gradient Boosting (XGBoost) algorithm proposed by Chen and Guestrin [11] is a novel implementation method for Gradient Boosting Machine and in particular, classification and regression trees. The algorithm uses the idea of boosting, which combined all the predictions of a set of weak learners for creating a strong learner through additive training strategies. XGBoost not only reduces over-fitting by simplifying the objective functions but also optimizes the computation resources b y parallel calculations.

5.3 Categorical Boosting (CatBoost)

Categorical Boosting (CatBoost) is a machine learning algorithm designed by Dorogush et al. [12] that uses gradient boosting on decision trees. CatBoost has the flexibility of giving indices of categorical columns so that it can be encoded as one-hot encoding using one-hot-max-size (Use one-hot encoding for all features with the number of different values less than or equal to the given parameter value).

If no features are passed in categorical features argument, CatBoost will treat all the columns as numerical variables. For remaining categorical columns which have unique number of categories greater than one-hot-max-size, CatBoost uses an efficient method of encoding which is similar to mean encoding but reduces overfitting. The process goes like this.

1. Permuting the set of input observations in a random order. Multiple random permutations are generated.
2. Converting the label value from a floating-point or category to an integer.
3. All categorical feature values are transformed to numeric values using the following formula:

$$\text{avg\_target} = (\text{countInClass} + \text{prior})/(\text{totalCount} + 1) \qquad (2)$$

where countInClass is how many times the label value was equal to "1" for objects with the current categorical feature value. Prior is the preliminary value for the numerator. It is determined by the starting parameters. totalCount is the total number of objects (up to the current one) that have a categorical feature value matching the current one. Mathematically, this can be represented using the below equation.

Assume that we are given a dataset of observations $D = \{X_i, Y_i\}_{i=1,\dots,n}$, where $X_i = \{x_{i,1}, \dots, x_{i,m}\}$ is a vector of m features, some numerical, some categorical, and $Y_i \in R$ is a label value. Let $\sigma = \{\sigma_1, \sigma_2, \dots, \sigma_n\}$ be the permutation, then $x_{\sigma_p,k}$ is substituted with

$$\left(\sum_{j=1}^{p-1} [x_{\sigma_j,k} = x_{\sigma_p,k}] Y_{\sigma_j} + a.P \right) \Big/ \left(\sum_{j=1}^{p-1} [x_{\sigma_j,k} = x_{\sigma_p,k}] + a \right) \qquad (3)$$

where P is prior value and parameter a, $a > 0$, which is the weight of the prior. Adding prior is a common practice and it helps to reduce the noise obtained from low-frequency categories [12].

5.4 Light Gradient Boosting Machine (LightGBM)

LightGBM, a machine learning algorithm developed by Ke et al. [13], is a fast, distributed, high-performance gradient boosting framework based on decision tree algorithm used for ranking, classification, and many other machine learning tasks.

Since it is based on decision tree algorithms, it splits the tree leaf wise as shown in Fig. 7a with the best fit whereas other boosting algorithms split the tree depth wise or level wise as illustrated in Fig. 7b rather than leaf wise. So when growing on the same leaf in LightGBM, the leaf-wise algorithm can reduce more loss than the level-wise algorithm and hence results in much better accuracy which can rarely be

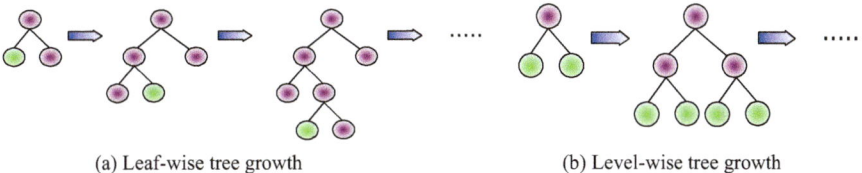

(a) Leaf-wise tree growth   (b) Level-wise tree growth

Fig. 7 Tree structure

achieved by any of the existing boosting algorithms. Also, it is often very fast, hence the word "Light" is used for this method.

Leaf-wise splits lead to increase in complexity and may lead to over-fitting and it can be overcome by specifying another parameter max-depth which specifies the depth to which splitting will occur.

6 Numerical Results

GBM, XGBoost, CatBoost, and LightGBM packages in Python are used as prediction models. After tuning parameters for prediction methods and considering 20% of samples as validation dataset, four methods were implemented on an Intel® Core i7 2.79 GHz PC with 8 GB of memory. Obtained results are reported in Table 1. As it can be seen, CatBoost and LightGBM perform better than GBM and XGBoost in terms of RMSE. The learning curves for CatBoost and LightGBM are depicted in Fig. 8a and b, respectively. Although LightGBM leads to more over-fitting, the results show that LightGBM outperforms CatBoost in terms of Learning RMSE, Test RMSE, and CPU run time. LightGBM gives better accuracy while its running time is approximately 9.5 times faster than CatBoost.

Figure 9 is the feature importance plot for LightGBM method. The top ten important features are listed in this figure. As it is shown in this figure, "totals.hits" is the most important feature far better than second important feature. Next important feature is "totals.pageviews". As expected, there is more chance for a transaction to happen when a customer insists to hit and view pages frequently. Among created features, "user_pageviews_sum" and "user_hits_sum" are the most important ones. Same conclusions about page views and hits are valid for these two features.

Table 1 Prediction results

Method	Learning RMSE	Test RMSE	Time (s)
GBM	1.533	1.548	709
XGBoost	1.788	1.790	796
CatBoost	1.452	1.521	822
LightGBM	**1.429**	**1.517**	**86**

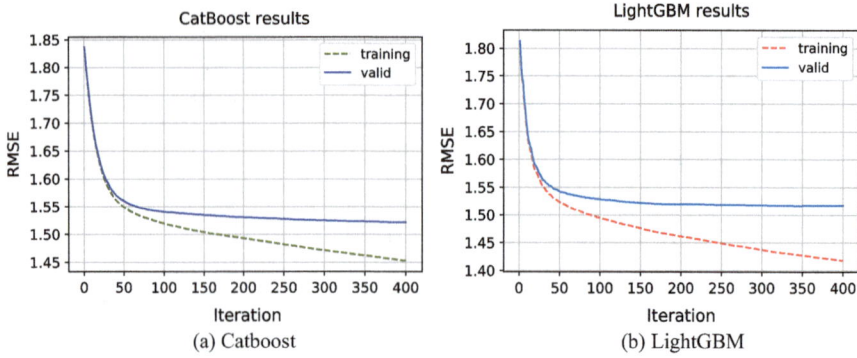

Fig. 8 Learning curve

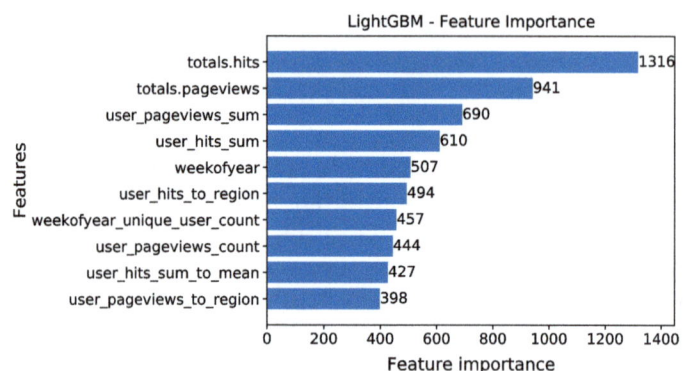

Fig. 9 Feature importance for LightGBM

7 Summary

In this paper, Google Merchandise Store customer dataset is used to predict revenue per customer. Feature engineering rules have been applied to keepuseful features and create new set of features. GBM, XGBoost, CatBoost, and LightGBM have been applied to predict revenue. Based on numerical results, LightGBM performs better than other prediction methods in terms of learning RMSE, test RMSE, and running time. For future research, additional machine learning methods such as logistic regression and random forests can be studied and compared. Ensemble methods can serve as another future direction to improve prediction accuracy.

References

1. Y. Li, Z. Yang, X. Chen, H. Yuan, W. Liu, A stacking model using URL and HTML features for phishing webpage detection. Futur. Gener. Comput. Syst. **94**, 27–39 (2019)
2. X. Ma, J. Sha, D. Wang, Y. Yu, Q. Yang, X. Niu, Study on a prediction of P2P network loan default based on the machine learning LightGBM and XGBoost algorithms according to different high dimensional data cleaning. Electron. Commer. Res. Appl. **31**, 24–39 (2018)
3. X. Sun, M. Liu, Z. Sima, A novel cryptocurrency price trend forecasting model based on LightGBM. Financ. Res. Lett. (2018)
4. B. Wang, Y. Wang, K. Qin, Q. Xia, Detecting transportation modes based on LightGBM classifier from GPS trajectory data, in *2018 26th International Conference on Geoinformatics*, (IEEE, 2018), pp. 1–7
5. F. Li, L. Zhang, B. Chen, D. Gao, Y. Cheng, X. Zhang, Y. Yang, K. Gao, Z. Huang, J. Peng, A light gradient boosting machine for remaining useful life estimation of aircraft engines, in *2018 21st International Conference on Intelligent Transportation Systems (ITSC)* (IEEE, 2018), pp. 3562–3567
6. Y. Cao, L. Gui, Multi-step wind power forecasting model using LSTM networks, similar time series and LightGBM, in *2018 5th International Conference on Systems and Informatics (ICSAI)*, (IEEE, 2018), pp. 192–197
7. Z. Mei, F. Xiang, L. Zhen-hui, Short-term traffic flow prediction based on combination model of XGBoost-LightGBM, in *2018 International Conference on Sensor Networks and Signal Processing (SNSP)*, (IEEE, 2018), pp. 322–327
8. X. Chen, X. Liu, A weighted bagging LightGBM model for potential lncRNA-disease association identification, in *International Conference on Bio-Inspired Computing: Theories and Applications*, (Springer, Singapore, 2018), pp. 307–314
9. J.H. Friedman, Greedy function approximation: a gradient boosting machine. Ann. Stat. 1189–1232 (2001)
10. H. Pham, S. Olafsson, Bagged ensembles with tunable parameters. Comput. Intell. **35**, 184–203 (2019). https://doi.org/10.1111/coin.12198
11. T. Chen, C. Guestrin, XGBoost: A scalable tree boosting system, in *Proceedings of the 22nd ACM SIGKDD international conference on knowledge discovery and data mining*, (ACM, 2016), pp. 785–794
12. A.V. Dorogush, V. Ershov, A. Gulin, CatBoost: gradient boosting with categorical features support, arXiv preprint arXiv:1810.11363 (2018)
13. G. Ke, Q. Meng, T. Finley, T. Wang, W. Chen, W. Ma, Q. Ye, T-Y. Liu, LightGBM: a highly efficient gradient boosting decision tree, in *Proceedings of the Advances in Neural Information Processing Systems* (2017), pp. 3146–3154

Predicting Metropolitan Crime Rates Using Machine Learning Techniques

Saba Moeinizade and Guiping Hu

Abstract The concept of smart city has been gaining public interests with the considerations of socioeconomic development and quality of life. Smart initiatives have been proposed in multiple domains, such as health, energy, and public safety. One of the key factors that impact the quality of life is the crime rate in a metropolitan area. Predicting crime patterns is a significant task to develop more efficient strategies either to prevent crimes or to improve the investigation efforts. In this research, we use machine learning techniques to solve a multinomial classification problem where the goal is to predict the crime categories with spatiotemporal data. As a case study, we use San Francisco crime data from San Francisco Police Department (SFPD). Various classification methods such as Multinomial Logistic Regression, Random Forests, Lightgbm, and Xgboost have been adopted to predict the category of crime. Feature engineering was employed to boost the model performance. The results demonstrate that our proposed classifier outperforms other published models.

Keywords Machine learning · Multinomial classification · Crime prediction

1 Introduction

Smart cities utilize multiple technologies to improve the performance of transportation, health, energy, public safety, and education leading to a better quality of life. Big data analytics is one of the recent technologies that has a huge potential to enhance smart city services. In a smart city, big data applications have the potential to serve many sectors such as providing better customer service, improving healthcare, optimizing route and schedules in transportation, and increasing safety [1, 2].

One of the key factors that impact the quality of life is the crime rate in a metropolitan area. Predictive policing applies analytical and quantitative techniques to identify potential criminal activities by making statistical predictions. Predictive policing contributes to the public safety and region planning decision-making process [3]. A

S. Moeinizade · G. Hu (✉)
Industrial and Manufacturing Systems Engineering, Iowa State University, Ames, IA 50014, USA
e-mail: gphu@iastate.edu

© Springer Nature Switzerland AG 2020   77
H. Yang et al. (eds.), *Smart Service Systems, Operations Management,*
and Analytics, Springer Proceedings in Business and Economics,
https://doi.org/10.1007/978-3-030-30967-1_8

crime usually happens at a certain place and time involving a perpetrator and a target. There are several questions that need to be answered: 1. who will commit a crime? 2. who will be offended? 3. what type of crime will occur? and 4. when and where a new crime will happen [4]?

In this study, we aim to address the third question of predicting crime categories given the time and location of a crime. In the case study, we use past data of San Francisco city which has the spatiotemporal information of 12 years (2003–2015). The dataset is part of a competition on Kaggle in 2016 to predict criminal occurrences in the city. Our goal is to predict the probability that a crime belongs to a certain category such as Larceny/Theft, Vandalism, and Robbery, given its time and location.

There has been a rich body of knowledge about predictive crime analytics. Watson used demographic and mobile phone data to predict whether a specific area in the city of London will be a crime hot spot or not [5]. Engel presented a support vector machine approach to predict the hot spot location of crimes [6]. Several classification methods such as decision tree, neural network, and Naïve Bayes are used by Bishop to predict the crime hot spot and then an ensemble of different models is proposed to stabilize the results [7]. Many other significant contributions in this field are related to crime pattern prediction using spatial and temporal data. Cortes and Vapnik use Naïve Bayes and decision tree to predict crime category for different states in USA [8]. Dietterich [9], Breiman [10], Freund and Schapire [11], Friedman [12], and Darekar [13] use the same data of San Francisco city to predict the crime category by implementing different methods including Naïve Bayes, decision tree, random forest, logistic regression, and neural network.

Our approach shares some similarities with the approaches mentioned above, in that, we predict category of crime given the spatiotemporal data. However, we focus more on feature engineering and implement some state-of-the-art ensemble methods that have not been studied by previous authors. Ensemble methods help to improve machine learning results by combining the merits of multiple models. We implement some decision tree-based ensemble models including two gradient boosting models and show that the new approach has better predictive performance.

This paper is organized as follows: Sect. 2 describes the dataset and feature engineering techniques. In Sect. 3, we discuss the methods and then, the experiments and results are presented in Sect. 4. Finally, Sect. 5 includes conclusions and discusses future work.

2   Data Exploration and Feature Engineering

A crime incident is a complex phenomenon which is hard to predict. However, it is closely associated with temporal and spatial factors. In this section, we explore the data and then propose feature engineering techniques.

The dataset contains incidents derived from SFPD Crime Incident Reporting system from across all of San Francisco's neighborhoods. The training set has 884,262 observations of nine variables. Table 1 demonstrates a description of these variables.

Table 1 Dataset description

Variable	Description
Dates	Timestamp of the crime incident which is in the format of "year-month-day hour: minute: second"
Category	Category of the crime incident. This is the target variable that should be predicted. It includes 39 distinct values such as Assault, Larceny/Theft, and Robbery, etc.
Descript	Detailed description of the crime incident
DayOfWeek	The day of the week with seven distinct values
PdDistrict	Name of the Police Department District which has 10 distinct values
Resolution	How the crime incident was resolved
Address	The approximate street address of the crime incident
X	The longitude coordinate of the crime
Y	The latitude coordinate of the crime

The test set has 884,262 observations of six variables which are the same variables as the training set excluding Descript, Resolution, and the target variable, Category. The data ranges from 1/1/2003 to 5/13/2015. The training and test set rotate every week which means that the odd weeks belong to the test set, and even weeks belong to the training set.

There are 39 categories of crime that should be predicted. Figure 1 shows the

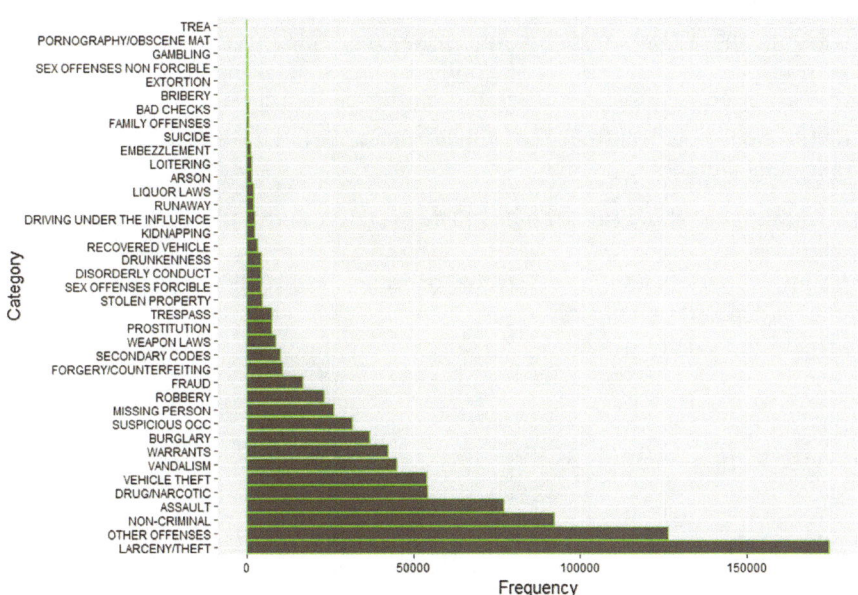

Fig. 1 Crime frequency for each category

frequency of each type of crime in the training data. We observe that the crime distribution is not even and some categories like TREA (Trespassing or loitering near posted industrial property), Gambling, Extortion, etc., are rare, whereas some other categories like Larceny/Theft, Other Offenses, NonCriminal, etc., are happening with a higher frequency. However, we decided to keep all the 39 categories because the evaluation is based on predicting the probability of each class.

To help facilitate the machine learning process, we performed a series of feature engineering techniques on the raw data before learning the algorithms. This can increase the predictive power of learning models and lead to a better classification. Here are some techniques used for this case study

- Dropping the Resolution and Descript features from the training set. This information is not available in the test set and furthermore it doesn't make sense to use the description of an incident or the way of resolving it to make a prediction before the crime's occurrence.
- Several features including Hour, Month, Year, and Min30 are extracted from the Dates feature. Then, the Dates column is dropped. The Min30 feature is calculated as the absolute value of the minute subtracted by 30 which reflects the fact of time distribution symmetry [1].
- The information in the address field has two different formats. Some crimes happened on the intersection of two streets (i.e., an address like TURK ST /JONES ST) while others happened on a specific block (i.e., an address like 300 Block of OFARRELL ST). Therefore, we add a new binary feature named "Block" which is extracted from the Address column to take the value of 1 when the address field contains the word "Block" and 0 otherwise.
- Dropping the Address feature as it's challenging to use characters such as street names and numbers in prediction. Moreover, the latitude and longitude data provide enough information about the location of an incident.
- Removing the outliers from Y (Latitude) and replacing it with the mean value of remaining latitudes.

Implementing the above steps results in a new set of features which are shown in the following table.

To predict the target variable, category of crime, the models will be trained based on the nine features provided in Table 2. In the next section, we will elaborate on classification models.

3 Methods

In this research, we are dealing with a multinomial classification problem where the goal is to classify observations into one of three or more classes. Multinomial classification problems are a form of supervised learning that aims at assigning a class label for each input observation. Given a training dataset of the form (x_i, y_i), where $x_i \in R^n$ is the ith example and $y_i \in \{1, \ldots, K\}$ is the ith class label, the objective

Table 2 Derived features

Variable	Description
Hour	The hour of crime incident
Month	The month of crime incident
Year	The year of crime incident
DayOfWeek	The day of the week with seven distinct values
PdDistrict	Name of the Police Department District which has 10 distinct values
Min30	Absolute value of the minute subtracted by 30
Block	The approximate street address of the crime incident
X	The longitude coordinate of the crime
Y	The latitude coordinate of the crime

is to learn a model F, such that $F(x_i) = y_i$ for new unseen examples [2]. Several algorithms have been developed based on decision trees [3], k-nearest neighbors [4], Naive Bayes [5], logistic regression [6], neural networks [7], and support vector machines [8] to address multinomial classification problems. To select the appropriate models, we need to explore the assumptions of each model and then select the ones which are more suitable given the problem and dataset we are dealing with. In the next section, we will discuss the pros and cons of each model in the context of this case study.

3.1 Classification Models

Naïve Bayes (NB)
The Naïve Bayes model is easy to build and can be useful for large datasets. However, this model assumes that features are independent, which may be a wrong assumption given the feature space of this case study (for example the PdDistrict feature is related to X and Y).

K-Nearest Neighbor (KNN)
K-nearest neighbor is a simple approach which assigns the unlabeled objects to the most common class of its k-nearest neighbors. However, objects tend to be classified into the most common classes when the training data is not uniformly distributed. In Fig. 1, we see that the distribution of classes is varied among the training dataset, thus we believe this algorithm will have a poor performance for this case study.

Multinomial Logistic regression (MLR)
Multinomial logistic regression is a classification method that generalizes the binary logistic regression to multinomial class problems. In this method, the results are converted to probability values in (0,1) using SoftMax function. MLR is an efficient

algorithm and the advantage is that the output can easily be interpreted as probabilities (we also want to predict the probability for each class). We think that good performance can be obtained with this classifier.

Support Vector Machines (SVM)

Support vector machine constructs a hyperplane in space, which can be used for classification or regression problems. The idea is to keep the hyperplane as far as possible from the nearest training data points. SVM usually results in high accuracy, but it might not be very efficient as the algorithm becomes very slow when the dataset is rather big.

Decision Trees (DT)

Decision trees are a powerful classification technique. A decision tree splits the training data based on the feature that gives the maximum information gain to produce a generalization for unseen data. Then a new observation can be classified by following a path from the root node to a leaf node. However, a single decision tree can suffer from high bias when the tree is too simple or it can suffer from high variance when it becomes too complicated. Ensemble learning [10], which combines several decision trees instead of using a single one can help overcome these challenges. Bagging and boosting are two techniques of ensemble learning which deal with reducing variance and bias, respectively.

In bagging, several decision trees are built on different subsets of training data chosen randomly with replacement and the final model is the average of all the predictions from different trees. Random Forest (RF) [11] is an extension to bagging where the selection of features is also random. The RF method can handle big datasets very well and has a good potential for this case study.

In boosting, many models with different structures are weighted together to reduce the bias. The idea is to combine many weak learners to create a powerful model. A weak learner is a predictor that is slightly correlated with the true labels. However, a strong learner is a predictor which is well-correlated with the true labels. In boosting, the models are built sequentially where the next model is focusing to reduce the errors of the previous models [12]. Gradient boosting is a special gradient-descent-based boosting technique which tries to reduce errors sequentially that can be applied to both regression and classification problems. This method was developed by Jerome H. Friedman in 2001 [13]. Two fast implementations of gradient boosting method are extreme gradient boosting (Xgboost) and light gradient boosting machine (Lgbm). These two methods have gained huge popularity among data mining competitions in terms of their performance and scalability. The main difference between these two methods is that the Xgboost algorithm grows trees level(depth)-wise, but the Lgbm does leaf-wise (best-first). We believe that these two algorithms can achieve a good performance on our dataset.

3.2   Model Selection and Evaluation

According to the discussion in Sect. 3.1, we select the MLR method as well as three state-of-the-art decision tree-based methods, RF, Xgboost, and Lgbm to tackle the San Francisco crime classification problem.

To evaluate each model, the multi-class logarithmic loss is calculated according to the following formula:

$$\log \text{loss} = -\frac{1}{N} \sum_{i=1}^{N} \sum_{j=1}^{M} y_{i,j} \log(p_{i,j}) \tag{1}$$

Here, log is the natural logarithm, N is the number of samples in the test set, M is the number of classes, $y_{i,j}$ is a binary variable that equals 1 i fobservation i is in class j and 0 otherwise, and $p_{i,j}$ is the predicted probability that observation i belongs to class j.

The dataset used for this case study is part of an open competition on Kaggle which was hosted in 2016. Each incident of the test set has been labeled with one true class by Kaggle. Therefore, we submit a set of predicted probabilities (one for every category) for each incident. Then, the submission will be evaluated using the log loss score.

4   Experiments and Results

The experiments are conducted in a supercomputer with a 128 GB memory at Iowa State University. Working programming language is R, and the integrated development environment is *RStudio*. The most used R libraries are *"caret"*, *"nnet"*, *"e1071"*, *"ranger"*, *"xgboost"*, and *"lightgbm"* for computations. Each algorithm has some specifications in terms of implementing and parameter tuning which are elaborated as follows:

MLR

The MLR method cannot deal with categorical features. Hence, we need to transform the categorical features to dummy variables by using the one-hot encoding technique. This increases the size of the feature space (i.e., the DayOfWeek column should be transformed to seven different features representing each day of the week) which increases the learning time. To avoid overfitting, we have used regularization. The final log loss on the test set is **3.073**.

RF

Random forests typically perform quite well with categorical variables in their original format. For this algorithm, two main parameters, number of trees and number of features selected for each tree, are tuned by conducting a search grid among different values and then the best parameters with the minimum out of bag error are selected. The final log loss score on the test set is **2.359**.

Xgboost

This algorithm manages only numerical values, so we need to create dummy variables from the categorical features. There are two main parameters that can help avoid overfitting. The first one is the maximum depth of the tree and the second one is the learning rate. After employing a four-fold cross-validation and tuning the two parameters, we get the log loss score of **2.951** on the test set.

Lgbm

The advantage of this algorithm is that it can easily handle the categorical features. Some important parameters including maximum depth of a tree, maximum number of leaves in one tree, and learning rate are tuned by conducting a search grid over different values with a cross-validation. The final log loss score is **2.276** which is better than what we achieved in the previous methods.

Table 3 summarizes the log loss scores for proposed four classification methods on SF crime dataset. We see that the two algorithms (RF and Lgbm) that are capable to handle categorical data are performing better. The results show that the Lgbm method is outperforming the other three methods and then RF, Xgboost, and MLR achieve a better score, respectively.

It should be noted that the Lgbm method covers a large number of parameters which makes the tuning process complicated. However, there are some parameters that are more important than others and can boost the algorithm performance after tuning. Therefore, we focus on the most important parameters and conduct a search grid over different possible values for each parameter to find the best values. The following table provides detailed information on the best parameter values used in our Lgbm model (Table 4).

To have a comparison, we have looked through the other sources that worked on SF crime prediction and got their performance of the best classification method. The following table gives a summary of results for these different sources (Table 5).

From Table 3, we see that the best result achieved from other sources is a gradient boosting approach with a log loss of 2.315 which indicates that our best classifier (Lgbm with log loss of 2.276) has achieved a better performance than published methods.

The contribution of our proposed method lies in the process of feature engineering, model selection, and parameter tuning. We observed that decision tree-based

Table 3 Performance of our classification algorithms on SF Crime data

Classification algorithm	Log loss
MLR	3.073
RF	2.359
Xgboost	2.951
Lgbm	**2.276**

Table 4 Parameter tuning of Lgbm

Parameter	Value	Description
Max_depth	7	Maximum depth of a tree, this parameter can handle overfitting
Num_leaves	80	Maximum number of leaves in one tree
Cat_smooth	15	Used for categorical features to reduce the effect of noises
Drop_rate	0.2	A fraction of previous trees to drop during the dropout
Learning_rate	0.1	Controls the magnitude of change in the estimate of each tree's impact on the outcome
Bagging_fraction Categorical_features	0.9 categoricals	The fraction of data used in each iteration, to speed up the learning and avoid overfitting used to specify categorical features

Table 5 Performance of classification algorithms from other sources on SF Crime data

Source	Log loss of best classifier	Approach
[1]	**2.315**	Gradient boosting
[14]	2.550	Naïve Bayes
[15]	2.394	Gradient boosting
[16]	2.537	Bayesian model
[17]	2.333	Deep neural network
[18]	2.45	Random Forest

ensemble models are strong predictors. In this case study, Lgbm method achieved the best performance. Producing much more complex trees by following a leaf-wise split approach rather than a level-wise approach is one of the main factors that made Lgbm achieve better prediction results. Furthermore, Lgbm can easily handle categorical features which is an advantage since most of the features are categorical in this case study.

5 Conclusion and Future Work

In this research, we used machine learning techniques to solve a multinomial classification problem. The problem we are tackling with is predicting new crimes given historical crime data. This is important because predicting crime patterns can enhance the efficiency of police departments by learning about crimes before they occur.

San Francisco crime data from SF Police Department is used as a case study. After exploring the data, feature engineering was employed to boost the model performance. Then, various classification methods such as Multinomial logistic regression, Random forests, Light gradient boosting machine, and Extreme gradient boosting machine were adopted to predict the category of crime. Our best classifier, Light gradient boosting machine achieved a log loss score of 2.276 which outperformed other previous published models.

In summary, we explored four different classifiers to predict the crime category given spatiotemporal data. The process of feature engineering, model selection, and tuning resulted in better prediction than existing reported results. However, this study is subject to a few limitations which point out future research directions: First, more information such as societal and ecological features can help enhance prediction results; second, classifying the data into larger groups with smaller number of classes can give more insights; third, future work can extend on employing more complicated classifiers such as deep neural networks to learn the structure of the data better.

References

1. G. Alperovich, *Multi-class Classification Problem: Crimes in San-Francisco* (2016), pp. 1–5
2. M. Aly, Survey on multiclass classification methods.pdf. no. November (2005), pp. 1–9
3. L. Breiman, J. Friedman, C.J. Stone, R.A. Olshen, *Classification and Regression Trees* (Taylor & Francis, 1984)
4. S.D. Bay, Combining nearest neighbor classifiers through multiple feature subsets, in *Proceedings of the Fifteenth International Conference on Machine Learning* (1998), pp. 37–45
5. T.J. Watson, An empirical study of the Naive Bayes classifier (2001)
6. J. Engel, Polytomous logistic regression. Stat. Neerl. **42**(4), 233–252 (1988)
7. C.M. Bishop, *Neural Networks for Pattern Recognition* (Oxford University Press Inc., New York, NY, USA, 1995)
8. C. Cortes, V. Vapnik, Support-vector networks. Mach. Learn. **20**(3), 273–297 (1995)
9. T.M. Choi, J. Gao, J.H. Lambert, C.K. Ng, J. Wang, *Optimization and Control for Systems in the Big Data Era: An Introduction*, vol. 252 (2017)
10. T.G. Dietterich, Ensemble methods in machine learning, in *Proceedings of the First International Workshop on Multiple Classifier Systems* (2000), pp. 1–15
11. L.E.O. Breiman, Random forest(LeoBreiman).pdf (2001), pp. 5–32
12. Y. Freund, R.E. Schapire, A decision-theoretic generalization of on-line learning and an application to boosting. J. Comput. Syst. Sci. **55**(1), 119–139 (1997)
13. J. Friedman, Greedy Function approximation: a gradient boosting machine. Ann. Stat. **29**(5), 1189–1232 (2001)
14. S. Darekar, *Predicting and Analysis of Crime in San Francisco* pp. 1–25
15. J. Ke, X. Li, J. Chen, *San Francisco Crime Classification*. no. November (2015), pp. 1–7
16. C. Hale, F. Liu, CS 229 Project Report : San Francisco Crime Classification.
17. G.H. Larios, Case Study Report San Francisco Crime Classification (2016)
18. P. Date, *UCLA UCLA Electronic Theses and Dissertations An Informative and Predictive Analysis of the San Francisco Police Department Crime Data* (2016)

Optimizing Ensemble Weights for Machine Learning Models: A Case Study for Housing Price Prediction

Mohsen Shahhosseini, Guiping Hu and Hieu Pham

Abstract Designing ensemble learners has been recognized as one of the significant trends in the field of data knowledge, especially, in data science competitions. Building models that are able to outperform all individual models in terms of bias, which is the error due to the difference in the average model predictions and actual values, and variance, which is the variability of model predictions, has been the main goal of the studies in this area. An optimization model has been proposed in this paper to design ensembles that try to minimize bias and variance of predictions. Focusing on service sciences, two well-known housing datasets have been selected as case studies: Boston housing and Ames housing. The results demonstrate that our designed ensembles can be very competitive in predicting the house prices in both Boston and Ames datasets.

Keywords Machine learning · Optimal ensemble · Bias-Variance trade-off · House price prediction

1 Introduction

The world's economies have been shifted toward the service sector in the recent decades. The service-related economy accounts for 65% of the world's GDP as of 2017, which has observed a rise from 61% in 2006. In addition, services sector is the leading sector in 201 countries and service-related economy consists of more than 80% of total GDP for 30 countries [1]. This has led to more research in the service sector. The term "Service science, management, and engineering (SSME)" was first used by IBM to describe service science as an interdisciplinary approach to the study, design, and implementation of service systems [2]. Service science is defined as "an abstraction of service systems in the same way that computer science

M. Shahhosseini · G. Hu (✉) · H. Pham
Industrial and Manufacturing Systems Engineering, Iowa State University Ames, Ames, IA 50011, USA
e-mail: gphu@iastate.edu

© Springer Nature Switzerland AG 2020

H. Yang et al. (eds.), *Smart Service Systems, Operations Management, and Analytics*, Springer Proceedings in Business and Economics, https://doi.org/10.1007/978-3-030-30967-1_9

is an abstraction of computer-based information systems" [3]. In general, SSME focuses on system design, industry services, software and hardware implementation of service, and etc. [4].

One of the many disciplines of service sciences focuses on information processing services. These services collect, manipulate, interpret, and transmit the data to create value for the end user. Issues such as representation, infrastructure, and self-service are the most significant problems in these services [3].

Machine learning (ML) has been used as one of the powerful tools to deal with the data. Due to the flexibility, machine learning models have been developed in a variety of application domains such as agriculture, bioinformatics, financial trading, fraud detection, and smart city management [5]. Several studies have used machine learning algorithms for housing price predictions. In addition, ML models have been implemented on housing datasets for various types of prediction. C4.5 Decision Tree, RIPPER, Naïve Bayes, and AdaBoost ML algorithms have been designed to predict Virginia housing prices [6]. A hybrid of genetic algorithm and support vector machines (G-SVM) was proposed in [7] to forecast China housing prices. In another study, SVM was combined with particle swarm optimization (PSO) to forecast real estate prices [8]. Artificial neural networks (ANN) and hedonic regression were compared in predicting housing prices in Turkey using a household budget survey data from 2004 [9]. In an empirical study for residential estate appraisal, it was shown that Random Forests perform better than several ML techniques such as CART, KNN, multiple linear regression, ANN, and Boosted trees [10].

Despite the prediction accuracy achieved by individual ML models, ensemble learning has been proposed to improve prediction accuracy by aggregating predictions of multiple base learners [11]. The ensemble is typically constructed by weighting (in the case of regression) or voting (in the case of classification) the predictions of base learners. The final resulting ensemble often achieves better predictions in comparison to any of single base learners [12]. For instance, the winners of famous real-world data analysis competitions, such as Netflix Prize and KDD Cup 2013, have chosen ensemble approaches as their prediction strategies [13]. The merits of ensemble learners have generated increasing interests to incorporate this method in research and practice. The results of a comparative evaluation of three ensemble learning methods including Bagging, Boosting, and Stacking for credit scoring show the advantage of ensemble learners over base learners [14]. Three financial datasets were chosen to analyze the performance of ensemble learners for classification problem of bankruptcy prediction and it was demonstrated that ensemble learners outperform the best stand-alone method which was multilayer perceptron neural network [15].

The objective of this paper is to optimize machine learning predictions of ensemble learners by finding the best weights for constructing ensembles for house price prediction. Two housing datasets for Boston and Ames have been chosen to demonstrate and validate the optimization model. Multiple learners including LASSO regression, Random Forests, Deep Neural Networks, Extreme Gradient Boosting (XGBoost), and Support Vector Machines with three kernels (polynomial, RBF, and sigmoid) have been chosen as base learners for prediction. The predictions made by base learners are used as inputs of the proposed optimization model to find the optimal weights. The objective is to minimize the mean squared error (MSE) of the predictions, which account for both bias and variance of the predictions.

The paper is organized as follows. Materials and methods are introduced in Sect. 2. Section 3 is dedicated to the results and discussions and the paper is concluded in the last section.

2 Materials and Methods

Although shown in various studies that ensemble learners outperform individual base models, designing the optimal method to combine base models remains a significant problem. In many data science competitions, the winners are the ones who could identify the best way to integrate the merits of different models and achieve superior performance.

It has been shown that the optimal choice of weights aims to achieve minimal prediction error by designing the ensembles for the best bias and variance balance. Every predictive model contains error from bias and variance with the amount of each determined by the interaction between the data and model choice. Bias is defined as a model's understanding of the underlying relationship between features and target outputs; whereas, variance is the sensitivity to perturbations in training data [16]. Mathematically, for a given dataset $(X, Y) = \{(x, y) : x \in \mathbb{R}^{n \times m}, y \in \mathbb{R}^n\}$, we assume there exists a function $f : \mathbb{R}^{n \times m} \to \mathbb{R}$ with noise ϵ such that $y = f(x_i) + \epsilon$ where $\epsilon \sim N(0, 1)$.

Using any myriad of supervised learning techniques, we approximate $f(x)$ with $\hat{f}(x)$ [17]. We define the following:

$$Bias\left[\hat{f}(x)\right] = E\left[\hat{f}(x)\right] - f(x) \tag{1}$$

and

$$Var\left[\hat{f}(x)\right] = E\left[\hat{f}(x)^2\right] - E\left[\hat{f}(x)\right]^2 \tag{2}$$

Taking the mean squared error as the metric of precision, the objective to be minimized can be defined as

$$E\left[\left(y - \hat{f}(x)\right)^2\right] = \left(Bias\left[\hat{f}(x)\right]\right)^2 + Var\left[\hat{f}(x)\right] + Var(\epsilon) \tag{3}$$

The third term in the above formula, irreducible error, is the variance of the noise term in the true underlying function which cannot fundamentally be reduced by any model [16].

Naturally, a model with low bias and low variance is desired but not always producible. One common approach to reduce variance among models is to create a bootstrapped aggregated ensemble. Whereas boosting models is used to reduce bias. Each strategy possesses its strength and weaknesses, and finding the optimal balance between the two remains a challenging problem [18. 19].

Taking both bias and variance into account, mean squared error (MSE) has been chosen as the objective function in the mathematical model for optimizing ensemble weights.

$$Min \, MSE \left(w_1 \hat{Y}_1 + w_2 \hat{Y}_2 + \ldots + w_k \hat{Y}_k, Y \right)$$

$$s.t.$$

$$\sum_{j=1}^{k} w_j = 1 \tag{4}$$

$$w_j \geq 0, \forall j = 1, \ldots, k.$$

where w_j is the weights corresponding to base model j ($j = 1, \ldots, k$), $\hat{Y}_j$ represents the vector of predictions of base model j, and Y is the vector of actual target values. This optimization problem can be formulated as a quadratic programming problem.

$$Min \, \frac{1}{n} \sum_{i=1}^{n} \left(y_i - \sum_{j=1}^{k} w_j \hat{y}_{ij} \right)^2$$

$$s.t. \tag{5}$$

$$\sum_{j=1}^{k} w_j = 1$$

$$w_j \geq 0, \quad \forall j = 1, \ldots, k.$$

in which, n is the number of observations, y_i demonstrates actual target values of observation i ($i = 1, \ldots, n$), and $\hat{y}_{ij}$ is the prediction of observation i by base model j.

This model is a nonlinear convex program. Since the constraints are linear, the convexity can easily be proved by computing the Hessian matrix of the objective function. Therefore, since a local minimum of a convex function on a convex feasible region is guaranteed to be a global minimum, we can conclude that the optimal solution achieves global optimality [20].

We use Python's SciPy optimization library [21] to solve this problem. This library contains numerous algorithms for constrained and unconstrained optimization. For this study, we apply a sequential least squares programming (SLSQP) algorithm, a special case of sequential quadratic programming [22]. SLSQP utilizes the Han–Powell quasi-Newton method with a BFGS update resulting in robust results to an optimal solution [23].

Three measures have been used to evaluate the model performance. First, mean squared error (MSE) that is a measure of difference between predicted and observed values; second, mean absolute percentage error (MAPE) which expresses accuracy as percentage; and third, the coefficient of Determination (R2) that is defined as the proportion of the variance in the response variable that is explained by independent variables. The R2 ranges from 0 to 1, where values near 1 indicate a perfect fit of predicted values to the observed data.

Validating the results has been done with 10-fold cross-validation to estimate the true prediction error. In addition, the hyperparameters in each of models have been tuned by conducting a grid search with five-fold cross-validation.

The proposed optimization model has been applied on two well-known housing datasets. Next two sections describe the details of Boston and Ames housing datasets.

2.1 Boston Housing

This dataset was collected by the U.S. Census Service regarding housing information in the Boston metropolitan area. The dataset was originally published by Harrison, D. and Rubinfeld, D.L. in a study investigating methodological problems associated with the willingness to pay for clean air using Boston housing dataset [24]. The original dataset is small in size with 506 cases. A description of the variables is presented in Table 1.

Table 1 Boston dataset variables

Variable	Type	Description
CRIM	Numeric	Per capita crime rate by town
ZN	Numeric	Proportion of residential land zoned for lots over 25,000 sq.ft.
INDUS	Numeric	Proportion of non-retail business acres per town
CHAS	Numeric	Charles River dummy variable (1 if tract bounds river; 0 otherwise)
NOX	Numeric	Nitric oxides concentration (parts per 10 million)
RM	Numeric	Average number of rooms per dwelling
AGE	Numeric	Proportion of owner-occupied units built prior to 1940
DIS	Numeric	Weighted distances to five Boston employment centers
RAD	Numeric	Index of accessibility to radial highways
TAX	Numeric	Full-value property-tax rate per $10,000
PTRATIO	Numeric	Pupil–teacher ratio by town
B	Numeric	$1000(Bk-0.63)^2$ where Bk is the proportion of blacks by town
LSTAT	Numeric	% lower status of the population
MEDV[*]	Numeric	Median value of owner-occupied homes in $1000's

[*]Target variable

In order to make better predictions, we have scaled the input data to be in the $(0,1)$ range. Then, seven machine learning models including LASSO regression, Random Forests, Neural Networks, XGBoost, and SVM with three kernels (polynomial, RBF, and sigmoid) were applied on the dataset.

2.2 Ames Housing

Ames housing dataset was presented by De Cock in 2011 as an alternative to the Boston housing dataset. It describes the sale of individual residential property in Ames, Iowa from 2006 to 2010. The original dataset contains 2930 observations and 80 variables. This dataset is used in an ongoing Kaggle data science challenge started in 2016. In this competition, the dataset is split into a train set with a size of 1460 observations and a test set of 1459 observations. In this study, only the train set of this dataset is used to make predictions. Some of the important variables of this dataset are shown in Table 2.

Preprocessing tasks and data cleanings have been done on this dataset before applying models. These tasks include but not limited to the following:

- Removing outliers observed with two variables ("GrLivArea" and "LotArea")
- Imputing missing values for many of the variables
- Log-transformation of the target variable
- Log-transformation of the numeric input variables
- Removing highly correlated input variables ("GarageArea", "1stFlrSF", and "TotRmsAbvGrd")
- Constructing three new features with existing variables
- Converting categorical variables to numeric with One-Hot encoding.

Afterward, the following ML models were applied on this dataset to prepare the inputs for proposed optimization:

1. LASSO regression
2. Random Forests
3. Deep Neural Network
4. Extreme Gradient Boosting (XGBoost)
5. Support Vector Machines with polynomial kernel
6. Support Vector Machines with RBF kernel
7. Support Vector Machines with sigmoid kernel.

3   Results and Discussion

The results of each of seven base machine learning algorithms on Boston and Ames housing datasets are presented in Table 3. Each of these models is tuned with a five-fold cross-validation and the error rates shown in the table are estimates of 10-fold cross-validation. Based on the results for Boston housing dataset, XGBoost and

Table 2 Ames dataset variables (some of the variables are shown here)

Variable	Type	Description
YearBuilt	Numeric	Original construction date
Neighborhood	Categorical	Physical locations within Ames city limits
Street	Categorical	Type of road access
BldgType	Categorical	Type of dwelling
MSSubClass	Categorical	The building class
Foundation	Categorical	Type of foundation
LotArea	Numeric	Lot size in square feet
RoofStyle	Categorical	Type of roof
Bedroom	Numeric	Number of bedrooms above basement level
FullBath	Numeric	Full bathrooms above grade
TotalBsmtSF	Numeric	Total square feet of basement area
1stFlrSF	Numeric	First floor square feet
TotRmsAbvGrd	Numeric	Total rooms above grade (does not include bathrooms)
GrLivArea	Numeric	Above grade (ground) living area square feet
GarageCars	Numeric	Size of garage in car capacity
GarageArea	Numeric	Size of garage in square feet
OverallQual	Categorical	Overall material and finish quality
ExterQual	Categorical	Exterior material quality
KitchenQual	Categorical	Kitchen quality
BsmtQual	Categorical	Height of the basement
SalePrice[*]	Numeric	The property's sale price in dollars

[*]Target variable

Table 3 Base model results for Boston and Ames housing datasets

Error measure	LASSO	Random forests	Neural network	XGBoost	SVM (poly)	SVM (RBF)	SVM (sigmoid)
	Boston housing dataset						
MSE	35.579	22.479	28.336	21.367	44.315	26.578	34.577
MAPE	20.06%	16.35%	19.99%	16.44%	20.87%	16.06%	18.83%
R^2	0.5785	0.7337	0.6643	0.7469	0.4751	0.6852	0.5904
	Ames housing dataset						
MSE	0.0132	0.0183	0.4549	0.0368	0.0275	0.0681	0.0196
MAPE	0.66%	0.77%	3.72%	1.16%	1.00%	1.39%	0.82%
R^2	0.9167	0.8842	-1.8796	0.7669	0.8258	0.5692	0.8758

Table 4 Optimal ensemble weights

	w^a_{LASSO}	w^b_{RF}	w^c_{NN}	w^d_{XGB}	w^e_{SVM-p}	w^f_{SVM-r}	w^g_{SVM-s}	obj^h
Boston	0	0.113	0.280	0.508	0.073	0.003	0.023	18.901
Ames	0.742	0.221	0	0	0.037	0	0	0.0126

[a]LASSO optimal weight
[b]Random forests optimal weight
[c]Neural network optimal weight
[d]XGBoost optimal weight
[e]SVM (polynomial) optimal weight
[f]SVM (RBF) optimal weight
[g]SVM (sigmoid) optimal weight
[h]Objective function value

Random Forests are the best algorithms predicting the median price of the houses with the least MSE and MAPE and highest R-squared values. In other words, not only these two models predict with highest accuracy but also they explain the variation in the target more than other chosen models. Moreover, prediction results of Ames housing dataset find LASSO and Random Forests as the models with the least MSE and MAPE. These models could explain most of the variations in the target variable with having R-squared values of 0.92 and 0.88, respectively.

The prediction vectors of each of the above models are used in the optimization model to find the optimal weight of constructing ensembles with the base learners. Table 4 shows the obtained optimal weights. As it can be seen from the weights in the table below, the optimization model assigns the weight of zero to some models which means that the ensemble excluding these base learners will perform better. Furthermore, the objective function which is the mean square error of the ensemble is less than the MSE of all base learners for both datasets, that shows the strength of the optimal ensemble in predicting the targets.

The ensemble with equal weights ($w_i = 1/7$), which is a common practice among data scientists in order to construct ensembles out of some base learners, is considered as a benchmark. The error measures for the ensembles with optimal weights are calculated and compared with the benchmark in Table 5.

The ensembles with optimal weights outperform the benchmark ensemble as well as each of the base models for both datasets. This is demonstrated in Fig. 1. Comparing the error measures, the optimal ensemble has lower MSE, lower MAPE, and higher R^2 value.

4 Conclusion

A new optimization framework has been proposed in this study which optimizes the mean squared error of multiple base learners to find the optimal weights in designing a new ensemble from them. The designed formulation can result in ensembles

Table 5 Error rates for optimal and benchmark ensembles

Error measure	Optimal ensemble	Ensemble with equal weights (1/7)
	Boston housing dataset	
MSE	18.901	21.95
MAPE	%15.25	%15.26
R^2	0.7761	0.7399
	Ames housing dataset	
MSE	0.0126	0.0281
MAPE	%3.49	%3.34
R^2	0.9199	0.8222

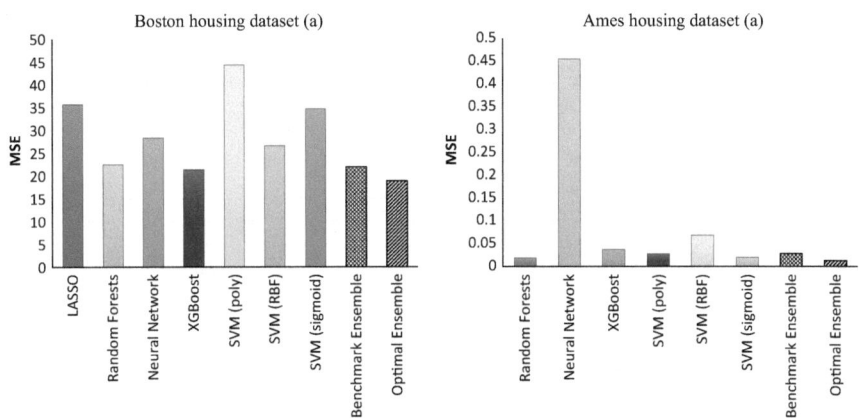

Fig. 1 Comparing optimal ensembles with benchmark ensembles and base learners

that minimize bias and variance of predictions. To validate the performance of the proposed methodology, two famous housing datasets, Boston and Ames datasets, were used as case studies. Seven machine learning algorithms including LASSO, random forests, neural networks, and XGBoost along with support vector machines with three kernels were considered as base learners. The created ensembles from the optimal weights found from our optimization model were compared to ensembles created from assigning equal weights to each individual learner and each of the base learners. The results showed that the designed ensemble can outperform the benchmark ensemble, as well as all the individual base learners.

The proposed methodology presented a systematic way to find the optimal weights of aggregating predictive learners to create better performing ensembles. This method performed better than each predictive learners in both housing datasets considered in this study. This methodology is generalizable to other datasets in other fields, given that the individual learners are accurate and diverse enough to effectively capture the structure of the data. This diversity in models is the reason for superiority of ensembles. Specifically, having different types of learners (e.g., linear and nonlinear learners) and aggregating these diverse models in a systematic way provides a way to represent different aspects of the data. Hence, this methodology is expected to be generalizable with the ability to predict better compared to initial single base learners.

For the future work, designing a methodology which incorporates finding best ensemble weights while tuning the hyperparameters of each base learner is recommended. This method can find the best hyperparameters and optimal weights of creating ensemble at the same time.

References

1. World development indicators. World Bank (1978)
2. B. Hefley, W. Murphy, Service Science, Management and Engineering: Education for the 21st Century. Springer Science & Business Media (2008)
3. H. Katzan, Foundations of service science concepts and facilities. J. Serv. Sci. 1(1) (2008)
4. G. Xiong, Z. Liu, X. Liu, F. Zhu, D. Shen, Service Science, Management, and Engineering: Theory and Applications. Academic (2012)
5. L. Breiman, Statistical modeling: the two cultures (with comments and a rejoinder by the author). Stat. Sci. 16(3), 199–231 (2001)
6. B. Park, J.K. Bae, Using machine learning algorithms for housing price prediction: the case of Fairfax County, Virginia housing data. Expert Syst. Appl. 42(6), 2928–2934 (2015)
7. J. Gu, M. Zhu, L. Jiang, Housing price forecasting based on genetic algorithm and support vector machine. Expert Syst. Appl. 38(4), 3383–3386 (2011)
8. X. Wang, J. Wen, Y. Zhang, Y. Wang, Real estate price forecasting based on SVM optimized by PSO. Opt.-Int. J. Light Electron Opt. 125(3), 1439–1443 (2014)
9. H. Selim, Determinants of house prices in Turkey: hedonic regression versus artificial neural network. Expert Syst. Appl. 36(2), 2843–2852 (2009)
10. E.A. Antipov, E.B. Pokryshevskaya, Mass appraisal of residential apartments: an application of random forest for valuation and a CART-based approach for model diagnostics. Expert Syst. Appl. 39(2), 1772–1778 (2012)
11. P.-N. Tan, M. Steinbach, V. Kumar, Introduction to data mining, addison, ed. by M.A. Boston, USA: Wesley Longman, Publishing Co., Inc (2005)
12. D. Talia, P. Trunfio, F. Marozzo, Data analysis in the cloud: models, techniques and applications. Elsevier (2015)
13. M. Sugiyama, Introduction to statistical machine learning. Morgan Kaufmann (2015)
14. G. Wang, J. Hao, J. Ma, H. Jiang, A comparative assessment of ensemble learning for credit scoring. Expert Syst. Appl. 38(1), 223–230 (2011)
15. L. Nanni, A. Lumini, An experimental comparison of ensemble of classifiers for bankruptcy prediction and credit scoring. Expert Syst. Appl. 36(2), 3028–3033 (2009)
16. J. Friedman, T. Hastie, R. Tibshirani, The Elements of Statistical Learning (no. 10). Springer series in statistics New York (2001)
17. L. Breiman, Bias, variance, and arcing classifiers (1996)
18. H. Pham, S. Olafsson, Bagged ensembles with tunable parameters. Comput. Intell. 35(1), 184–203 (2019)
19. H. Pham, S. Olafsson, On Cesaro averages for weighted trees in the random forest. J. Classif. (2019)
20. S. Boyd, L. Vandenberghe, Convex Optimization. Cambridge university press, Cambridge (2004)

21. E. Jones, T. Oliphant, P. Peterson, others. SciPy: Open source scientific tools for Python (2001). http://www.scipy.org
22. D. Kraft, A software package for sequential quadratic programming, Forschungsbericht-Deutsche Forschungs- und Versuchsanstalt fur Luft- und Raumfahrt (1988)
23. A. Wendorff, E. Botero, J.J. Alonso, Comparing different off-the-shelf optimizers' performance in conceptual aircraft design, in *17th AIAA/ISSMO Multidisciplinary Analysis and Optimization Conference* (2016), p. 336
24. D. Harrison Jr., D.L. Rubinfeld, Hedonic housing prices and the demand for clean air. J. Environ. Econ. Manag. **5**(1), 81–102 (1978)

How Do Pricing Power and Service Strategy Affect the Decisions of a Dual-Channel Supply Chain?

Houping Tian and Chaomei Wu

Abstract With the rapid development of e-commerce, an increasing amount of manufacturers are implementing a dual-channel strategy, i.e., introducing an online direct sale channel on the existing traditional retail channel. However, it is noted that for the small and medium-sized manufacturers, they may face the resistance of the dominant retailer caused by this strategy. It is well known as channel conflict, and some challenging issues remain to be further explored, e.g., How to price and coordinate the dual-channel under different pricing power structures? And, if possible, when should the retailer offer the value-added service to enhance his competitiveness and mitigate the channel conflict? The paper compares the decision models under a different power structure (i.e., Manufacturer-Stackelberg, Retailer-Stackelberg, and Vertical Nash) and service strategy (providing the value-added service or not); the analyses show two important findings. First, for the manufacturer or the retailer, they both prefer to be the leader to enjoy the power advantage and higher profits. However, for the entire supply chain and the consumers, they prefer equal power structure to the other two power arrangements. Second, the retailer will get more profits while providing value-added service. Interestingly, it is also beneficial to the manufacturer.

Keywords Pricing power · Service strategy · Dual-channel supply chain · Supply chain coordination · Game theory

1 Introduction

With the rapid development of the Internet, advances in Internet technologies and third-party logistics have attracted a growing number of manufacturers to engage in the online direct sale channel along with their preexisting traditional retail channel. Obviously, this dual-channel strategy is increasingly popular in the e-age.

H. Tian · C. Wu (✉)
School of Economics and Management, Nanjing University of Science and Technology, Nanjing 210094, People's Republic of China
e-mail: 1172507762@qq.com

© Springer Nature Switzerland AG 2020
H. Yang et al. (eds.), *Smart Service Systems, Operations Management, and Analytics*, Springer Proceedings in Business and Economics,
https://doi.org/10.1007/978-3-030-30967-1_10

Meanwhile, retailers also start to explore new sale models to enhance their competitiveness. They can attract customers by providing the value-added service, where the advantage is that retailers have direct contact with customers. In the PC industry, Dell has earned a reputation for providing high-quality customer support for software installation, maintenance, and hardware troubleshooting. However, some companies have not provided consumers with more attractive value-added service. Companies such as Wal-Mart, JD, etc., use electronic tags (online and offline price synchronization), which is convenient for consumers to purchase online or offline. Therefore, the retailer needs to decide whether to offer value-added service to consumers based on product types, consumer preferences, etc. Many scholars have done relevant researches on it. Yao and Liu (2005) explored the dual-channel pricing problem when the retailer provides the channel service [1]. Zhang et al. (2018) studied the dominant retailer to cope with the disadvantage of information inequality by deciding whether to provide services or not [2]. From the above studies, it can be seen that most of them take the service as the premise or treat the service as an exogenous variable. However, in this paper, we use the service as a decision-making variable and study the impacts on the decision-making of supply chain members in the case of the retailer providing services or not.

In a marketing channel, both the manufacturer and the retailer have "power", defined by EI-Ansary and Stern (1972) as "the ability of one channel member to control the decision variables in the marketing strategy of another member in a given channel at a different level of distribution" [3]. Different power structures also appear in real-world supply chains. Apple has a strong influence and can dominate the decision-making of the supply chain. Conversely, Wal-Mart can dominate supply chain decision-making when trades with most small manufacturers. When P&G trades with Wal-Mart, both sides have equal channel power and neither side can dominate supply chain decision-making. In this paper, we introduce three typical power structures: Manufacturer-Stackelberg, Retailer-Stackelberg, and Vertical Nash. Karray and Sigué (2018) studied the situation where the manufacturer dominated the market [4] and Chen et al. (2017) et al. analyzed the dominated retailer's optimal pricing decision [5]. Many scholars have combined the three kinds of power structures to make some management insights. Wei et al. (2013) explored the pricing issues for two suppliers and a retailer that produce complementary products [6]. Chen et al. (2016) indicated that for the channel members of the supply chain, the more power, the more profits. However, for the whole supply chain, the balanced distribution of power will bring more profits [7]. Luo et al. (2017) studied the optimal pricing strategy of differentiated brands under different power structures [8]. Through the above research, we can see that the power structure has a great impact on the dual-channel pricing, which has great inspiration for this paper.

This study makes two significant contributions to our understanding of supply chain management. One is to consider the effects of the value-added service on optimal pricing and profits of supply chain members. The other is to research the impacts of power structure on dual-channel pricing and service strategy.

The remainder of this paper is organized as follows. Section 2 introduces the game model with three power structures without services: M model (manufacturer-leader's Stackelberg game), R model (retailer-leader's Stackelberg game), and N model (Nash game). Then we give the key results. In Sect. 3, we consider the retailer who provides the value-added service, the same three power structures: M-RS model, R-RS model, and N-RS model. In Sect. 4, we conduct a comparative analysis of the power structure and the value-added service and draw relevant managerial insights. Section 5 will give some numerical examples to explore the above results. We make the conclusion in Sect. 6.

2 Game Model Without Service

Consider a dual-channel supply chain consisting of a manufacturer and a retailer. The manufacturer sells through the traditional supply chain of retailer and a direct channel through the Internet. The manufacturer produces a single product at a unit cost c. The retailer will resell the product through her own traditional channel. The online and offline products are homogeneous. Customers may use either the retail channel or the direct channel to purchase the product.

About the demand function, linear demand functions have been adopted in Chiang et al. (2003) and Huang and Swaminathan (2009), et al. [9, 10]. The corresponding demand functions to the manufacturer and the retailer are described as follows:

$$
\begin{aligned}
D_d &= \gamma a - b p_d + \theta(p_r - p_d) \\
D_r &= (1 - \gamma)a - b p_r + \theta(p_d - p_r)
\end{aligned}
\tag{1}
$$

D_d denotes the consumer demand from the direct channel, D_r denotes the consumer demand from the retail channel, and a represents the total basic demand size of the market. We can see that the total base demand is divided into two parts by the retail channel and the direct channel (γa and $(1 - \gamma)a$). γ represents consumers' loyalty to the online direct channel, where $\gamma \in [0, 1]$. When homogeneous customers perceive that there is a price difference between the two channels, they will switch to the low price channel to purchase products at the transfer rate θ.

2.1 M Model

The M model that refers to the manufacturer dominates the market and acts as a leader, the retailer is the follower. The manufacturer first decides the wholesale price and direct sale price to maximize his profit based on retailer's response. The retailer subsequently decides the retail price after observing the manufacture's decisions. The M model can be described as follows:

$$\max_{w,p_d} \pi_m = (p_d - c)D_d + (w - c)D_r$$

$$s.t.\ p_r^* \in \arg\max_{p_r} \pi_r = (p_r - w)D_r \qquad (2)$$

Employing the backward induction, we can obtain the optimal decisions and the profits as follows:

Proposition 1

$$w^M = \frac{c}{2} + \frac{ab(1-\gamma)+a\theta}{2b(b+2\theta)};\ p_d^M = \frac{c}{2} + \frac{a(\theta+b\gamma)}{2b(b+2\theta)};$$

$$p_r^M = \frac{4c\theta - 3a\gamma}{4(b+2\theta)} + \frac{3a}{4(b+\theta)} + \frac{a\theta(2\theta-b\gamma)+b^3c}{4b(b+\theta)(b+2\theta)};$$

$$\pi_r^M = \frac{[a(1-\gamma)-bc]^2}{16(b+\theta)};$$

$$\pi_m^M = \frac{a^2(b+\theta+2b\gamma^2)}{8b(b+2\theta)} + \frac{a^2[b^2\gamma(\gamma-2)+\theta^2]}{8b(b+\theta)(b+2\theta)}$$
$$- \frac{ac(b+2\theta+b\gamma)}{4(b+\theta)} + \frac{bc^2(3b+4\theta)}{8(b+\theta)}.$$

2.2 R Model

The R model represents a dual-channel supply chain where there exist one larger retailer and one relatively smaller manufacturer. The retailer controls the market and acts as the leader, the manufacturer is the follower. In this part, let $p_r = w + \mu$ (Liu et al. [11]; Shi et al. [12]), the retailer first decides the margin profit μ, then the manufacturer subsequently decides the wholesale price and direct sale price to maximize his profit after observing the decisions of the retailer. The R model is obtained as follows:

$$\max_{\mu} \pi_r = \mu D_r$$

$$s.t.w^*,\ p_d^* \in \arg\max_{w,p_d} \pi_m = (p_d - c)D_d + (w - c)D_r \qquad (3)$$

Like Proposition 1, we can obtain Proposition 2 as follows:

$$w^R = \frac{b(b+2\theta)(a+2c\theta+3bc)+a(2\theta^2-b^2\gamma)}{4b(b+\theta)(b+2\theta)};\ p_d^R = \frac{c}{2} + \frac{a(\theta+b\gamma)}{2b(b+2\theta)};$$

$$p_r^R = \frac{4c\theta - 3a\gamma}{4(b+2\theta)} + \frac{3a}{4(b+\theta)} + \frac{a\theta(2\theta-b\gamma)+b^3c}{4b(b+\theta)(b+2\theta)};$$

$$\pi_r^R = \frac{[a(1-\gamma)-bc]^2}{8(b+\theta)};$$

$$\pi_m^R = \frac{a^2\gamma(\gamma+2)}{8(b+2\theta)} + \frac{a^2(b+\theta)}{16b(b+2\theta)} + \frac{3a^2\theta^2}{16b(b+\theta)(b+2\theta)} + \frac{3a^2b\gamma(\gamma-2)}{16(b+\theta)(b+2\theta)}$$

$$- \frac{ac(b+4\theta+3b\gamma)}{8(b+\theta)} + \frac{bc^2(5b+8\theta)}{16(b+\theta)}.$$

2.3 N Model

In this section, the manufacturer and the retailer move simultaneously when making decisions. The N model can be described as

$$\begin{cases} \max_{\mu} \pi_r = \mu D_r \\ \max_{p_d,w} \pi_m = (p_d - c) + (w - c)D_r \end{cases} \tag{4}$$

Taken the partial derivative of the profit function of manufacturer and retailer with respect to the decision variable, we can get the following results:

Proposition 3

$$w^N = \frac{2a + 3c\theta + 4bc}{6(b+\theta)} - \frac{\gamma a}{b(b+2\theta)} + \frac{a\theta(3\theta - b\gamma)}{6b(b+\theta)(b+2\theta)}; \quad p_d^N = \frac{c}{2} + \frac{a(\theta + b\gamma)}{2b(b+2\theta)};$$

$$p_r^N = \frac{2a}{3(b+\theta)} + \frac{3c\theta - 2\gamma a}{3(b+2\theta)} + \frac{b^2c(2b+\theta) + a\theta(3\theta - b\gamma)}{6b(b+\theta)(b+2\theta)};$$

$$\pi_r^N = \frac{[a(1-\gamma)-bc]^2}{9(b+\theta)};$$

$$\pi_m^N = \frac{a^2(4b + 9\theta + 2b\gamma + 8b\gamma^2)}{36b(b+2\theta)} + \frac{5a^2(b\gamma^2 - 2b\gamma - \theta)}{36(b+\theta)(b+2\theta)}$$

$$+ \frac{bc^2(13b + 18\theta) - 2ac(4b + 9\theta + 5b\gamma)}{36(b+\theta)}.$$

3 Game Model with Service

In the previous chapter, we considered the manufacturer to redesign his traditional channel structures by engaging in direct sales. The mixed channel indeed has its advantages, which can attract more consumers for the manufacturer. However, it also leads to channel conflict, with the retailer feeling threatened by direct competition. One way of mitigating this channel conflict, where the retailer is allowed to add value to the products to differentiate its offering to the customers, is proposed in this paper. The level of the retail service is denoted by $v(v > 0)$. The retail services refer to all forms of demand-enhancing services provided by the retailer, which include the presale services (consulting services, product advertising, etc.), in-sale services (delivering products on time, etc.), and after-sales services (follow-up tracking services, etc.). These value-added services provided by the retailers can have a significant impact on channel choice and the loyalty of the customers.

The demand function is similar to the previous chapter. The difference is that the retailer provides value-added service, so the net price of traditional channel is $p_r - v$. We use a quadratic cost function for the service cost, $c_R = c_r v^2/2$, where c_r is an efficiency parameter for the retailer's value-added cost. The function is convex, and it is also used successfully by Tsay and Agrawal (2000), Yao et al. (2008), etc. [13, 14]:

$$D_d = \gamma a - b p_d + \theta(p_r - v - p_d)$$
$$D_r = (1 - \gamma)a - b(p_r - v) + \theta(p_d - p_r + v) \qquad (5)$$

3.1 M-RS Model

In this part, the manufacturer, as the leader, first announces the wholesale price and the online direct sale price. The retailer, as the follower, decides the offline retail price and service level. The problem is described as follows:

$$\max_{w, p_d} \pi_m = (p_d - c)D_d + (w - c)D_r$$

$$s.t. p_r^*, v^* \in \arg\max_{p_r, v} \pi_r = (p_r - w - c_R)D_r \qquad (6)$$

The following results can be obtained:

Proposition 4

$$p_d^{M-RS} = \frac{c}{2} + \frac{a(\theta + b\gamma)}{2b(b + 2\theta)}; \ w^{M-RS} = \frac{1}{4c_r} + \frac{c}{2} + \frac{a\theta + (1 - \gamma)ab}{2b(b + 2\theta)};$$

$$v^{M-RS} = \frac{1}{c_r}; \ p_r^{M-RS} = \frac{7}{8c_r} + \frac{3a}{4(b + \theta)} + \frac{4c\theta - 3a\gamma}{4(b + 2\theta)} + \frac{a\theta(2\theta - b\gamma) + b^3 c}{4b(b + \theta)(b + 2\theta)};$$

$$\pi_r^{M-RS} = \frac{[b + \theta + 2c_r((1 - \gamma)a - bc)]^2}{64c_r^2(b + \theta)};$$

$$\pi_m^{M-RS} = \frac{a^2(b + \theta + 2b\gamma^2)}{8b(b + 2\theta)} + \frac{a^2[b^2\gamma(\gamma - 2) + \theta^2]}{8b(b + \theta)(b + 2\theta)} - \frac{ac(b + 2\theta + b\gamma)}{4(b + \theta)}$$

$$+\frac{a(1-\gamma)-bc}{8c_r}+\frac{bc^2(3b+4\theta)}{8(b+\theta)}+\frac{b+\theta}{32c_r^2}.$$

3.2 R-RS Model

In this section, the retailer, as a leader, first determines the service level and offline margin profit. The manufacturer, as a follower, then decides the wholesale price and the online direct sale price

$$\max_{\mu,v} \pi_r = (\mu - c_R)D_r$$

$$s.t.\, w^*, p_d^* \in \arg\max_{w,p_d} \pi_m = (p_d - c)D_d + (w - c)D_r \qquad (7)$$

The following results can be found:

Proposition 5

$$p_d^{R-RS} = \frac{c}{2}+\frac{a(\theta+b\gamma)}{2b(b+2\theta)}; w^{R-RS}=\frac{1}{8c_r}+\frac{a+2c\theta+3bc}{4(b+\theta)}+\frac{a(2\theta^2-b^2\gamma)}{4b(b+\theta)(b+2\theta)};$$

$$v^{R-RS}=\frac{1}{c_r}; p_r^{R-RS}=\frac{7}{8c_r}+\frac{3a}{4(b+\theta)}+\frac{c\theta}{(b+2\theta)}+\frac{2a\theta^2+b^3c-abr(3b+4\theta)}{4b(b+\theta)(b+2\theta)};$$

$$\pi_r^{R-RS}=\frac{[b+\theta+2c_r((1-\gamma)a-bc)]^2}{32c_r^2(b+\theta)};$$

$$\pi_m^{R-RS}=\frac{a^2\gamma(\gamma+2)}{8(b+2\theta)}+\frac{a^2(b+\theta)}{16b(b+2\theta)}+\frac{3a^2\theta^2}{16b(b+\theta)(b+2\theta)}+\frac{3a^2b\gamma(\gamma-2)}{16(b+\theta)(b+2\theta)}$$
$$-\frac{ac(b+4\theta+3b\gamma)}{8(b+\theta)}+\frac{a(1-\gamma)}{16c_r}+\frac{bc^2(5b+8\theta)}{16(b+\theta)}-\frac{bc}{16c_r}+\frac{b+\theta}{64c_r^2}.$$

3.3 N-RS Model

The manufacturer and the retailer act simultaneously to optimize their respective profits

$$\begin{cases} \max_{\mu,v} \pi_r = (\mu - c_R)D_r \\ \max_{p_d,w} \pi_m = (p_d - c)D_d + (w - c)D_r \end{cases} \qquad (8)$$

So, we can get

Proposition 6

$$p_d^{N-RS} = \frac{c}{2} + \frac{a(\theta + b\gamma)}{2b(b + 2\theta)}; \; w^{N-RS} = \frac{1}{6c_r} + \frac{2a + 3c\theta + 4bc}{6(b + \theta)} - \frac{\gamma a}{6(b + 2\theta)}$$

$$+ \frac{a\theta(3\theta - b\gamma)}{6b(b + \theta)(b + 2\theta)};$$

$$v^{N-RS} = \frac{1}{c_r}; \; p_r^{N-RS} = \frac{5}{5c_r} + \frac{2a}{3(b + \theta)} + \frac{3c\theta - 2\gamma a}{3(b + 2\theta)} + \frac{b^2 c(2b + \theta) + a\theta(3\theta - b\gamma)}{6b(b + \theta)(b + 2\theta)};$$

$$\pi_r^{N-RS} = \frac{[b + \theta + 2c_r(a(1 - \gamma) - bc)]^2}{36c_r^2(b + \theta)};$$

$$\pi_m^{N-RS} = \frac{a^2(4b + 9\theta + 2b\gamma + 8b\gamma^2)}{36b(b + 2\theta)} + \frac{5a^2(b\gamma^2 - 2b\gamma - \theta)}{36(b + \theta)(b + 2\theta)}$$

$$+ \frac{bc^2(13b + 18\theta) - 2ac(4b + 9\theta + 5b\gamma)}{36(b + \theta)} + \frac{a(1 - \gamma) - bc}{9c_r} + \frac{b + \theta}{36c_r^2}.$$

4   Comparisons and Managerial Implications

In this section, we make comparisons and analyses of performances in different market power structures and whether value-added service is provided or not. After that, we derive some managerial insights.

(a)   Analyses of channel power structures

After comparison, it is found that the conclusions about power structures are similar whether the retailer provides the service or not. Therefore, we only select the retailer who provides the service to make an explanation. Here, we have two Corollaries.

Corollary 1

(1) $v^{M-RS} = v^{R-RS} = v^{N-RS} \equiv 1/c_r$.
(2) $p_d^{M-RS} = p_d^{R-RS} = p_d^{N-RS}, w^{M-RS} > w^{N-RS} > w^{R-RS}, p_r^{M-RS} = p_r^{R-RS} > p_r^{N-RS}$.

It can be seen that the service level provided by the retailer is the same whether the manufacturer or the retailer occupies a dominant position in the market or not. In terms of the service, it is the retailer's unique competitive advantage. Regardless of the channel status, the service level decision only depends on the service efficiency, it increases with the service efficiency. In Corollary 1(2), for the manufacturer, the direct sale price is the same despite the power status of supply chain members. This result is similar to Corollary 1(1). The wholesale price decreases as the power shifts from the manufacturer to the retailer. It can be seen that the manufacturer is very sensitive to the power structure when deciding the wholesale price. For the retailer, the optimal retail price in M-RS model is equal to that in R-RS model. It means that the retailer with dual-channel has more flexibility in different power structures.

In addition, the retailer's optimal retail price is lowest in N-RS model. It indicates that consumers are generally better off in a symmetric power structure. This can be explained by the fact that when the manufacturer and the retailer have a more balanced power in the N-RS model, there is a more intense competition which drives the prices down.

Corollary 2

$$\pi_r^{R-RS} > \pi_r^{N-RS} > \pi_r^{M-RS}, \pi_m^{M-RS} > \pi_m^{N-RS} > \pi_m^{R-RS}, \pi^{N-RS} > \pi^{M-RS} = \pi^{R-RS}.$$

It can be seen from above that a member can get the highest profit by relying on his own leadership, but the profit of his competitors is the lowest at that time. The profit conflict will intensify the competition among members and the loss of the total profits of the supply chain. When the power is symmetric, neither party occupies the leading position. The profit of each member in a middle level and the competition among members be less fierce. Therefore, from the perspective of the total profits of the supply chain and consumers, the balanced power is the optimal decision.

(b) Analyses of the channel service

By comparing, we found that the impacts on the value-added service are similar across different power structures. Therefore, we only select N model and N-RS model to make an explanation.

Corollary 3

$$w^{N-RS} > w^{N}, p_d^{N-RS} = p_d^{N}, p_r^{N-RS} > p_r^{N}.$$

From Corollary 3, we can find that the manufacturer's wholesale price is higher when the retailer offers the value-added service. Because the retailer will become more competitive after providing services and attract more consumers. In order to gain more offline profit, the manufacturer can raise the wholesale price. The manufacturer's online direct sale price is the same regardless of whether the retailer provides services or not. The online channel is the manufacturer's unique channel which is his advantage. The retailer provides services or not has no effect on the pricing of the direct channel. It is obvious that when the retailer offers the value-added service, which can improve the value of products, the retailer price rises.

Corollary 4

$$\pi_r^{N-RS} > \pi_r^{N}, \pi_m^{N-RS} > \pi_m^{N}, \pi^{N-RS} > \pi^{N}.$$

From Corollary 4, we can see that when the retailer provides the value-added service, both the manufacturer and the retailer enjoy the higher profits. It is a win-win situation, so the manufacturer is also willing when the retailer chooses to offer value-added service.

5 Numerical Study

In this section, we analyze the impacts of system parameters on optimal pricing decisions through numerical studies. Then we illustrate our analytical results about the power structure and the value-added service and find subtler explanations through the table data. We select to illustrate the optimal decisions in N-RS model because their performances are moderate among all the situations we have considered. The other models can be analyzed similarly, so we omit them for saving space.

(a) The impacts of service efficiency.

Consider the following parameters: $a = 100, b = 1, c = 10, \gamma = 0.5, \theta = 0.1, c_r \in [0.1, 0.5]$. Figure 1 shows the impacts of the service efficiency on optimal prices and profits of supply chain channel members. In the left side of Fig. 1, it shows that with the increase of c_r, the optimal retail price, the wholesale price and the value-added service are decreasing, while the direct sale price remains unchanged. The reason, as the retailer's service efficiency shrinks, the value-added service level provided to consumers will eventually decrease, so will the corresponding the retail price. For the manufacturer, he lowers the wholesale price in order to encourage the retailer to provide the value-added service. The online direct channel is the proprietary channel for the manufacturer so that the optimal price is not related to the service efficiency. From the right side of Fig. 1, it is obvious that profits of all supply chain members decrease as the service efficiency shrinks.

(b) The impacts of consumers' channel loyalty.

Here we set $c_r = 0.25, \gamma \in [0, 1]$. We can get Fig. 2, which indicates the influences of consumers' channel loyalty on optimal prices and profits. In the left, with the increase of γ, the optimal retail price and the wholesale price are decreasing, the direct sale price is increasing, while the value-added service remains unchanged. As consumers become more loyal to the online channel, consumers are more dependent on the online channel, resulting in the increasing of the direct sale price. Inversely,

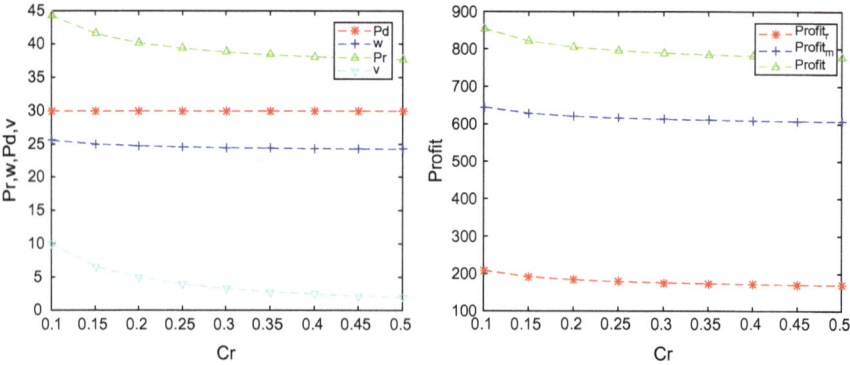

Fig. 1 Variations of optimal prices and profits with c_r in N-RS model

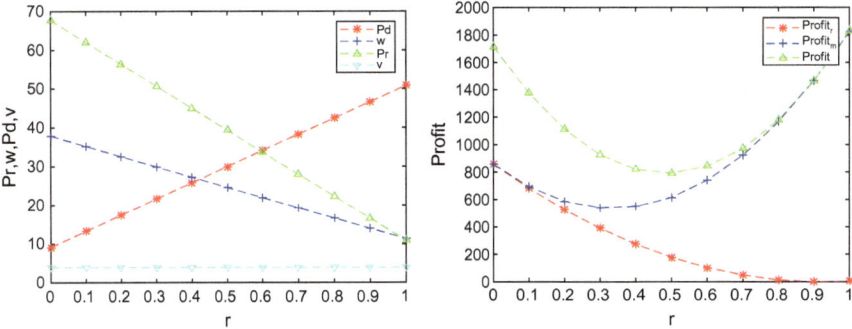

Fig. 2 Variations of optimal prices and profits with γ in N-RS model

the retailer needs to lower the offline retail price to retain consumers. Similar to the above, the manufacturer needs to encourage the retailer to provide the value-added service owing to the low offline channel loyalty, thus reducing the wholesale price. In the right, with the increase of consumers' loyalty to the online direct channel, the retailer will lose part of consumers, resulting in lower profit. The profit of the manufacturer and the total profits of supply chain are in a convex shape. At the beginning, the direct channel is not mature enough and the majority of consumers still trust the offline channel more. Although the online channel demand is rising, the loss of the offline channel is greater, involving the decreasing profits. Later, the online advantage gradually increased with the maturity of the online channel, so the profits are increasing.

Next, we briefly discuss the results of our extensive numerical to verify some of the analytical findings and gain more managerial insights. The parameters are set as $\gamma = 0.5$, $c_r = 0.25$. The results are shown in Table 1.

Table 1 indicates that both the manufacturer and the retailer can be benefited with the value-added service than without it. In addition, it shows that both of them prefer to be the leader to enjoy the higher profits. As to the total profits of the supply chain, N model and N-RS model have the highest total profits among the different power and service arrangements. So a balanced power structure is beneficial to the supply chain. These data are consistent with the conclusions above.

Table 1 Profits of the different models

	π_r	π_m	π
M model	90.91	618.18	709.09
R model	181.82	527.27	709.09
N model	161.61	597.98	759.60
M-RS model	101.18	638.73	739.92
R-RS model	202.37	537.55	739.92
N-RS model	179.88	616.25	796.13

6 Summary

In this paper, we consider a dual-channel supply chain with a manufacturer and a retailer. The manufacturer sells through the traditional retailer and the online direct channel to the customer. First, we established three power games for the supply chain members in the case that the retailer doesn't provide the value-added service. Then, the value-added service is considered in the three power games. Lastly, we have made a comparative analysis and obtained relevant management insights in the above cases.

The main conclusions of this study are summarized as follows. First, the direct channel, the value-added service is the advantage of the manufacturer and the retailer, respectively. Second, whether for the manufacturer or the retailer, who is intended to be the leader due to the higher profit. But a balanced power structure is best for the entire supply chains and the consumers. Third, for each member of the supply chain, providing the value-added service will generate more profits. Moreover, the manufacturer will encourage the retailer to provide the value-added service when the service efficiency is low or the retailer is in a weak power position.

Our paper can be extended in the following two aspects. First, we only consider the value-added service provided by the retailer. However, it can also be provided by the manufacturer. So further study can compare the cases that which one is better to provide the service. Second, we investigated only three extreme power structures (i.e., M-Stackelberg, R-Stackelberg, and equal pricing power), the further research can model the decisions of the players under a general pricing power structure.

Acknowledgements This research was supported by the National Natural Science Foundation of China Grant 71402074, 71472089, and Fundamental Research Funds for the Central Universities Grant 30920130111011.

References

1. D.Q. Yao, J.J. Liu, Competitive pricing of mixed retail and e-tail distribution channels. Omega **33**(3), 235–247 (2005)
2. S. Zhang, J. Zhang, G. Zhu, Retail service investing: an anti-encroachment strategy in a retailer-led supply chain. Omega (2018)
3. A.L. El-Ansary, L.W. Stern, Power measurement in the distribution channel. J. Mark. Res. **9**, 47–52 (1972)
4. S. Karray, S.P. Sigué, Offline retailers expanding online to compete with manufacturers: strategies and channel power. Ind. Mark. Manage. **71**, 203–214 (2018)
5. X. Chen, H. Zhang, M. Zhang et al., Optimal decisions in a retailer Stackelberg supply chain. Int. J. Prod. Econ. **187**, 260–270 (2017)
6. J. Wei, J. Zhao, Y. Li, Pricing decisions for complementary products with firms' different market powers. Eur. J. Oper. Res. **224**(3), 507–519 (2013)
7. X. Chen, X. Wang, X. Jiang, The impact of power structure on the retail service supply chain with an O2O mixed channel. J. Oper. Res. Soc. **67**(2), 294–301 (2016)

8. Z. Luo, X. Chen, J. Chen et al., Optimal pricing policies for differentiated brands under different supply chain power structures. Eur. J. Oper. Res. **259**(2), 437–451 (2017)
9. W.Y.K. Chiang, D. Chhajed, J.D. Hess, Direct marketing, indirect profits: a strategic analysis of dual-channel supply-chain design. Manage. Sci. **49**(1), 1–20 (2003)
10. W. Huang, J.M. Swaminathan, Introduction of a second channel: implications for pricing and profits. Eur. J. Oper. Res. **194**(2), 258–279 (2009)
11. B. Liu, R. Zhang, L. Ma, et al., Pricing decisions in a dual channels system with different power structures, in *IEEE*, vol. 29, No. 2, pp. 523–533 (2012)
12. R. Shi, J. Zhang, J. Ru, Impacts of power structure on supply chains with uncertain demand. Prod. Oper. Manag. **22**(5), 1232–1249 (2013)
13. A.A. Tsay, N. Agrawal, Channel dynamics under price and service competition. Manuf. Serv. Oper. Manag. **4**(2), 372–391 (2000)
14. D.Q. Yao, X. Yue, J. Liu, Vertical cost information sharing in a supply chain with value-adding retailers. Omega **36**(5), 838–851 (2008)

Designing Value Co-creation for a Free-Floating e-Bike-Sharing System

Christoph Heitz, Marc Blume, Corinne Scherrer, Raoul Stöckle and Thomas Bachmann

Abstract Value co-creation requires a system that links actors together for mutual value creation. In our paper, we describe the development of such a system in the context of the new free-floating e-bike-sharing system (BSS) in Zurich, Switzerland. This BSS is based on the idea that users of the BSS co-create value by adapting their usage behavior such that the overall service level is maximized. This creates value for other users and reduces the provider's costs for redistribution, but requires some kind of incentive system for influencing the user behavior. We describe a systematic approach of designing such a system by operationalizing the concept of value and value generation for the different actors: What exactly is the value that is to be created, and how can it be measured? By which activities is value created, and what are the options for stimulating these activities? Which design options maximize value creation? We found that this required combining two different research approaches: Empirical social research was necessary to understand user needs, value perception, motivational patterns in response to incentives, and communication needs. Operational research was necessary for assessing different options for the incentive system with respect to the value creation both for provider and users. By interlinking both research activities, we were able to design an incentive system that allows reducing the number of bikes by 30% without diminishing the service level. Users are offered a reward for dropping-off their bikes in dynamically changing reward zones whose locations are determined based on the bike distribution and the future demand pattern. These incentives lead to two distinct behavioral responses which were assessed

C. Heitz (✉)
Institute for Data Analysis and Process Design, Zurich University of Applied Sciences, Winterthur, Switzerland
e-mail: heit@zhaw.ch

M. Blume · C. Scherrer
Institute for Marketing Management, Zurich University of Applied Sciences, Winterthur, Switzerland

R. Stöckle
smide KmG, Uster, Switzerland

T. Bachmann
La Mobilière, Bern, Switzerland

© Springer Nature Switzerland AG 2020   113
H. Yang et al. (eds.), *Smart Service Systems, Operations Management, and Analytics*, Springer Proceedings in Business and Economics,
https://doi.org/10.1007/978-3-030-30967-1_11

and quantified in an extensive real-life field test during a period of 13 weeks. The impact of the measured behavioral change on service level and the required number of bikes was modeled via simulation.

Keywords Value co-creation · Bike sharing · Operations research · Social research · User co-production · Operationalizing of value creation · User behavior modeling · Simulation · Design of service systems

1 Introduction

Free-floating bike-sharing systems (BSS) are relatively new. In such systems, bikes can be dropped off at any location within a specified area, and no docking stations are needed. Bikes can be located and unlocked via smartphone. In recent years, free-floating BSS have been rapidly introduced worldwide, for example, by *Call a Bike* (Germany) or *LimeBike* (USA). In Switzerland, the first free-floating e-bike system has been launched in 2017 in Zurich by the insurance company La Mobilière and is now being operated by the company *smide* (www.smide.ch).

Free-floating systems avoid expensive docking stations. On the other hand, the redistribution of bikes from areas where they are dropped off to areas where they are needed is much more costly than for station-based systems, because the bikes are distributed over the complete area rather than concentrated at few stations. This is a particular critical issue since redistribution is typically the largest cost factor of a BSS even for station-based BSS [1]. In order to reduce the redistribution costs, several authors have suggested to transfer at least a part of the redistribution task to the users, which is called user-based redistribution [2, 3]. Nearly, all of the literature, however, is focused on station-based systems, and to our knowledge, there are no publications on user-based redistribution for free-floating systems.

In this paper, we study the problem of designing a system implementing user-based redistribution in the context of a free-floating BSS. We frame this problem as a problem of value co-creation: Users contributing to redistribution are creating value for other users by making bikes available where they are needed. At the same time, they are creating value for the operator because they save costs. We operationalize the concept of value and value generation for the different actors by defining quantitative measures for the created value, and linking value creation to behavior change of users which may be stimulated by incentives. With simulation methods, different options for incentive systems were assessed in terms of value creation, revealing possible value generation mechanisms and their potential. With empirical social research, we evaluated user needs and behavioral patterns, value perception in the context of a drop-off incentive system, response to incentives, and communication needs. By combining both results, we were able to design an incentive system that allows reducing the number of bikes by about 30% without diminishing the service level. Users are offered a reward for dropping-off their bikes in dynamically changing reward zones which are based on the bike distribution and the future demand pattern.

2 Operationalizing Value Co-creation for User-Based Redistribution

The basic value proposition of a BSS is that it satisfies the transportation demands of users. For a free-floating BSS, users are not forced to predefined stations but can drop-off their bike wherever they want (and are legally allowed to park a bike), which increases the service value compared to station-based BSS. Operationally, however, value can only be created if a bike is available at the place where and at the time when a transportation demand arises. So, the value creation depends on the service level β, which we define as the probability that a user finds a bike within maximum walking distance of about 300–500 m [3]. The higher the service level, the more value is created. The optimum situation is given for $\beta = 1$, where all transport demands can be met. Of course, many other variables might influence the value perception, for example the state of the bikes, but in the context of studying user-based redistribution, the service level is a natural and well-defined measure for the value creation of the BSS.

The BSS operator's goal is to maximize the service level at minimum costs. There are two different options of improving the service level: One is to increase the number of bikes in the serviced area: By increasing the number of bikes, it is always possible to achieve a high service level. This, however, has a direct effect on the costs. The other way is to redistribute bikes: If it were possible to move a dropped bike immediately to the place where the next demand event will occur, a perfect service level could be realized even with few bikes. From the perspective of the BSS operator, user-based redistribution in a BSS creates value by allowing to provide a given service level with less bikes, or increasing the service level with a given fleet size.

User-based redistribution requires the stimulation of specific user behavior in a controlled way. In the context of our research, we focused on mechanisms for influencing the dropping behavior, i.e., for motivating users to drop their bike at specific locations, by offering incentives. These incentives depend on the current bike distribution and the future spatiotemporal demand pattern, as the purpose is to make sure that bikes are dropped at locations where they will be needed in the (near) future (see [4] for details). Thus, rewards are dynamic and constantly changing. Empirical evidence shows that offering such incentives might lead to two different behavioral responses: Users might change their drop-off location at the end of a ride, or the offered reward might lead to additional rides. In the latter case, users might, for example, choose to use the BSS instead of public transport, because they can earn a reward for this ride.

The effect of an incentive system in terms of value creation depends on (a) when and where dropping incentives are offered, and (b) how users react to those incentives. While (a) is the basic design problem that the service provider has to solve, (b) can be assessed by empirical research. For our approach, we solved these questions independently: The question of how to design the incentive system was approached via stochastic simulation, i.e., by Operations Research methods. This allowed us to understand in detail how much the service level can be influenced by different behavioral changes of users, and how the dynamical incentive system should work. The question of how users respond to incentives and on which factors the reaction depends was solved by methods borrowed from empirical social research. This helped us to understand which kind of incentives should be offered, which motivational structures could be addressed, and on which factors the perceived value depends.

The link between the two approaches is a model of how users respond to drop-off incentives. As the simplest approach, we modeled the users' behavioral changes with two parameters: The probability p_1 that a user is willing to drop-off the bike not at his exact destination but at some other place near the destination (within maximum walking distance), and the proportion p_2 of additional trips that are generated by the incentives. While the simulation environment allowed to study the positive effect of incentives as a function of p_1 and p_2 and study different options of when and where incentives are offered, the empirical research with users allowed to estimate the quantitative level of response and to determine which factors influence this level.

3   Operations Research for Design of Incentive System

We developed a simulation model for analyzing the effect of user incentivation in BSS. We study a quadratic area A of 3 km × 3 km with a total number of N bikes. The dynamics of the bike distribution is given by the following process: Demand occurs as a Poisson process in a random cell. If a bike in the environment is available, it is removed. The destination cell is chosen randomly and the bike is dropped off, while incentives might change the drop-off location. The details of the simulation model can be found in [4]. Similar to [3] and [2], we define the service level β as the probability that a demand is met, i.e., that a bike is available within the walking distance. The studied dynamical system is balanced: pick-up rates and drop-off rates are equal in each cell. This is unrealistic, of course, but we chose this configuration because it allows to study the effect of user-based redistribution directly, without the need for explicitly modeling the operator's redistribution activity. The modeled situation mimics a redistribution strategy where the operator focuses his redistribution activity on counterbalancing macroscopic and systematic flows, but ignores the stochastic fluctuations of the bike distribution.

Depending on the current bike distribution, the future expected demand pattern and the future dropping behavior, for each location the value of dropping a bike here is calculated. The earlier the dropped bike would be needed to avoid a service failure, the higher is this value contribution [4]. We assume that the resulting value map is mapped into a reward map such that higher rewards are offered at locations where a drop would create more value. The reward function is communicated continuously to all users via their smartphone. Rewards are offered in advance, and the reward only depends on the dropping location [4].

We assume that offering a location-based drop-off incentive has two different behavioral effects: (a) With probability p_1, users choose to drop-off their bike some-where close to, but not directly at, their destination, thus maximizing their reward. (b) With probability p_2, an additional trip is generated. For example, $p_2 = 0.1$ means that 10% additional trips are generated by offering the incentives, compared to not offering incentives at all. For these additional trips, we assume that the pick-up distri-bution is the same as before, but the drop-off location is the location with the highest reward within the whole area. This is motivated by our empirical findings with users of *smide*, indicating that additional trips are mainly generated by high rewards.

In Fig. 1, the resulting service level for different combinations of p_1 and p_2 are shown. It can be seen that both mechanism of behavioral change leads to substantial increase in the service level, and for achieving a given service goal, the number of needed bikes can be strongly reduced. For a service level goal of 0.95, without user incentives, around 250 bikes would be needed for the simulated area of 9 km^2. It can be seen that, even with a modest response to the incentives, a substantial reduction of bikes can be obtained. This nonlinear behavior is consistent with earlier findings (e.g., [3]). As an example, with 10% additional trips and a 30% user response for drop-off displacement of 300 m, the number of bikes could be reduced by 50%.

The simulations show the impact of user incentivation as a function of the user response p_1 and p_2. They clarify the value creation both for the users (increase of service level) and for the provider (potential for reducing the bike fleet) and the mutual relation between these two quantities. In addition, the influence of the basic parameters p_1 and p_2 of the users' reponse to the incentives can be assessed. These results yield a fundamental understanding of how value co-creation by user-based redistribution can be achieved. A crucial element, of course, are the response parameters p_1 and p_2. In the next section, we describe our empirical social research for understanding which factors determine the user response to incentives. In Sect. 5, results of a field test are shown that allow a first quantification of these parameters.

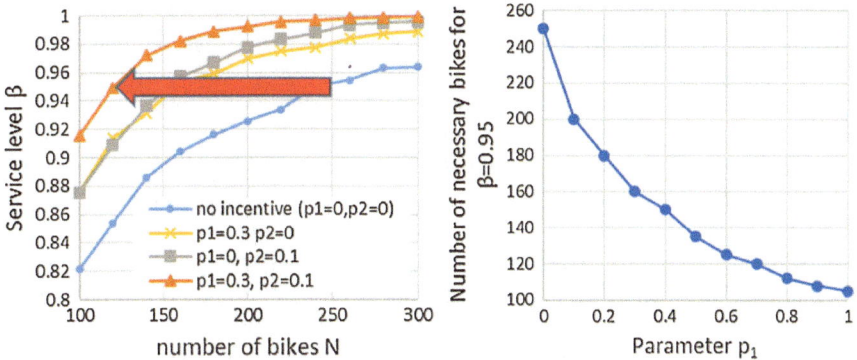

Fig. 1 Left: Service level as a function of number of bikes for different parameter values of p_1 and p_2. For a service level goal of 95%, the number of necessary bikes can be reduced by 50% with even modest participation of users. Right: Sensitivity analysis for p_1 (with $p_2 = 0$)

It must be noted that the shown simulation model is still rather simplified. It models only the main features of the incentivation system (offer higher reward for locations where a drop is more likely to avoid a service failure in the future), but not yet the detailed way of how exactly the reward map is defined. Different implementations might be possible. For the user research described in Sect. 4, we assume that the reward map consists of so-called *bonus zones*, defined as small circular areas with a radius of about 50 m. A reward is given if a bike is dropped with this zone. Bonus zones are set in areas where the value is high, while two reward zones have a minimum distance of at least 500 m. It can be shown that this minimizes wastage (i.e., paying rewards for users that did not change their behavior), thus minimizes incentive costs. The optimization of number and spatial distribution of such bonus zones, as well as their reward levels, is a complicated optimization problem by itself that we did not investigate in-depth and that requires further research. Note that the response parameters p_1 and p_2 are a necessary ingredient for any control logic setting bonus zones.

4 Empirical Social Research: Understanding the Motivation of Co-create Value

While the OR research (Sect. 3) was focused on understanding how, technically, users might co-create value in a free-floating BSS, the second stream of research activities focused on the optimum design of an effective incentive system in this context. To this aim, we tried to understand motivation patterns of bike-sharing users, in particular with respect to changing their usage behavior, in order to design an incentive system that is aligned with users' needs and creates optimal value for them. This is the classical domain of service design, covering value perception, value proposition design, and communication issues. Methodologically, we followed classical service design approaches [5–7], but with specific emphasis on value co-creation.

As described above, value co-creation happens if users drop-off their bike at locations where it will be needed in the near future. However, the biggest advantage of a free-floating BSS compared to a station-based system is the possibility to drop-off a bike *directly at the destination* as opposed to somewhere else. In terms of value perception, asking for parking a bike *not at the destination* but somewhere else is thus somewhat contradictory. It is of upmost importance for the success of the co-creation idea to solve this contradiction by an adequately improved value proposition.

In a first step, we investigated the basic personal and situational characteristics that influence the willingness of users to adapting their drop-off location. Three focus groups [8] were conducted in October and November of 2016 to pinpoint these characteristics. Each focus group consisted of eight participants, distinguished by their current usage pattern: (a) *People who rarely* ride bicycles (0–3 days per week), and are interested in a BSS, (b) *People who often* ride bicycles (4–7 days per week), and are interested in a BSS, and (c) People who were already *active users* of the *smide* e-bike-sharing service in Zurich. We conducted focus group discussions, moderated by a member of the project team and covering three main topics: (a) Individual pattern of bike usage, (b) Free-floating BSS: understanding and appeal of a free-floating BSS, price sensitivity, and presumptive (or for focus group #3, actual) use of a free-floating BSS, and (c) Incentive system: Appraisal of bonus zone concept, personal and situational willingness to adapt the drop-off location of a bike, and judgement of different types of rewards.

The gathered data shows that the most important factor for perceived value of a BSS is the availability of a bike within walking distance of about 300–400 meters. In the eyes of the participants, it was considered a service failure if the nearest e-bike is hard or time consuming to access, increasing the likelihood of abstaining from using the service. This result justifies the choice of the service level as the main optimization criterion (see Sect. 3). Participants also confirmed a perceived value reduction if a drop-off location different to their destination was to be chosen. The possible socially grounded motivation of "helping other users to minimize their time and effort by providing a bike where it is needed" was acknowledged by most participants, but was not judged as sufficient for a behavioral change, and the need for a direct reward in exchange of choosing an inferior location for drop-off became clear.

We also investigated how the drop-off incentive system could be motivated and explained to users, and how this influenced acceptance and behavioral response. We found that the most appealing way of framing the bonus zone concept was to emphasize the *combination* of personal benefit (reward) and shared social benefit for all users by the improved service level. An online survey with 246 *smide* users in December of 2017 later confirmed this early finding. In an open-ended question why they liked the bonus zones, 109 out of 286 positive comments stressed the "Win-win"-situation of earning free rewards in combination with doing good for the fellow *smide* users.

In the three focus groups, we presented basic alternatives of the type of reward. Social rewards like virtual batches [9], or privileges like reservations of bikes, did evoke some limited sympathy. However, the incentive that attracted by far the most interest was the option to use the reward for future rides. To our surprise, receiving free minutes for future rides was seen more attractive than the corresponding monetary value, especially if the number of offered minutes reached or exceeded the total duration of the current trip. For example, receiving 8 min for the next ride was seen as more attractive than receiving 2.00 Swiss Francs—the corresponding cost of buying 8 min of use. This finding can be explained by the concept of value perception: Value is not an objective measure, but a uniquely defined personal and psychological judgment [10, 11]. A compelling value communication is based on highlighting concrete benefits that are relevant for the customer [12]. It seems that, in the context of a BSS, free minutes have a higher relevance than money, and are thus judged more valuable.

Pronounced individual differences with respect to the willingness to co-create value by adapting the drop-off location showed up. Some participants of the focus groups profusely rejected the idea of contributing to an improved service level by parking their bike at a location suggested by the system. In their eyes, any thought of parking the bike away from their individual destination contradicts to two expectations: (a) since they paid for this service, they felt it was not their job to take care of the service quality and (b) since the option to park the bicycle right at their destination was one important reason why they consider bicycling as superior to other means of transportations, sacrificing this option made the service worthless. In contrast, to other users, the idea of parking their bikes at a location that is ideal for the next user was not a strange thought. They acknowledged the benefit of helping others to easily reach a bike. Like holding the door open for a stranger, they saw not an immediate benefit, but a midterm and indirect benefit for the good of the entire social group of bike-sharing users, including themselves.

These results suggested that there are different types of users, distinguished in their motivation to co-create value by changing their drop-off location. In fact, the results of the field test (see Sect. 5) show a pronounced difference of response to the offered drop-off incentives. In order to understand the underlying motivational motives, we separated the users into three distinctive groups according to their self-reported percentage of consciously choosing a bonus zone for dropping-off their bike. The first group consisted in persons ignoring the bonus zones entirely, the second group consisted of users that occasionally chose a bonus zone, and the third group were users that said that they took extensive advantage of bonus zones. We conducted a series of personal, semi-structured in-depth interviews with open-ended questions. Fourteen interview partners were selected that matched the above defined criteria of the three user groups according to the *theoretical sampling* strategy of the Grounded Theory research approach [13, 14]. The first group was represented by four interviewees (no rides into bonus zones), the second and third by five persons each, with 30 and 58% of their trips ending in bonus zones.

Based on these interviews, we were able to distinguish three different personas [15–17], i.e., constructed representatives of those user groups, while the description was focused on the motivational patterns underlying their usage behavior and their response to drop-off incentives.

- Users that ignore bonus zones are represented by "Johannes, the hunter": Johannes does only care about reaching his destination fast—every minute counts. He is ignoring bonus zones and is deliberately unwilling to park his e-bike apart of his destination to help others. He feels that, since he pays for this service, he should not be bothered with anything that is not in his immediate personal interest.
- Users occasionally using bonus zones are represented by "Marcel, the opportunist": Marcel decides on a case-by-case basis if he reacts to a nearby bonus zone or not. He weighs the required personal effort against the number of bonus minutes offered, taking the current situation into account (such as available time, necessary deviation from his actual destination and weather conditions).
- Users with a large percentage of bonus zone drops were found to have a different motivational pattern. This group is represented by "Timo, the gatherer": Timo likes to scan the booking app to spot attractive opportunities for earning free minutes. He uses the bonus zones frequently not only for personal gain, but to enjoy the

pleasant feeling that he has contributed to a well-functioning service and he has helped others to find and use bikes more easily.

It should be noted that the main behavioral motives of the three depicted user groups—saving time, earning free minutes, and helping others—are not mutually exclusive, but are on an additive basis. That is, virtually *all* users like to save time by using the e-bike-sharing service, *some* like to earn free minutes for their own benefit, and for *a few*, the fact that they also support other users and the e-bike-sharing service, in general, is an additional relevant aspect. This overlapping structure is in contrast to most literature where motives of personas are mutually exclusive (e.g., [15–17]).

Four important conclusions for the design of the incentive systems could be drawn from our results:

1. The reward offered for dropping-off a bike in a bonus zone should be made in the form of free minutes.
2. The entire incentive system, including the display of bonus zones in the booking app, should not interfere with the main purpose of the BSS of allowing users to reach their destination as fast as possible. Offering bonus zones must be done in such a way that the largest group, represented by "Johannes the hunter" can easily ignore it.
3. Users that take advantage of the bonus zones do this differently. The group represented by "Marcel the opportunist" does this on a case-to-case basis. Bonus zones have to be displayed clearly but not too dominantly. The most susceptible group, represented by "Timo, the gatherer", needs an easy way to scan for opportunities.
4. With respect to the bonus zone concept, there are motives for using and for deliberately not using them. For maximizing user satisfaction, a group-specific communication strategy seems most appropriate, taking into account the motivational structure of the three identified user groups. For example, for users of type "Johannes, the hunter", the bonus concept should not be emphasized, and the focus should be on the efficiency of the free-floating bike sharing as such. In contrast, for users of type "Timo, the gatherer", the advantage of the bonus zone concept as a means of improving the situation for everybody should be stressed.

5 Field Test of Incentive System

A field test for assessing the bonus zone concept was performed during 13 weeks in the fall of 2017. The goal of this field trial was to test and to quantify the response of users to offering rewards for parking a bike in a designated bonus zone. At this time, *smide* was already introduced in Zurich and had more than 6,000 registered users. The total user base at the beginning of the field test was divided into a control group and an experimental group of about equal size by random assignment. Users who registered during the field test were assigned alternately to each group. At the end of the experiment, the control group and the experimental group consisted of 1445 and 1556 users, respectively, where only users were considered who made a least one ride during the observation period. The total number of rides in control group and experimental group was 15,856 and 18,793, respectively.

The mobile app for the experimental group was extended with the incentive system. The city map in the app not only displayed the location of the bikes, but in addition showed different bonus zones distributed over the city, offering rewards of 5, 10, or 15 free minutes, added to the otherwise prepaid minute account. The bonus zones were positioned at locations with low dropping rate in the past (i.e., unattractive areas). They were changed daily in order to avoid habituation effects. Methods of design of experiments were used to ensure a balanced distribution of the bonus zones with respect to location and reward height. The number of bonus zones offered at each point of time was set to 30, such that about 70% of the dropping density was in walking distance to one of the bonus zones, i.e., for about 70% of all rides at least one bonus zone was offered near the destination. Note that the bonus zones were not optimized for improving the service level, but for studying the response of the users to the incentives.

The average number of rides per user per week for the experimental group was 0.93, compared with 0.84 for the control group. Thus, the incentive system led to a net increase in usage by 10%. Interestingly, the number of rides of an average user *not* ending in a bonus zone was nearly identical for users of the experimental group and the control group (see Fig. 2, left). So, offering rewards seem not to cannibalize the trips that users paid for. The additional trips are ending in a bonus zone. In terms of redistribution, this means a substantial potential, since 10% of the total traffic can be redirected to the bonus zones. Thus, the dropping rate in the bonus zones can be substantially increased, which is exactly what is necessary for the incentive system to be effective.

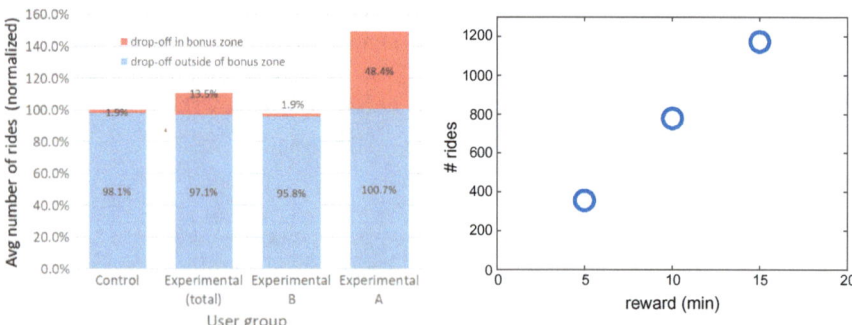

Fig. 2 Left: Average number of rides into a bonus zone (red) or non-bonus area (blue) for users of control group, experimental group (total), experimental B, and experimental A. The number of rides is normalized: 100% corresponds to the average total number of rides of users of the control group. Right: Number of bikes ending in a bonus zone as a function of bonus level

Bonus rides may be generated by users who choose to drop-off their bike somewhere near, but not directly at their destination, or by users who make a ride which they would not make otherwise (see Sect. 3). The quantity of our gathered data was not sufficient to distinguish between these effects. However, the fact that the number of rides *not* ending in bonus zones is identical for control group and experimental group indicates that the dominating behavioral effect of the incentive system is the emergence of additional rides, i.e., rides that would not have been made without the offered incentives. In the extreme case, this would correspond to $p_1 \approx 10\%$, and $p_2 \approx 0$ (see Sect. 3). From Fig. 1, it can be concluded that, based on our findings of the field test, a saving potential of about 30% can be expected.

As suggested by our empirical research (Sect. 4), we expected the experimental group to consist of three subgroups. For analyzing the behavior of the subgroup that is susceptible to the incentive system, we split the experimental group into two subgroups A and B, where A was meant to coincide with the subgroup of users who did take advantage of the bonus zones (combining the two personas "Marcel the opportunist" and "Timo the gatherer"), and B corresponds to the users that ignored the bonus zones completely ("Johannes the hunter"). Subgroup B should have a riding behavior that is statistically identical to the experimental group. In particular, the percentage of rides ending in a bonus zone should be the same. Note that the control group did not see the bonus zones on their app, nor could users of this group gain any rewards. Nevertheless, 1.91% of the rides of the control group ended in a bonus zone by chance. This should also be the case for a user of the experimental group who ignored the bonus zones. As a proxy for belonging to subgroup A or B, we used the percentage p of rides ending in a bonus zone, evaluated for each user of the experimental group. If $p > p_0$, the user was allocated to subgroup A, otherwise to subgroup B. Setting $p_0 = 12.5\%$ led to about 2% of rides ending in a bonus zones for subgroup B, and was thus used for the subgroup assignment.

Users of subgroup A are the ones that are co-producing value. As can be seen in Fig. 2 (left), the average number of rides into non-bonus zones (blue) is similar to the control group for both subgroups "Experimental A" and "Experimental B". However, users of group A have about 50% more rides than users of the control group, where the additional rides are rides into a bonus zone. Thus, the users that are susceptible for the incentive system show a substantial change in their usage behavior: On the average, they choose to drop-off their bike in a bonus zone instead of somewhere else for about a third of their trips.

In Fig. 2 (right), the number of rides into bonus zones for the different levels of offered rewards is shown. It can be seen that increasing the reward increases the attractiveness of the bonus zone. Surprisingly, this increase is nearly perfectly linear, and no saturation effect can be seen.

6 Conclusion

Designing value co-production in services needs a clear understanding of how value is generated for both the client and the provider, and how the activities of both actors are related to value creation. In the context of user-based redistribution in a free-floating bike-sharing system, we tried to understand and model value creation both qualitatively and quantitatively. With a simulation of the logistic system, we could assess quantitatively the potential of behavioral change in terms of value creation, quantified by the basic quantities of service level (value for users) and bike fleet reduction (value for provider). The simulation also clarified quantitatively the trade-off between these two parameters. Two basic parameters p_1 and p_2 were used for modeling the user response on incentives. Sensitivity analysis showed that even a modest participation of users results in a significant improvement in terms of service level or costs. Qualitative social research was necessary to understand the motivational patterns underlying the willingness to co-produce. Different motivational patterns could be distinguished and were described in the form of personas. These results allowed an optimization of the design of the incentive system (framing, reward design, user communication, and user interface design). Usage data collected in an extensive field test helped to further quantify the response of the different user groups to the offered incentives. In particular, it was possible to estimate the values of the response parameters p_1 and p_2, obtained in the setting of the field test. Fed back into the simulation model, this helped to calibrate the model and to increase the prediction accuracy of the simulation model.

Methodologically, an important learning of this research was the fact that neither classical OR methods, nor classical user research methods, would have been sufficient to design such a system. The OR results were necessary to understand the complete system, to assess the overall effects of implementing a user incentive scheme for value co-creation, and to identify the most relevant user-related behavioral features in this context. On the other hand, empirical social research including a field test was necessary to create a realistic behavioral model of the users, and to optimize the design of the incentive system for maximizing perceived value of the incentive system. While both approaches contributed important insights about how to design user-based redistribution into a BSS, it was the *combination* of both research streams that was found to be the crucial element. This confirms the often-heard call for interdisciplinary research in the field of service innovation and research.

Acknowledgements This research was supported by the Swiss Commission of Technology and Innovation CTI (grant 25040.1 PFEN-ES).

References

1. H. Ferrando, E. Anaya, I. Arauzo, Guía metodológica para la implantación de sistemas de bicicletas públicas en España (Intituto Para Diversif. Ahorro Energ., 2007)
2. J. Pfrommer, J. Warrington, G. Schildbach, et al., Dynamic vehicle redistribution and online price incentives in shared mobility systems. IEEE Trans. Intell. Transp. Syst. **15**(4), 1567–1578 (2014)
3. A. Singla, M. Santoni, G. Bartók, P. Mukerji, M. Meenen, A. Krause, Incentivizing users for balancing bike sharing systems, in *AAAI* (2015)
4. C. Heitz, R. Etschmann, R. Stöckle, T. Bachmann, M. Templ, User-based redistribution in free-floating bike sharing systems, in *Operations Research Proceedings 2018, Brussels, Belgium* (2018)
5. L. Leifer, M. Lewrick, P. Link (eds.), *Das Design Thinking Playbook: Mit traditionellen, aktuellen und zukünftigen Erfolgsfaktoren* (Zürich, Versus, 2017)
6. H. Plattner, C. Meinel, U. Weinberg, Design Thinking: Innovation lernen—Ideenwelten öffnen (mi, München, 2009)
7. F. Uebernickel, W. Brenner, B. Pukall, T. Naef, B. Schindlholzer, *Design Thinking: das Handbuch*, Erste edn. (Frankfurter Allgemeine Buch, Frankfurt am Main, 2015)
8. A. Kuss, R. Wildner, H. Kreis, Marktforschung: Grundlagen der Datenerhebung und Datenanalyse, 5., vollst. überarb. u. erw. Aufl (Springer Gabler, Wiesbaden, 2014)
9. K. Werbach, D. Hunter, *For the Win: How Game Thinking can Revolutionize Your Business* (Wharton, Philadelphia, 2012)
10. S.L. Vargo, P.P. Maglio, M.A. Akaka, On value and value co-creation: a service systems and service logic perspective. Eur. Manag. J. **26**(3), 145–152 (2008)
11. G. Steinhardt, *Market-value Pricing: Definitions, Concepts, and Processes for Market-value Centric Pricing* (Springer, Cham, 2019)
12. T.T. Nagle, *The Strategy and Tactics of Pricing: a Guide to Growing More Profitably*, 6th edn. (Routledge, New York, 2018)
13. Barney G. Glaser, Anselm L. Strauss, *Grounded Theory: Strategien qualitativer Forschung, 3*, unveränd edn. (Huber, Bern, 2010)
14. K. Charmaz, *Constructing Grounded Theory: A Practical Guide Through Qualitative Analysis, Report* (Sage Publications, London, 2011)
15. A. Cooper, R. Reimann, D. Cronin, C. Noessel, J. Csizmadi, D. LeMoine, *About Face: The Essentials of Interaction Design*, 4th edn. (Wiley, Indianapolis Indiana, 2014)
16. K. Goodwin, *Designing for the Digital Age: How to Create Human-centered Products and services* (Wiley & Sons, 2011)
17. J.S. Pruitt, T. Adlin, *The Essential Persona Lifecycle: Your Guide to Building and Using Personas* (Elsevier, Amsterdam, 2010)

Research on Electricity Falling Accident Based on Improved Bode Accident Causation Model

Qian Yuanyuan, Xu Jie, Mi Chuanmin and Peng Qiwei

Abstract In view of the applicability of accident causation models to different types of accidents, this paper comprehensively considers various kinds of accident causation theories, compares, and analyses seven kinds of accident causation models, and points out the limitations of existing models. Then, according to the mechanism of electricity falling accident, combining the theory of Bode accident causality chain with the theory of man–machine–environment system, this paper presents an accident causation model suitable for the analysis of electricity falling accident, which has a strong guiding significance for the prevention of electricity falling accident. Finally, the management improvement measures are put forward from the aspects of safety operation standard, safety production responsibility system, safety education, and training, in order to prevent the recurrence of electricity falling accident.

Keywords Accident causation model · Electricity falling accident · Preventive measures

1 Introduction

There are many high-altitude operations in power generation, power supply, and infrastructure construction of electric power system, such as the installation and maintenance of boilers, installation and maintenance of some generator equipment or lifting equipment, and installation of lighting equipment at high places, tower erection and line stringing of power transmission and transformation companies or power supply enterprises. If the operators at high altitudes do not pay attention to or

Q. Yuanyuan (✉) · M. Chuanmin
School of Economics and Management, Nanjing University of Aeronautics and Astronautics, Nanjing 210016, China
e-mail: 352356316@qq.com

X. Jie
State Grid Wuxi Power Supply Company, Wuxi 214062, China

P. Qiwei
NARI Group Corporation or State Grid Electric Power Research Institute, Nanjing 211000, China

© Springer Nature Switzerland AG 2020 127
H. Yang et al. (eds.), *Smart Service Systems, Operations Management, and Analytics*, Springer Proceedings in Business and Economics,
https://doi.org/10.1007/978-3-030-30967-1_12

comply with the relevant regulations of construction safety at high altitudes, accidents such as falling at high altitudes and being hit by objects may occur. Statistics show that falling accident is one of the most common accidents in the electric power industry, and is also an event type with high casualty.

In order to avoid serious casualties and property losses, electric power enterprises need to pay attention to safety management. In 2005, Sherill innovatively proposed that safety management is to carry out safety culture education from the aspects of management, system, technology, and employees, so as to minimize the potential safety hazards brought by unsafe factors to electric power companies [1]. In order to prevent high-altitude falling accidents in electric power industry and carry out effective safety management, it is necessary to analyze the causes of high-altitude falling accidents, and take measures against relevant factors to prevent the occurrence of accidents. A large number of research and practice have proved the importance of accident causation model in the research of safety science theory and accident prevention. Accident causation model is not only the theoretical basis of accident prevention and control, but also the tool of accident investigation and analysis [2].

At present, there are three basic models for accident analysis: sequential accident model, epidemiological accident model, and systematic accident model. Among them, the sequential accident model interprets the causes of accidents as a series of discrete-time chain reactions [3]. In the epidemiological accident model, accidents are analogous to the spread of diseases, and generated by the combination of multiple dominant and recessive factors happening at the same time and in the same space [4]. These two models describe accidents as simple causal event chains, which are obviously inadequate and insufficient for capturing the nonlinear interaction and relationship between components. In the systematic accident model, Clarkson et al. mapped the factors of accidents in the social-technical system by using causality diagram to analyze the Australian air accident [5]. HFACS is an accident analysis method based on Swiss cheese model [6]. Leveson proposed the system theory accident model and process (STAMP) model, which was applied to analyze the China–Jiaoji railway accident in 2008 and the serious China–Yongwen railway accident in July 23, 2011 [7, 8]. Zhang Hong et al. used the "2–4" model of accident cause to analyze the falling accident in building construction, and classified all kinds of reasons by using statistical analysis [9].

In the field of electric power, many scholars have studied the theory and method of human factor analysis in power accidents. For example, Yang Shaojun et al. believe that human unsafe behavior is the main factor of accidents, because 80% of accidents are caused by human unsafe behavior in the electric power industry [10]. Through sorting out the relevant research, it is not difficult to find that there is a lack of overall analysis of accident unsafe factors in the domestic electric power industry. In addition, the study of accident causes in the electric power safety started late and the speed of its development is relatively slow, lagging behind the study of coal mine production [11], construction, and other fields. Therefore, this paper studies and compares different accident causation theories and analyses the mechanism of electricity falling accident. On this basis, we construct an accident causation model of electricity falling accident, select typical cases for analysis, and put forward countermeasures to deal with falling accident. This model will play a guiding role in formulating prevention and control measures for electric power enterprises, and help to realize scientific and systematic safety management.

2 Accident Causation Theory

Accident causation theory has developed from single factor to multifactor, from chain reaction to system integration, resulting in a variety of theoretical models. As early as 1936, Heinrich expounded the causal chain theory of industrial accidents in his book *Industrial Accident Prevention* [12]. The core idea of this theory is that the occurrence of casualties is not an isolated event, but the result of a series of causes and events. This theory provides a valuable method for later researchers to study the mechanism of accidents and promotes the development of accident causation theory such as Bode causal chain theory and Adams model, which has become the forerunner of scientific accident research, occupying an important historical position [13].

On the basis of Heinrich causal chain theory, Bode put forward the modern causal chain theory of accidents. He believes that although the unsafe behavior of human and the unsafe state of substances are important reasons for the accident, but in-depth investigation, the root cause of accident is management failure. The model of Bode also contains five factors, but the meaning of each factor is different from Heinrich's [14]: (a) Insufficient control, including control of unsafe behavior and unsafe state, which is the core of safety management. (b) Basic causes, including personal reasons and working reasons. Among them, personal reasons are physical and mental problems, such as lack of knowledge, skills and incorrect motivation; working reasons include inappropriate operating procedures, abnormal use methods, and so on. (c) Direct causes, including the unsafe behavior of human and the unsafe state of substances, is the appearance of management defects and basic causes. (d) Accidents, meaning energy beyond its threshold or the contact of human body with substances that hinder normal physiological activities. (e) Injury loss, refers to the result of an accident, including casualties and financial losses.

For the aspect of the indispensable energy in the production process, two theories have emerged. One is the theory of accidental release of energy, which refers to the accidental escape or release of energy beyond the constraints set by human [15]. The second is two types of hazards, which holds that accidents result from the interaction of two types of hazards. The first type of hazard is the energy body with energy, which may lead to the accidental release of energy, is the premise of the second type of hazard and determines the severity of accident. The second type of hazard is a variety of factors leading to energy restraint or shielding, including material failure, human error and environmental factors, and determining the possibility of the accident [16].

In addition, many scholars believe that the main cause of accident is human error. This kind of theory mainly includes Surrey model, Haier model, Wigglesworth's general human error model, Lawrence model, and so on. Wigglesworth model constitutes the basis of human error theory. He pointed out that human errors constitute the basis of all types of accidents, and defined human errors as "human erroneously or inappropriately responding to an external stimulus" [17]. However, Skiba considered that only when both production operators and mechanical equipment occur at the same time, accidents can occur. This theory is called trajectory intersection theory.

Through the study of various accident causation models, this paper points out the limitations of some models (see Table 1), in order to play a guiding role on the construction of accident causation model in electric power industry.

Table 1 The limitation of electricity causation model

Accident Causation Model	Limitation
Heinrich causal chain theory	The description is too absolute and simplified and completely attributes the causes of unsafe human behavior and unsafe state of substances to human shortcomings
Bode causal chain theory	Totally attributes the causes of accidents to management failures
The theory of accidental release of energy	Management and other factors can not be measured from the viewpoint of energy; not reveal the deep-seated causes leading to accidental release of energy
Two types of hazards	Not consider the effect of organizational management factors on energy shielding; difficult to guide enterprises to control the second type of hazards effectively
Epidemiological theory	Must have sufficient samples for statistics and evaluation, and in these areas, the theory lacks clear guidance
The human error model	Attributes the accident entirely to human error, does not consider the factors of machine and environment and is difficult to make qualitative analysis of special types of accidents
Trajectory intersection theory	Not reflect the deep-seated reasons that lead to unsafe behavior of human and unsafe state of substances

3   Causation Model of Electricity Falling Accident

3.1   Mechanism of Electricity Falling Accident

By analyzing the characteristics of electric falling accident and studying its evolution, we divide the causes into four categories: management system, human, substance, and environment. In the process of electric power production, there are unstable factors in human, substance, environment, and management. Under certain time and certain space, accidents will occur after the interaction of unstable factors. Specific analysis is as follows:

(1) Management defects refer to the factors related to safety control in the process of power production, i.e., the lack of management function. Management not only refers to safety technical measures, operating procedures, operating standards, rules, and regulations, but also includes the understanding and attention to safety production, scientific safety management methods, and the implementation of management responsibilities. Management is the basis for the occurrence of hidden dangers leading to accidents.

(2) Human's unsafe behavior refers to the abnormal behavior of those who have caused or may cause accidents. As the protection object of safety management, human is also an unsafe factor and the focus of safety management. In the electricity falling accident, human's unsafe behavior is mainly divided into three categories: abnormal physiology and psychology of operators; violations; operation errors.

(3) The unsafe state of substances refers to the state of substances (including machine, materials, objects of production, and other factors of production) that have caused or may cause accidents. The unsafe state of objects can also be divided into three categories: insufficient material strength, poor installation, wear, and aging; defects of labor protective equipment; unqualified protective facilities and device failure.

(4) Unsafe environment refers to the production and operation environment that has an impact on safety production, i.e., potential risk factors caused by air quality, lighting, temperature, humidity, bad weather, and workplace defects (narrow space, rugged terrain, slippery ground, disorderly placement of goods, etc.).

3.2   Construction of Model

In the process of high-altitude operation, management errors result in defects of management, human, substance, etc. Human defects lead to unsafe behavior of human beings, and material defects lead to unsafe state of equipment and facilities, together with unsafe production environment caused by defects of external environment. In this way, human's unsafe behavior, the unsafe state of substances and the unsafe

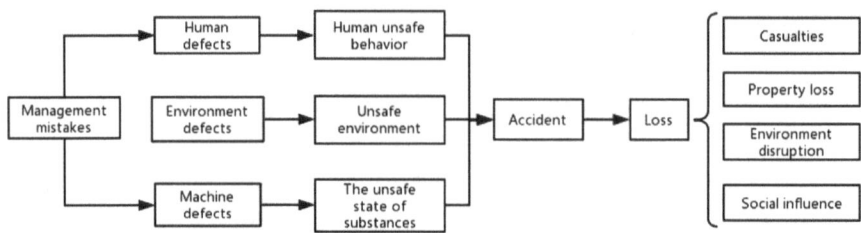

Fig. 1 An improved Bode accident causation model

environment work together, which leads to the occurrence of electric falling accidents, and ultimately results in the loss of accidents, mainly for casualties, property losses, environmental damage, and social impact.

From this point of view, the occurrence of electric accidents is a process in which human, machine (including tools, machines, computers, etc.) and environment are interrelated and interacted. Qian Xuesen, a famous scientist, first put forward a new concept in 1981: man–machine–environment system. In this system, man is the main body of work, such as operators or decision makers; machine is all objects controlled by human, such as automobiles, aircraft, ships; environment refers to the specific working conditions of human and machine, such as external working space, physical environment, biochemical environment, social environment [18].

Therefore, considering the advantages and limitations of various accident causation theories, this paper combines Bode causal chain theory with man–machine–environment system theory, and proposes an improved Bode accident causation model suitable for electricity falling accidents as shown in Fig. 1.

In order to prevent accidents and strengthen supervision and guidance, we need in-depth analysis of electricity falling accident to find its deep-seated causes. The causes of the accident can be divided into essential causes, direct causes and indirect causes. In this model, the essential cause of electric falling accident is management failure, which can be divided into two aspects: external management failure and internal management failure. The direct reason is human's unsafe behavior, the unsafe state of machines and equipment, and the unsafe environment. Among the three main direct causes, unsafe human behavior is the most important and direct cause of accidents. Indirect causes include three aspects: human defects; material defects; environmental defects.

4 Case Studies

4.1 The Description of Accident

According to the work plan, the transmission overhaul center of a maintenance company carries out live detection of zero-value porcelain insulators at five intervals of

A, B, C, D, and E in a substation of XXX kilovolt. The live shift of the transmission maintenance center surveyed the work site the day before.

On the day of the accident, S, the person in charge of the work and Z (deceased, male, 29 years old, undergraduate, working in 2010), M, and W, the deputy director of the power transmission maintenance center, arrived at the station and went through the work ticket permit formalities for live operation. After carrying out the safe confession and checking the protective appliances, S is in charge of guardianship, Z and M are in turn to carry out work.

At 9:40, M completed the detection of A, B, and C intervals.

At 9:45, Z climbed the ladder on the south side of D-interval and started work. In the process of operation, Z, without the safety belt or the reserve protection rope, transferred the working position by hands supporting on the upper crossbeam and feet stepping on the lower crossbeam. S, the person in charge of the work, and W, the cadre, were about 6 meters below the working portal frame, but did not stop violations.

At 10:05, Z completed the D-interval detection work and prepared to cross the column to carry out the next interval detection. When he was about 0.5 meters away from the column, he suddenly fell from the D-interval beam (18 meters from the ground) to the ground. Z was immediately sent to the hospital. After rescue, Z died.

4.2 Causes for Accidents

The accident exposed many weaknesses and management loopholes in the safety production of the responsible units, and there were many problems in safety management and technical management from the leadership, management to operation.

Direct causes. Human's unsafe behavior includes employees not wearing safety belts, not hanging back-up protective ropes, which belongs to serious violations in operation, and also includes the insufficient supervision. First, the operator violated the safety regulation that workers at high places should not lose safety protection when transferring their working positions. He transferred working positions without fastening safety belts and reserve protective ropes. Second, the responsible person and the on-site personnel are not in place to perform their duties, and serious violations in the movement of workers were not stopped in time. Their safety awareness is weak, and there are serious loopholes in the safety management. The unsafe factors of substances include the lack of compulsory measures to prevent personnel from violating regulations.

Indirect causes. From the aspect of human, responsible person's safety awareness is weak, safety education is insufficient, staff quality is low, people have mental state by luck.

Essential causes. The implementation of safety rules and regulations is not strict, the daily safety management of teams and groups is not sufficient, the organization of production operations and on-site supervision and inspection are not solid, and the work of safety education is not carried out in-depth.

5 Accident Prevention Countermeasures

5.1 *Establish Safety Operation Standards*

Safety operation standards are safety measures that operators should implement in each stage of operation. According to the identified hazard factors, safety operation standards can be formulated from both technical and managerial aspects [19]. The basic requirements for each stage of the operation process include the following aspects:

(1) Pre-operation. In order to ensure that the site meets the safety working conditions, the standard of this stage is mainly to clarify the implementation of the relevant basic preparatory work before production operation in order to eliminate the unsafe state of substances. Including (a) Risk assessment; (b) Whether the intact condition of objects and environmental conditions meet the requirements, and whether the safety protection measures are perfect; (c) Whether the production preparation of materials, tools and other substances is implemented and meets the requirements; (d) Whether the safety operation measures are formulated and implemented.

(2) Operation. The main content of this stage is to standardize human's operation behavior and prevent accidents caused by human's unsafe behavior, including (a) standardization of operation procedures; (b) safety operation behavior norms; and (c) on-site supervision requirements.

(3) Post-operation. Similar to the preparatory stage before operation, the main content of this stage is to formulate clearance and confirmation measures for the site, put forward inspection requirements, avoid the unsafe factors left on the site.

5.2 *Improve the Responsibility System for Safety Production*

(1) The main body of responsibility for safety production should be clarified. It mainly includes three levels of responsibility subjects: first, to clarify the most fundamental responsibility subject of the project department; second, to clarify the responsibilities of the construction team and subcontractors, so as to achieve "each is responsible for its responsibilities"; and finally, to clarify the responsibilities of the company leadership and functional departments.

(2) Institutionalization of responsibility incentive and restraint. Do not leave blind areas in the implementation of responsibilities, do assessment according to their respective responsibility objectives and implementation of responsibilities, and

implement quantitative scoring. Projects, departments, and individuals with outstanding safety work and no accidents shall be commended, given certain material rewards; for those subjects resulting in accidents shall be checked to the end, and give full play to the warning role of the responsibility system [20].

5.3 Strengthen Safety Education and Skills Training

Electric power enterprises should regularly organize grassroots units to enrich and update their safety knowledge. In view of different positions, we should carry out the education of safety knowledge, safety regulations, and discipline-abiding in a planned and focused manner, so as to enable employees to understand the connotation of post safety in a deeper level and know what to do is safe, what to do is unsafe. At the same time, we should creatively carry out safety debates, safety speeches, and other various, popular safety education activities, so as to effectively improve staff's safety awareness and safety skills. Every employee should receive safety education regularly, firmly establish the consciousness of "safety first" and form a good atmosphere of "safety at all times" in the whole collective, so as to sublimate the safety ideology of all employees.

In addition, it is necessary to strengthen the skills training of operators so as to improve their abilities qualitatively and eliminate unsafe human behavior. We can use typical cases of falling accident as safety teaching material to make operators aware of potential risk factors, make them pay more attention to the operation, thus eliminate human behavior errors.

5.4 Cut Down or Discontinue Operations in Unsafe Environment

Under special weather conditions such as freezing, fog, thunderstorm, and gale, we should take effective measures or stop operations. Especially at high places (such as chimneys, water towers), we must have reliable communication devices and signals to prevent contact out of control.

Minimize the amount of work at high altitudes as much as possible. We can raise the equipment combination rate and prefabrication rate, and some high-altitude operations will be moved to the ground to do. This is the most fundamental and effective way to reduce the hazards of falling accidents at high altitudes.

6 Conclusion

Based on the analysis of various accident causation theories and the occurrence mechanism of electric falling accidents, this paper synthesizes Bode causal chain theory and man–machine–environment system theory, then puts forward an improved Bode accident causation model, which conforms to the general law of electric falling accidents and plays a positive role in perfecting and developing accident causation theory.

After analyzing the accident with the model, it is concluded that the key to prevent the accident at this stage is to strengthen safety management. Strengthening the implementation of the responsibility system for production safety and improving the safety quality of the whole staff can standardize human's safety behavior and improve the level of safety management. Strengthening the safety management of construction site can eliminate the unsafe state of substances. The unsafe behavior of human and the unsafe state of substances have been controlled, and the safe production of electric power has been guaranteed.

References

1. W.H. Sherill, C.M. Tam, C.F. Tung Karen, S.K. Man Ada, Safety culture divergences among management, supervisory and worker groups in Hong Kong construction industry. Int. J. Project Manag. **23**(7), 504–512 (2005)
2. W.U. Chao, Some advances in safety science fundamental theories of China in the recent ten years. Chin J Nonferrous Metals **26**(8), 1675–1692 (2016)
3. T.S. Ferry, *Modern Accident Investigation and Analysis*, 2nd edn. (Wiley, New York, 1988)
4. E. Hollnagel, *Barriers and Accident Prevention* (Ashgate, Hampshire, 2001)
5. J. Clarkson, A. Hopkins, K. Taylor, Report of the board of inquiry into F-111(Fuel tank) deseal/reseal and spray seal programs. R. Aust. Air Force (2001)
6. D.A. Wiegmann, S.A. Shappel, A human error approach to aviation accident analysis. The human factors analysis and classification system (2003)
7. M. Ouyang, L. Hong, M.H. Yu, Q. Fei, STAMP-based analysis on the railway accident and accident spreading: taking the China-Jiaoji railway accident for example. Saf. Sci. **48**(5), 544–555 (2010)
8. Tian Song, Deming Zhong, Hang Zhong, A STAMP analysis on the China-Yongwen railway accident. Comput. Saf. Reliab. Secur. **7612**, 376–387 (2012)
9. H. Zhang, Y. Gong, G. Fu, Causes classification and statistical analysis on falling accidents on construction sites based on "2–4" model. J. Saf. Sci. Technol. **13**(09), 169–174 (2017)
10. Y.A.N.G. Shaojun, L.U.O. Wanting, Analysis and countermeasure of human unsafe behavior. Electr. Saf. Technol. **12**(11), 1–5 (2010)
11. L.I.N. Yongming, On the cause tracing of the coal mine safety accidents based on the hierarchical topic model. J. Saf. Environ. **18**(02), 619–623 (2018)
12. H.W. Heinrich, *Industrial Accident Prevention* (McGraw-Hill, 1979)
13. L.I. Wanbang, X.I.A.O. Dongsheng, A review of accident causation theory. J. Univ. South China (Soc. Sci. Ed.) **01**, 57–61 (2007)
14. Q.I.N. Rong, P.E.N.G. Dongzhi, Discussion on accident causation theory. J. North China Inst. Sci. Technol. **03**, 1–10 (2005)
15. M. Zhong, Y. Wei, W. Fan, et al., Overview on accident causing theories. Fire Saf. Sci. (03), 38–44 (1999)

16. Baozhi Chen, *Hazard Source Identification, Control and Evaluation* (Sichuan Science and Technology Press, Chengdu, 1996)
17. C.R. Asfahl, *Industrial Safety and Health Management* (Prentice Hall, 1999)
18. B.G. Wang, W. Wang, W.G. Huang, et al., Application of H.S. Tsien's systems science in man-machine-environment system engineering. J. North China Inst. Sci. Technol. **11**(08), 1–18 (2014)
19. G.U.O. Hanjun, Method and application of hazard identification and evaluation for power plant operation. Electr. Saf. Technol. **19**(02), 1–5 (2017)
20. Baocheng Song, Safety management of electric power construction enterprises. Electr. Saf. Technol. **11**(09), 7–9 (2009)

Crop Yield Prediction Using Deep Neural Networks

Saeed Khaki and Lizhi Wang

Abstract The world's population is on the rise and in order to feed the world in 2050, food production will need to increase by 70% [1]. As a result, it is of great importance to construct powerful predictive models for phenotype prediction based on Genotype and Environment data (so-called G by E problem). The objective of the G by E analysis is to understand how genotype and the environment jointly determine the phenotype (such as crop yield and disease resistance) of plant or animal species. In this research, deep neural networks are trained and used as predictive models. Deep neural networks have become a popular tool in supervise learning due to considerable ability in training nonlinear features [5]. Recent articles have stated that the network depth is a vital factor in decreasing classification or regression error. But, deeper networks have a so-called vanishing/exploding gradients problem which makes the training and optimizing deeper networks difficult. He et al. proposed residual learning method which alleviates this problem very well and showed that deep residual networks are significantly better and more efficient than previous typical networks [5]. As a result, residual training has been used in this research to prevent gradient degradation and ease the optimization process. Finally, since it is difficult to predict the yield difference directly, two separate residual neural networks have been trained to predict yield and check yield. After training the networks, the RMSE for check yield and yield are 8.23 and 10.52, respectively, which are very good because of considerable amount of missing values, uncertainty, and complexity in the datasets.

Keywords G-by-E interaction analysis · Supervised learning · Machine learning · Deep neural networks

S. Khaki · L. Wang (✉)
Industrial and Manufacturing Systems Engineering, Iowa State University, Ames, IA 50014, USA
e-mail: lizhi.wang.2016@gmail.com

© Springer Nature Switzerland AG 2020
H. Yang et al. (eds.), *Smart Service Systems, Operations Management, and Analytics*, Springer Proceedings in Business and Economics,
https://doi.org/10.1007/978-3-030-30967-1_13

1 Introduction

As the earth population is growing, it is important to use new and more efficient techniques to enhance food production so as to feed the world in future. One of the ways of increasing food production is to use genotype and environment information to predict the phenotype of crops or animal species. In order to process genetic and environment data, predictive models are needed. In supervised learning literature, neural networks are very powerful models since they can train many nonlinear features which are very useful especially in complex datasets.

Neural Networks are made up of neurons that have learnable weights and biases. Each neuron receives some inputs, performs a dot product and optionally follows it with a nonlinearity [2]. Deep neural networks have the ability to achieve record-breaking results in analyzing data. They have been recognized for the groundbreaking results in numerous fields. The proven ability of deep learning has been a key factor too, why i t ḃ being used more and more for many datasets nowadays [2–4].

Previous researches have stated that as the number of hidden layers of a network increases, classification error might decrease. It is true, and it may seem if number of hidden layers is increased by any number, classification error will decrease continuously. However, the only problem with this approach is that deeper networks are hard to optimize and train. They have a so-called vanishing gradient problem [2, 5].

For solving this problem, data normalizing in first and intermediate layers of network is used to help deep networks to converge using stochastic gradient descent (SGD). He et al. stated that a deeper model should always produce less training error than its shallower counterparts. They proposed that the problem with deeper networks is optimization problem. Therefore, they proposed a new structure for neural networks called residual neural networks in which identity blocks or residual training have been used to prevent gradient degradation and make optimization task easier [5].

In this research, we have trained two deep neural networks for prediction of yield and check yield separately using residual learning approach. A different neural network has also been trained for 2017 weather prediction.

In the end, it is worth mentioning that Syngenta Contest poses more challenges this year since there exist many genetic markers and they also have many missing values. So, the dataset requires a thorough preprocessing before doing any further analysis.

This research consists of the following sections: In Sect. 2, data preprocessing is explained. In Sect. 3.1, deep neural network structure is elaborated for yield prediction. Section 3.2 presents 2017 weather prediction approach.

2 Analysis of Data

It is essential to give high-quality input data to any predictive model in order to get more accurate results. Data preprocessing is a data mining technique that consists of transforming raw data into a more transparent format.

Real-world data like Syngenta is often incomplete, inconsistent, and lacking in certain behaviors or trends, and is likely to produce many errors. So, it is a crucial task to do preprocessing on this type of data.

We preprocessed the data by first imputing all missing values in Genetic data. There are nearly 19,500 unique genetic markers provided in this dataset. However, there are many missing values in each genetic marker. As a result, a threshold as a maximum number of missing values allowed for each genetic marker is defined. If a genetic marker has missing values more than the defined threshold, it is thrown away. In the literature of data mining, it is often recommended not to have a big threshold since it increases the error and noise inside the dataset resulting in a very poor prediction. In this research, 23% threshold is used. We also used Minor Allele Frequency (MAF) of 0.01 as well for additional refinement of all genetic markers.

After deletion of genetic markers with excessive missing values based on defined threshold, all missing values are imputed. In statistics, imputation is the process of replacing missing data with substituted values instead of deletion of instances that have missing values. Imputation preserves observations by replacing missing data with an estimated value based on other available information. Once all missing values have been imputed, the dataset can then be analyzed using standard techniques for complete data [7]. There are many approaches for missing value imputation such as mean, median, most frequent, and stochastic regression methods. Since the genetic dataset consists of just categorical data with only three values $(-1, 0, 1)$, we tried both median and most frequent approaches which median method proved to work better on the validation set. Therefore, median approach is used for imputation.

After the above preprocessing, the remaining number of genetic markers was 4005. We did not do any special preprocessing on environment dataset except normalizing them.

3 Methodology and Results

In this research, deep neural network has been used for prediction since this method has proven to be very effective approach especially when a high amount of complexity and nonlinearity exist in data. As the complexity of data increases, the number of hidden layers is typically increased to form a powerful enough model to represent all the existing complexity and nonlinearity in the data. Due to having great number of genetic markers and weather variables, we have adopted deep learning method to create a powerful model for prediction to fully capture and represent the interaction between genetic and environment data.

In this research, we have trained three different neural networks for prediction of yield, check yield and 2017 weather variables, respectively. The network architectures are the same for yield and check yield predictions, but different architectures have been used for weather prediction. The network architectures are as follows.

3.1 Yield Prediction Neural Networks

In the training dataset, there are only 2,267 distinct hybrids. It shows same hybrids with same genetic markers produced different yields, so the only difference among them was environmental conditions. As a result, we should construct a powerful neural network which is able to train nonlinear features to fully represent the effect of environmental conditions on genetic information. In a neural network, each neuron receives some inputs, performs aggregation and follows it with an activation function. Figure 1 shows a simple neuron and Fig. 2 presents a neural network with one hidden layer. A neural network is simply a function mapping input data to outputs with trainable weights and biases which applies nonlinearity in each neuron.

A neural network usually has many weights and biases to be trained. The process of training is loop containing the following steps:

- Forward-propagation: In this phase, input data is fed into network and it is propagated forward through the network, layer by layer, until it reaches the output layer [6].
- Computing the Loss: In this phase, the output is compared with actual target values to see how accurate the current model is.

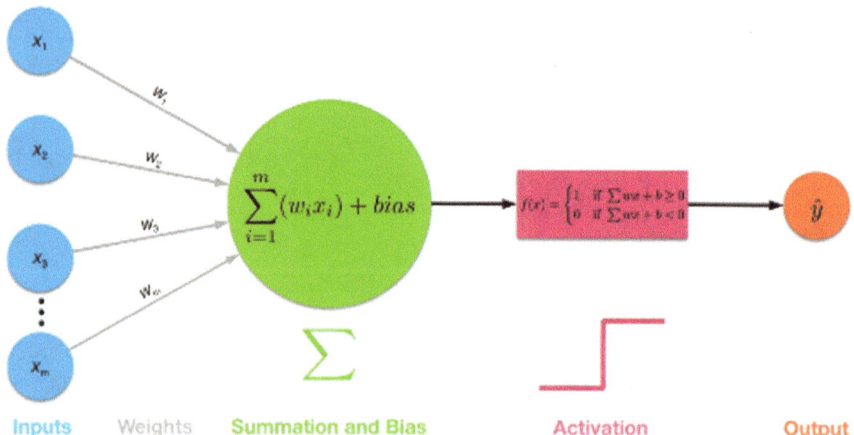

Fig. 1 Simple neuron

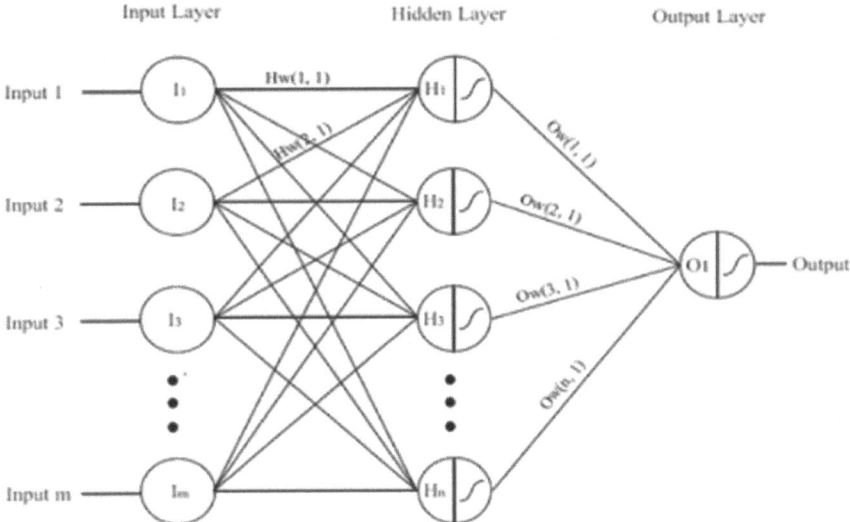

Fig. 2 One hidden layer neural network

- Backpropagation: In this phase, the gradient of loss function is computed with respect to all weights and biases for all layers in the network using chain rule from calculus.
- Update: In this phase, we update all weights and biases using gradient-based optimization methods like stochastic gradient descent (SGD), Adam, Adagrad optimization algorithms.

As the network becomes deeper and deeper, more non-linear features can be trained and as a result, the classification error might decrease. He et al. stated that a deeper model should always produce no higher training error than its shallower counterpart [5]. However, there is a serious problem with the deeper networks which is they are harder to train, or they have so-called vanishing gradient problem. Through backpropagation, gradient flows back from the output layer to the very beginning layers of network which includes many multiplications of usually small values. Therefore, it sometimes causes the gradients to become almost zero when it reaches to the first layers of neural network. Consequently, there will be no updates for weights of first layers resulting in high error.

To solve this problem, many methods have been proposed such as normalizing input data before each layer which helps deeper networks converge using stochastic gradient descent (SGD) [2, 5].

In 2016, He et al. proposed a new neural network structure which can significantly alleviate the gradient degradation problem. He et al. have addressed the degradation problem by introducing a deep residual learning framework. Instead of fitting a desired underlying mapping denoted as H(x), they let the stacked nonlinear layers fit a residual mapping of F(x): = H(x) − x.

Fig. 3 Residual shortcut
connection [5]

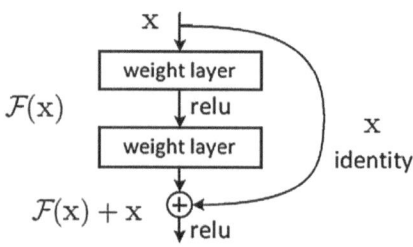

They have hypothesized that it is easier to optimize the residual mapping than to optimize the original network [5]. Figure 3 shows a residual building. In residual networks, there is a shortcut connection for every two stacked layers. As it is shown in Fig. 3, the shortcut connection is simply an add gate. So, in backpropagation, the add gate preserves the gradient and acts like a gradient super highway throughout the network. As a result, we make sure that all weights in the network will be sufficiently updated through the cycle of forward and backwardpropagation.

The interesting thing about these shortcuts is that they add neither extra parameters nor computational complexity to the network. The entire network can still be trained end-to-end by SGD using backpropagation, and can be easily implemented using common libraries like Tensorflow and Pytorch without modifying the solvers [5].

In this paper, we have trained a deep 21-layer neural network for both yield and check yield prediction. We have used residual shortcuts after every two stacked layers of the network. Batch normalization is also used for every intermediate layer for getting better convergence. Finally, the network is optimized using Adam optimizer. The whole process is coded in Python 3.5 using Tensorflow open-source software library.

Following implementation has been used:

- 21 Hidden Layers
- 40 neurons in each layer
- Adam optimizer with learning rate 0.0003
- Exponential decay with rate of 0.95 for the learning rate
- Batch size of 64
- Batch normalization for all hidden layers
- 300,000 Iterations
- Residual shortcuts
- 143,999 training samples
- 4453 samples as validation set (0.03 of training set)
- Relu family activation functions
- Xavier initialization for all weights

After training the model for both yield and check yield, the results are as follows (Table 1).

Figures 4 and 5 show the performance of the proposed deep residual neural network for yield prediction on training and validation set for 100 randomly chosen

Table 1 Model performance

	Training RMSE	Validation RMSE
Yield	10.3	12.15
Check yield	7.42	8.02

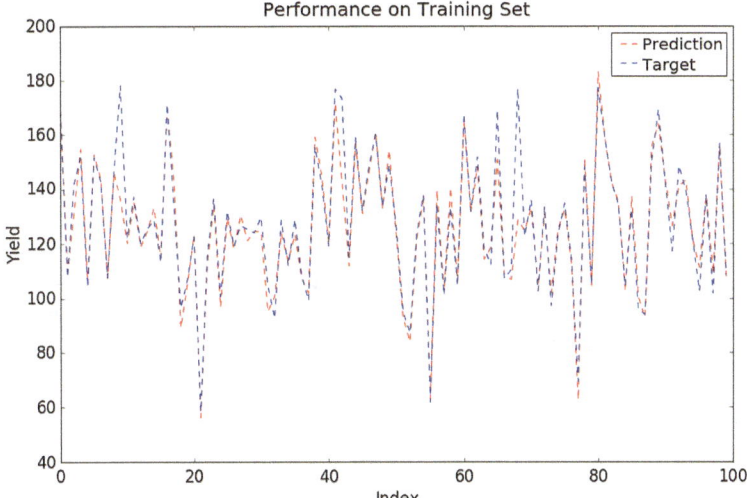

Fig. 4 Performance on training set

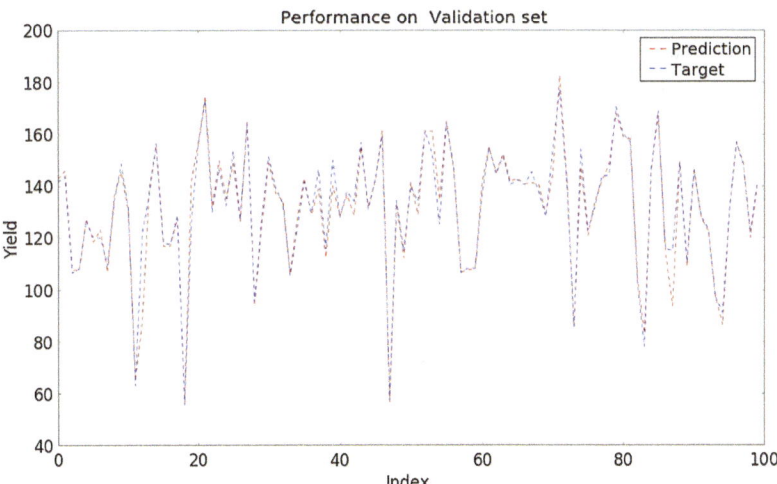

Fig. 5 Performance on validation set

crops from two sets. Finally, we subtract yield and check yield for getting the yield difference which is the final target value.

3.2 Weather Prediction Neural Networks

As we know, part of this year Syngenta Challenge is the prediction of 2017 weather variables. So, before final step, we need to predict the weather variables for 2017 year. There exist 315 distinct locations and for each of them, 72 weather variables should be predicted.

Weather data from 2001 to 2016 are available as training data. So, we can look at each variable for any location as single time series. For doing time series analysis, one hidden layer neural network has been trained. But, before anything else, training data should be created. Weather data are usually periodic data, so value of one specific variable in the next year may depend on just certain amounts for previous years. For instance, amount of weather precipitation in the Ames, IA may depend on just last 3 years. Since no information on the name of variables is given in the crop challenge weather dataset, all dependencies (lag) ranged between 1 and 13 years are considered for all 72 weather variables, and the one with minimum mean squared error (MSE) is selected.

For instance, let assume $x_1, x_2, \ldots, x_{16}$ are weather values for a specific variable and lag of 4 years is considered for them. Then, the training samples are as follows (Table 2).

For each variable, there are 315 locations and all locations together for one specific variable create the training dataset based on above method. Finally, from 2001 to 2015 weather data are used as training set and 2016 weather data are used as validation set. The Average MSE for all 72 variables on validation was 0.18. Then, results from this section are used in the yield prediction process.

Table 2 Training sample creation

Input variables	Target value
x_1, x_2, x_3, x_4	x_5
x_2, x_3, x_4, x_5	x_6
x_3, x_4, x_5, x_6	x_7
$\vdots$	$\vdots$
$x_{12}, x_{13}, x_{14}, x_{15}$	x_{16}

4 Conclusion

For In this paper, we proposed a machine learning method which showed a superior performance in 2018 Syngenta Crop Challenge using deep neural networks approach. The method used state-of-the-art modeling and solution techniques to predict yield, check yield, and yield difference.

References

1. http://www.fao.org/news/story/en/item/35571/icode/
2. A. Krizhevsky, I. Sutskever, G.E. Hinton, Imagenet classification with deep convolutional neural networks. Adv. Neural Inf. Proc. Syst. 1097–1105 (2012)
3. C. Szegedy, W. Liu, Y. Jia, P. Sermanet, S. Reed, D. Anguelov, A. Rabinovich, Going deeper with convolutions, in *Proceedings of the IEEE Conference on Computer Vision and Pattern Recognition*, pp. 1–9 (2015)
4. Y. LeCun, Y. Bengio, G. Hinton, Deep learning. Nature **521**(7553), 436–444 (2015)
5. K. He, X. Zhang, S. Ren, J. Sun, Deep residual learning for image recognition, in *Proceedings of the IEEE Conference on Computer Vision and Pattern Recognition*, pp. 770–778 (2016)
6. https://en.wikipedia.org/wiki/Artificial_neural_network
7. https://en.wikipedia.org/wiki/Imputation_(statistics)

Cloud-Based Life Sciences Manufacturing System: Integrated Experiment Management and Data Analysis via Amazon Web Services

Pei Guo, Raymond Peterson, Paul Paukstelis and Jianwu Wang

Abstract A vital need in the life sciences industry is software that manages large amounts of fast-moving data for manufacturing quality assurance, clinical diagnostics, and research. In the life sciences industry and research labs, lab information management systems (LIMS) are often used to manage expensive lab instruments. We propose a new software architecture for cloud-based life sciences manufacturing system through the following two advances: (1) full life cycle support of life science experiment through cloud services, (2) workflow-based easy and automatic experiment management and data analysis. This paper discusses our software architecture and implementation on top of Amazon Web Services by utilizing its services including Lambda architecture, API gateway, serverless computing, and Internet of Things (IoT) services. We demonstrate its usage through a real-world life sciences instrument and experimental use case. To our best knowledge, it is the first work on supporting integrated experiment design, experiment instrument operation, experiment data storage, and experiment data analysis all in the cloud for the life sciences.

Keywords IoT · Cloud manufacturing · Workflow · Lab information management system

P. Guo · J. Wang (✉)
Department of Information Systems, University of Maryland, Baltimore County, Baltimore, MD 21250, USA
e-mail: jianwu@umbc.edu

P. Guo
e-mail: peiguo1@umbc.edu

R. Peterson
Granite Point Ventures, Greenbelt, MD 20770, USA
e-mail: raymond.john.peterson@gmail.com

P. Paukstelis
Department of Chemistry and Biochemistry, University of Maryland, College Park, MD 20742, USA
e-mail: paukstel@umd.edu

© Springer Nature Switzerland AG 2020
H. Yang et al. (eds.), *Smart Service Systems, Operations Management, and Analytics*, Springer Proceedings in Business and Economics, https://doi.org/10.1007/978-3-030-30967-1_14

1   Introduction

Nowadays, cloud computing has been adopted by more and more scientific disciplines and industries as a viable way for better resource utilization and budgetary
cost management. Servitization is a key technology used in cloud computing to provide XaaS (Anything as a Service), which includes IaaS (Infrastructure as a service),
SaaS (Software as a service), and DaaS (Data as a Service). Following the same
principle, we explore how cloud computing and servitization could be used to help
Life Sciences Manufacturing Systems (LIMS), especially the following challenges:
(1) many instruments in life sciences manufacturing are expensive but might not be
used very often once purchased; (2) it takes a lot of efforts to train instrument operators such as students and technicians on how to use a specific instrument for their
experiments; (3) each conducted experiment requires extensive log taking in order to
be able to reproduce the result in the future; (4) data analysis results obtained from
experiments are often saved in a separate local computer which is difficult to share.

   To deal with the challenges, we propose a new generation of software architecture for cloud-based life sciences manufacturing system through the following two
advances: (1) the support for life science experiment full life cycle through cloud
services, (2) workflow-based easy and automatic experiment management and data
analysis. We design our architecture on how to utilize existing cloud services for
the above capabilities. To evaluate its feasibility and advantages, we implemented a
system on top of Amazon Web Services (AWS) based on the architecture and applied
it with a real-world use case. To our knowledge, this is the first report of serverless
and edge computing for the life sciences that integrates the primary functions of
instrument control, data storage, and data analysis.

   The rest of the paper is organized as follows. Section 2 walks through existing
works on LIMS and cloud manufacturing. Section 3 introduces our proposed cloud-
based LIMS architecture. Section 4 is an overview of the current implementation
with Sect. 5 explaining a specific use case with AWS. In Sect. 6, we conclude our
paper with some future work discussion.

2   Related Work

Grand View Research estimates that the LIMS market will grow 9.1% to $2.22 bil-
lion by 2025 [1]. Many current LIMS utilize client–server model with a database
and involve workflow for data management. First LIMS, explained in [2], imple-
ments laboratory automated mutation screening workflow to record data step-by-
step. It is a high throughput genetic platform with a client/server architecture and a
MySQL database aimed at candidate gene mutation screening. The LIMS connects
lab instruments and has database and web application. Another LIMS is ISPyB [3]
for synchrotron macromolecular crystallography, which also contains a web server,

a Model-View-Controller (MVC) front-end, and a MySQL database with an underlying data model. ISPyB's functionalities include sample management, experiment recording and planning, and searching and reporting. The particular strength of ISPyB relates to sample management and the recording of information during an experiment. A third LIMS, described in [4], is a data analysis and management pipeline for Liquid Chromatography–Tandem Mass Spectrometry (LC-MS/MS) proteomics, which is a web-based tool whose features include experiment annotation, protein database searching, protein sequence management, as well as data mining of the peptides and proteins identified. Most of current existing LIMS focus more on data storage and data analysis. But these LIMS have some limitations: (1) they are not cloud-based, thus execution of experiments could not be easily controlled remotely; (2) the experiment data is only stored locally; (3) data analysis only happens on one machine, which means it's not easy to scale up when dealing with large data sets.

In its highest form, Cloud Manufacturing (CMfg) is a customer-centric model that combines supply chain, manufacturing services, and software using Internet of Things (IoT) technology and Cloud Computing (CC) [5]. CMfg supplies manufacturing services from the manufacturing resources and capabilities. It is feasible to manage and to operate manufacturing software and hardware resources in a unified way to provide better efficiency and effectiveness in manufacturing process. For instance, CMfg supports deploying and distributing manufacturing software on cloud, which makes it easier for global marketing and for meeting the market business trends. Current CMfg frameworks such as Cloud Computing and Internet of Things-Based Cloud Manufacturing Service System (CCIoT-CMfg) [6] and Interoperable Cloud-based Manufacturing System (ICMS) [7] focus more on production life cycle from business perspective. Whereas in our proposed architecture, we focus on applying cloud services from the life sciences laboratory perspective to manage and operate through the experiment life cycle from design, execution to data capture, storage, and analysis.

3 Proposed Architecture for Cloud-Based Life Sciences Manufacturing System

There are four main stages for a life science experiment: experiment design, experiment instrument operation, experiment result data storage, and experiment result data analysis. As in Fig. 1, we will explain how to support the full life cycle of the experiment through cloud services.

Stage 1: Experiment Design. Every experiment involves several arranged steps and only a successful design results in effective and convinced outputs. Sometimes a user will frequently edit the experiment workflow to handle different experiment requirements, such as changing input parameters or the order of experiment steps. In this stage, the user can design his/her own specific experiment by dragging and dropping experiment step blocks on cloud-based user interface instead of re-editing

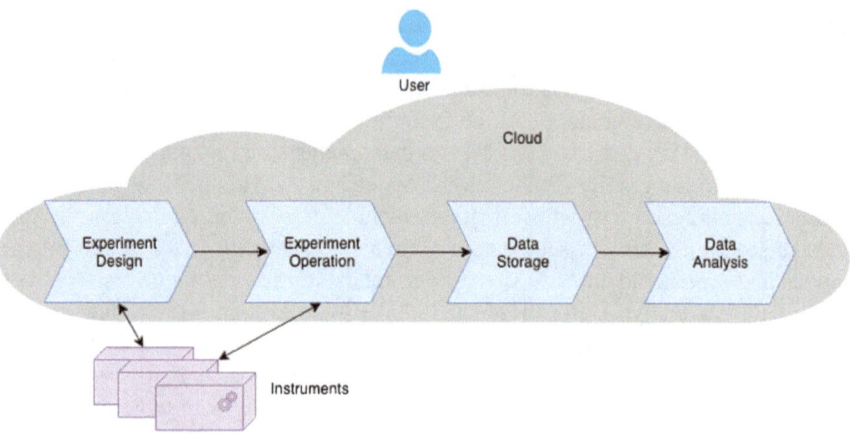

Fig. 1 Proposed Architecture for Cloud-based Life Sciences Manufacturing System

the code line by line. The possible cloud service solutions include workflow services such as Amazon Simple Workflow Service (SWF) [8] or Microsoft Flow [9].

Stage 2: Experiment Operation. After the experiment design, the experiment should be started to operate on instrument. In most cases, the user needs to control one instrument step-by-step or multiple instruments simultaneously. For convenience and efficiency, in this stage, scientific experiment instruments are connected to cloud service that user can start, stop, repeat experiments or even automatically run same experiment for several times on cloud by following a certain schedule. A solution is to utilize Internet of Things (IoT) cloud services, which can connect the instruments to support following steps such as data storage and data analysis for the integration of the architecture. Possible cloud IoT services solutions are AWS IoT [10], Azure IoT [11], etc.

Stage 3: Data Storage. Retention of raw data is crucial for research and quality assurance of industrial production. In our proposed architecture, as a solution in data storage stage, the raw data is directly uploaded to the cloud with both timestamp and the corresponding instrument tag. It makes sure that the experiment raw data could be traced back with keeping records of its time and source. In data format perspective, the data can be saved in raw data files format (txt, csv) or json format. Data can then be uploaded to cloud storage services such as Amazon Simple Storage Service (S3) [12], IBM Cloud Storage [13], and Azure Cloud Storage [14]. Or, the data can be directly uploaded and saved into a structured cloud database, for instance Amazon Relational Database Service (RDS) [15], Azure Cosmos DB [16], etc.

Stage 4: Data Analysis. The final stage of running experiments is not to store overwhelmed raw data, but to provide understandable and useful results and knowledge to the user by utilizing data analysis to transform the experiment raw data to intuitive numeric results or plots. The last but not least stage, in our proposed architecture, data analysis, finally applies machine learning technologies to process the raw data and to provide numerical and visualized result back to user. Workflow is also playing a role in this stage to support user needs such as essential experiment repeats. The data analysis service could run on a cloud computing platform as Amazon Elastic Compute Cloud (Amazon EC2) [17], or through a serverless compute cloud service such as AWS Lambda [18].

The advantages of our architecture are (1) it supports the full life cycle of life science experiments from experiments design and executing to result data storage and analysis; (2) it remains highly flexible that all the components are pluggable; (3) it is not vendor locked since the cloud services are interchangeable; (4) the workflow in experiment design stage and data analysis stage provides better user experiences.

4 Early Implementation for Cloud-Based Life Sciences Manufacturing System

The early implementation of our proposed framework includes three main functional modules, experiment operation, data storage, and data analysis as in Fig. 2, which are later three stages with same names corresponding to our proposed architecture in Fig. 1.

Experiment Operation Module. In this module, the user can manipulate instrument to conduct designed experiments. The raw data generated from experiment instrument is automatically published to cloud. The involved cloud technologies are AWS IoT and AWS Lambda.

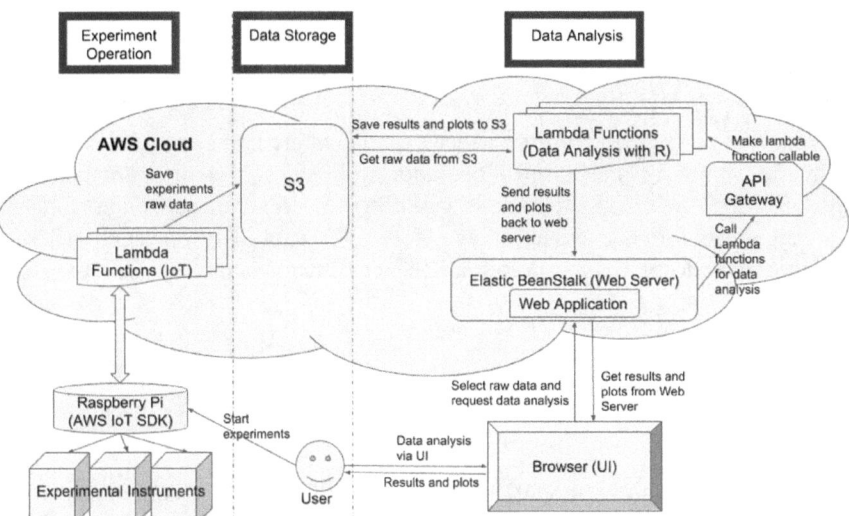

Fig. 2 Early implementation for Cloud-based Life Sciences Manufacturing System. Left: Experiment Operation module. Middle: Data Storage module. Right: Data Analysis module

Data Storage Module. In this module, the raw data published by AWS IoT service is subscribed by AWS Lambda function and saved to AWS S3 to fulfill data storage function. Each record saved in S3 is in text format, which is retrievable by data analysis module for running machine learning jobs. It also supports data analysis module to save the final data analysis results. The involved cloud technology is AWS S3 and AWS Lambda.

Data Analysis Module. In this module, experiment raw data is retrieved by an AWS Lambda function from AWS S3. After that, a [19] program running on AWS Lambda serverless service processes data analysis jobs and then outputs the regression results and plots. The data analysis lambda function is RESTful [20] callable by utilizing API Gateway [21]. User can interact with UI through web browser including selecting experiment data, viewing the data shape, selecting regression model, running regression analysis, and finally, getting the results and plots. Moreover, the results and plots are saved back to S3. The involved technologies are AWS S3, Lambda, API Gateway, R, AWS Elastic Beanstalk [22], AngularJS [23], and Node.js [24].

One focus of our system implementation is that we use serverless services as much as possible for better budget cost control. The limitations of our current implementation are (1) user can only start experiment locally; (2) all the components always have to stay connected to the Internet; (3) packages and codes have to be deployed manually; (4) current experiment design is not very easy to change (not enough flexibility).

5 Cloud-Based Life Sciences Manufacturing System Use Case with AWS

The details of our use case of life science experiment are introduced and explained in three subsections: (1) experiment operation section describes the equipment, the hardware connection, and experiment workflow; (2) data storage section contains the detail of data format and data storage process; (3) data analysis section includes implementation detail of its workflow and the introduction of the web UI for the end user.

5.1 Experiment Operation

The experiment operation module utilizes AWS IoT service, which connects the legacy lab instrument to AWS IoT cloud platform and executes the instrument following designed experiment steps to process the experiment and upload raw data.

The legacy instrument used in our implementation use case is a UV spectrometer (Beckman Coulter DU 640 Spectrophotometer) from a Biochemistry lab. It

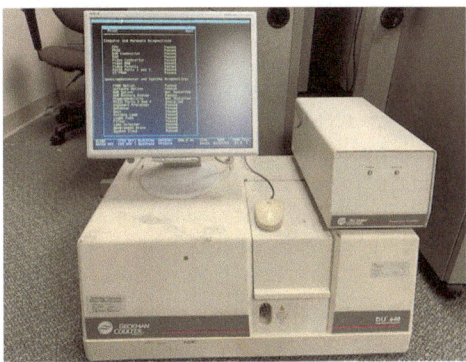

Fig. 3 Cloud connection of legacy instrument via Raspberry Pi. Left: Beckman Coulter DU 640 Spectrophotometer. Right top: Raspberry Pi. Right bottom: RS232 Serial to USB adapter

is connected with a display, a mouse, and a thermistor as in Fig. 3. Spectrophotometers measure the amount of light passing through a clear container that holds a liquid sample. Comparison of the light intensity before and after the sample provides absorption measurements at particular wavelengths. This light absorption correlates directly with concentration of material in the sample and can be highly informative about the physical properties of the molecule of interest. Spectrophotometers have many applications and as such are workhorses in chemistry, biochemistry, other life sciences laboratories, and quality assurance departments.

The Beckman DU640 has the capacity for six specialized containers called cuvettes. In the present experiment, we fill the cuvettes with clear liquid solution and a precise amount of short, complementary DNA strands. The instrument has thermal control from refrigerated temperature to near boiling. At low temperature, the DNA strands are hybridized into helical duplexes. These duplexes absorb a certain amount of light. As the sample temperature is raised in the instrument, the DNA strands bind less strongly to each other, and eventually separate, leading to increased UV absorbance. The amount of light absorbance measured at each temperature is proportional to the amount of DNA that is in duplex or single-strand. It is these two values, temperature and absorbance, that constitute the experimental data that we wish to generate and analyze. Over the temperature range, the temperature-absorbance relationship is S-shaped, with lower and upper asymptotes. The shape of this curve is informative of certain biophysical properties of use in biotechnology applications. One such application is amplification of patient DNA samples for accurate and rapid medical diagnostics. For this application, the use of DNA melt properties is to predict the behavior of DNA to design an optimal assay.

To manipulate the instrument by computer, the traditional setup is to connect the instrument by Recommended Standard 232 (RS232) [25] cable to a local desktop computer which has an embedded RS232 card and port or a USB port. One may communicate via RS232 protocol on USB port by use of an RS232/USB adapter. This desktop computer runs user interface software that enables the user to select experimental conditions and to execute packaged programs for instrument control. The instrument's firmware has an API architecture that accepts instructions from this computer and executes the instructions on the instrument's physical mechanisms and signal detectors. The instrument then sends generated raw signal data back to the local computer via this interface. DU 640 has instrument API interfaces which enable a scientist to control the instrument in custom fashion by writing a local Python script that communicates with the instrument by way of RS232 port. This is what we have done previously, to demonstrate the ability to control the instrument with a local Python script executing on a desktop computer.

Modern instruments have Ethernet network capability, which makes it possible that the RS232 cable connection could be replaced with either an Ethernet cable or wireless network connection. In our upgraded setup, we replace the desktop computer with a Raspberry Pi. The Raspberry Pi has a USB port but no RS232 port, necessitating a USB to serial RS232 adapter cable as in Fig. 3. On the Raspberry Pi, the Python script sets port configuration and instrument connection and manipulates instrument action by way of command line execution of individual statements or entire script.

To connect our instrument to the cloud, the Raspberry Pi is registered as a Thing in AWS IoT Console. In the meanwhile, AWS IoT SDK is installed on the Raspberry Pi. During the experiment, each record is published by the Raspberry Pi through AWS IoT infrastructure. And the published record is captured and handled by AWS lambda functions to be uploaded to AWS S3 for record keeping. In this manner, we effectively convert the legacy, nonnetworked Beckman instrument into an IoT device. The experiment execution critically relies on AWS IoT service, which connects the nonnetworked legacy laboratory instrument to AWS IoT web service. These data are then available to any Amazon web service that may subscribe to AWS IoT event-driven model.

Experiment Workflow Fig. 4 describes our current designated experiment workflow. The user logs into Raspberry Pi via SSH to start experiments through command line. The set of measurements starts by selecting a start set temperature (*Temp*), for instance, 17.5 °C. The program waits for 30 s before retrieving the actual temperature (*temp*) from the thermistor. Then the program compares *temp* to *Temp* and checks if they are equal. If not, the program waits for another 30 s to ensure it gets to *Temp*. When the temperature set point has been reached, the program records current temperature *temp*, sample absorbance *abs*, and timestamp *time* as a triplet (*time, temp,*

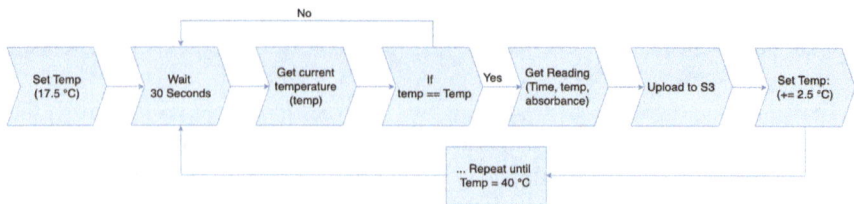

Fig. 4 Experiment design and execution workflow

abs). Each record triplet is published under each temperature setting. Then the temperature will increase by 2.5 °C and previous steps all execute as a loop. The program stops when the temperature arrives at 40 °C.

5.2 Data Storage

In the data storage module, AWS S3 is utilized for raw data record keeping. It ensures that every single recorded experiment result is orderly uploaded and stored on cloud with a timestamp, which makes sure that compared to user manually inputting records to spreadsheets, the experimental records will not lose by storing in a structural and retrievable format to make it feasible to trace back to the specific instrument that produces the corresponding raw data. AWS S3 provides the integration of AWS Lambda functions to support AWS IoT uploading the raw data as in previous module and R program on AWS Lambda retrieving the raw data for following data analysis module.

5.3 Data Analysis

In the data analysis module, experiment raw data saved in AWS S3 is retrieved by AWS Lambda functions, then it is concatenated to a full experiment data set. R is utilized on top of AWS Lambda to use the full data set to process data analysis by generating different regression models to output summaries and corresponding plots. In current data analysis implementations, data could be used to fit four different regression models. Moreover, the fitted regression model and the plots are saved back to AWS S3 for record keeping or future use.

A Lambda function can be executed directly or through an API Gateway. API Gateway is chosen here because it provides a RESTful API for the integration with our web application. Also, API Gateway provides built-in request throttling and supports robust error handling via HTTP status codes.

The data analysis workflow is designed as in Fig. 5. A user first selects the data in .csv file format following the instructions on the web page, then the data shape plot is displayed as the first glance for user. In next step, the user selects proper regression model with intuitively comparing the data shape and example fitted curve shapes.

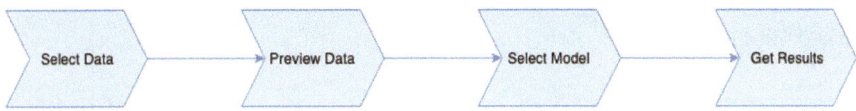

Fig. 5 Data analysis workflow

The final step is that the user gets the regression results summary with the plot. In the meanwhile, the plot and regression results are saved back to S3.

A web application is created as UI for the user to execute the steps in data analysis workflow. In front-end, AngularJS is used to build the user interface using a single page setting. We chose it because it is an MVC framework and it is testable with Karma or other testing frameworks. In the server-side, a Node.js web server hosts the front-end pages and services. Moreover, AWS Elastic Beanstalk is chosen as our cloud web hosting service. Compared to AWS Elastic Cloud Computing (EC2), our choice of AWS Elastic Beanstalk is based on the features of Elastic Beanstalk: (1) a layer beyond AWS EC2 with EC2 instances, databases, load balancer, auto-scaling, etc.; (2) auto managing the components for future upgrading; (3) version control.

6   Conclusion and Discussion

With the need of life science industry and research labs requiring managing lab instruments, executing experiments, keeping records and analyzing experiment results, we propose a new cloud-based life sciences manufacturing system architecture. The architecture could support cloud-based automatic instrument operation and data storage, data analysis from instrument experiment results, and workflow-based process management for instrument operation and data analysis. Our architecture utilizes existing cloud service to achieve our goals and it is not vendor locked. We evaluate the feasibility by implementing a real-world use case with cloud services provided by AWS.

For future works, we will extend our architecture from the following two aspects. First, we will study how to deploy our system on a hybrid cloud environment so that some parts of the life science experiment life cycle stages could be done at premises for better security control. Our implementation experimented how to utilize OpenCPU [26] as an external resource for our data analysis stage. Second, we plan to extend our architecture for edge computing. For instance, AWS Greengrass service [27] could be deployed on the Raspberry Pi to support data analysis on the edge and corresponding instrument control based on data analysis results.

Acknowledgements This work is supported in part by a Maryland Industrial Partnerships (MIPS) grant: the Low-Code Workflow Software for Life Sciences.

References

1. PR NEWSWIRE: Lims market size worth $2.22 billion by 2025—cagr: 9.1%: Grand view research, inc. http://news.sys-con.com/node/4372573. Accessed 21 Feb 2019
2. A. Thomas, C. Voegele, D. de Silva, F. Le Calvez-Kelm, S. Cuber, S. Tavtigian, A Laboratory Information Management System (LIMS) for a high throughput genetic platform aimed at

candidate gene mutation screening. Bioinformatics **23**(18), 2504–2506 (2007). https://doi.org/10.1093/bioinformatics/btm365

3. A.W. Ashton, D. Spruce, E.J. Gordon, G.A. Leonard, J. Gabadinho, K.E. Levik, L. Launer, M.A. Walsh, M. Nanao, O. Svensson, P. Brenchereau, R. Leal, S.D. Jones, S.M. McSweeney, S. Delagenire, S. Monaco, S. Veyrier, ISPyB: An information management system for synchrotron macromolecular crystallography. Bioinformatics **27**(22), 3186–3192 (2011), https://doi.org/10.1093/bioinformatics/btr535, https://dx.doi.org/10.1093/bioinformatics/btr535

4. A. Droit, J.M. Hunter, M. Rouleau, C. Ethier, A. Picard-Cloutier, D. Bourgais, G.G. Poirier, Parps database: A lims systems for protein-protein interaction data mining or laboratory information management system. BMC Bioinform. **8**(1), 483 (2007), https://doi.org/10.1186/1471-2105-8-483

5. D. Wu, M.J. Greer, D.W. Rosen, D. Schaefer, Cloud manufacturing: Strategic vision and state-of-the-art. J. Manuf. Syst. **32**(4), 564–579 (2013). https://doi.org/10.1016/j.jmsy.2013.04.008, http://www.sciencedirect.com/science/article/pii/S0278612513000411

6. F. Tao, Y. Cheng, L.D. Xu, L. Zhang, B.H. Li, Cciot-cmfg: Cloud computing and internet of things-based cloud manufacturing service system. IEEE Trans. Industr. Inf. **10**(2), 1435–1442 (2014). https://doi.org/10.1109/TII.2014.2306383

7. F. Tao, Y. Zuo, L.D. Xu, L. Zhang, Iot-based intelligent perception and access of manufacturing resource toward cloud manufacturing. IEEE Trans. Industr. Inf. **10**(2), 1547–1557 (2014). https://doi.org/10.1109/TII.2014.2306397

8. Amazon Simple Workflow Service (SWF). https://aws.amazon.com/swf/. Accessed 21 Feb 2019

9. Microsoft flow: Automate processes + tasks. https://flow.microsoft.com/en-us/. Accessed 21 Feb 2019

10. Amazon IoT. https://aws.amazon.com/iot/. Accessed 21 Feb 2019

11. Azure IoT hub. https://azure.microsoft.com/en-us/services/iot-hub. Accessed 21 Feb 2019

12. Amazon S3. https://aws.amazon.com/s3/. Accessed 21 Feb 2019

13. IBM: IBM cloud object storage. https://www.ibm.com/cloud/object-storage. Accessed 21 Feb 2019

14. Azure Storage. https://azure.microsoft.com/en-us/product-categories/storage/. Accessed 21 Feb 2019

15. Amazon Relational Database Service (RDS). https://aws.amazon.com/rds/. Accessed 21 Feb 2019

16. Azure Cosmos DB. https://azure.microsoft.com/en-us/free/cosmos-db/. Accessed 21 Feb 2019

17. Amazon Elastic Compute Cloud (Amazon EC2). https://aws.amazon.com/ec2/. Accessed 21 Feb 2019

18. Amazon Lambda. https://aws.amazon.com/lambda/. Accessed 21 Feb 2019

19. R Core Team (2019) R: A Language and Environment for Statistical Computing. R Foundation for Statistical Computing, Vienna, Austriahttp://www.R-project.org/

20. R.T. Fielding, Architectural styles and the design of network-based software architectures. Ph.D. thesis, University of California, Irvine (2000)

21. Amazon API Gateway. https://aws.amazon.com/api-gateway/. Accessed 21 Feb 2019

22. AWS Elastic Beanstalk. https://aws.amazon.com/elasticbeanstalk/. Accessed 21 Feb 2019

23. Google: Angularjs. https://angularjs.org/. Accessed 21 Feb 2019

24. Node.js Foundation: Node.js. https://nodejs.org/en/. Accessed 21 Feb 2019

25. J.L. Axelson, Serial port complete: programming and circuits for RS-232 andRS-485 links and networks with disk. Lakeview Res. (1999)

26. OpenCPU: Opencpu home. https://www.opencpu.org/. Accessed 21 Feb 2019

27. AWS IoT Greengrass. https://aws.amazon.com/greengrass/. Accessed 21 Feb 2019

Matching Anonymized Individuals
with Errors for Service Systems

Wai Kin (Victor) Chan

Abstract Data privacy is of great importance for the healthy development of service systems. Companies and governments that provide services to people often have big concerns in sharing their data. Because of that, data must be preprocessed (e.g., anonymized) before they can be shared. However, without identification, it is difficult to match data from different sources and thus the data cannot be used together. This paper investigates how the performance of two simple individual matching methods was affected by errors in the similarity scores between individuals. The first method is a greedy method (GM) that simply matches individuals based on the maximum similarity scores. The second method is an optimal assignment problem (AP), which maximizes the total similarity scores of the matched individuals. Consistent with the literature, we found that GM outperforms AP in most situations. However, we also discovered that AP could be better in fixing errors.

Keywords Data matching · Data correlation · Service systems

1 Introduction

Value co-creation is one core principle of the development of service systems [13]. A number of studies have demonstrated the benefits of value co-creation and how it fosters the development of many business and social sectors [9, 16, 17]. Many innovations in service science require the use of user data to enable or create new services. However, privacy is a big concern nowadays due to the raising of big data. As people (e.g., customers) are using more and more services, their digital trails could be collected and used by unauthorized individuals or organizations. As such, it is important to protect the privacy of customers while still allowing data to be used for service innovations.

W. K. V. Chan (✉)
Environmental Science and New Energy Technology Engineering Laboratory, Tsinghua-Berkeley Shenzhen Institute, Tsinghua University, Shenzhen 518055, People's Republic of China
e-mail: chanw@sz.tsinghua.edu.cn

© Springer Nature Switzerland AG 2020   161
H. Yang et al. (eds.), *Smart Service Systems, Operations Management, and Analytics*, Springer Proceedings in Business and Economics, https://doi.org/10.1007/978-3-030-30967-1_15

One application is data matching. In data matching, two or more databases from different owners (e.g., service organizations) are matched based on individual identity. The combined dataset is a larger database that contains much richer information about each individual and the whole population in the dataset. This allows more accurate services to be provided to customers. Unfortunately, due to privacy issue, owners of the databases cannot share their data unless at least identifications of individuals are encrypted or removed.

Privacy-preserving data matching (or record linkage) has been an active research area [2, 6, 10, 14]. One of the earliest work in data matching was done in [4]. Data matching has a number of applications [12], create an architecture for social genome databases by connecting social data from various sources [7, 8], use census data to match records at a household level. A graph-based approach was used to capture the structural relationship between household members.

There are a number of data matching methods, based on blocking and indexing [3, 5]. There are two basic types of protocols governing how data is exchanged between database owners: (1) two-party protocol and (2) three-party protocol. The three-party protocol requires a third party that is trusted by the database owners to do the matching of the data. The two-party protocol does not rely on the use of a third party. As such, sensitive personal information is removed or encrypted by using some encryption or encoding methods mutually agreed between the database owners. Many models have been studied [10] and a number of advanced encryption approaches have also been proposed, such as secure hash algorithm, secure secret function evaluation, and message digest function [11, 15].

The procedure of data matching involves the following steps in sequence: (1) data preprocessing, (2) indexing, (3) comparison, (4) classification, (5) clerical review, and (6) evaluation. Data preprocessing is to prepare and clean the data. Indexing is to filter out unlikely matches to reduce the complexity of matching. The comparison step compares records of the matching databases and assigns each pair of records a similarity score. The classification step matches the records based on the similarity scores. If the matching method classified records are possible matches, the clerical review is needed to make the decision. Finally, evaluation can be carried out to assess the matching quality.

This paper focuses on analyzing how matching quality is affected by errors, which could occur during the calculation of the similarity scores or simply due to data entry mistakes. To rule out influence from other factors, this paper will not apply any classification operations. In addition, we assume that the two matching databases are of the same size and correspond to the same n distinct individuals, that is, the matching is a one-to-one matching. Therefore, we assume that the $n \times n$ similarity scores have been obtained during, for example, the comparison step by using some similarity functions [18]. However, errors can occur either in record entries or during score calculation. Therefore, it is important to evaluate how such errors influence the matching quality. The assumption of two databases being equal size can be realized if the n individuals are fixed and the two databases contain data only for these n individuals.

As this paper only considers one-to-one matching, we focus on the two methods used in [2] and [6]: (1) greedy method (GM) and (2) assignment problem (AP).

Although [6] also allow errors in their study, the present paper focuses on examining the impact of the error rate on the matching quality. In addition, besides considering precision and recall, the present paper divides the errors into four types and explains why AP performed worse than GM. This shows counterintuitively that optimization is not always better. We aim at extending the study in [2, 6] by providing a detailed analysis of how errors affect matching quality.

Section 2 defines the problem and outlines the two approaches, GM and AP. Section 3 presents the main results and performs an analysis of the experimental results. Section 4 draws a conclusion and offers several future works.

2 Problem Definition and Two Data Matching Methods: Greedy Method and Assignment Problem

The original problem of privacy-preserving data matching (or linkage) is to find matching between two or more databases owned by different organizations in such a way that no sensitive information is shared across organizations.

This paper considers two post-matching methods: greedy method (GM) and optimal assignment problem (AP). As introduced in [2], the GM simply matches records based on the highest available similarity scores; that is, the pair of records with the highest similarity score is matched first and their associated links are removed. Then, the next pair of records with the highest similarity score (after removing links with the first pair) is matched and their associated links are removed. This process repeats until all records are matched. Note that because we do not employ any classification operations (i.e., no threshold is used), all records between the two databases are fully connected (i.e., no missing similarity score). Therefore, GM will eventually obtain a complete match between databases.

AP is a special case of the network flow problem [1]. Like many network flow problems, AP also satisfies the network property, meaning that linear relaxation solution is also an integer solution. As such, existing network flow solution algorithm can solve this problem efficiently.

As explained in the previous section, this paper assumes that the matching is conducted between two databases of equal size and that the databases are deduplicated. Precisely, let D_A and D_B be the two databases to be matched and $|D_A| = |D_B| = n$, where n is the size (number of records) in each database. Let binary variable $\delta_{ij} = 1$, if Record i in D_A and Record j in D_B are a true match and $\delta_{ij} = 0$, otherwise. It is also assumed that there exists a true perfect matching between the two databases, that is, for each Record i in D_A, there exists one and only one Record j in D_B that corresponds to the same individual as Record i in D_A. However, this true perfect matching is unknown.

A premise of this study is that similarity scores between the records of the two databases are available. In other words, the comparison step will result in a set of similarity scores, $\mathbf{S} = \{s_{ij}, \forall i \in D_A, \forall j \in D_B\}$. In the literature, it is commonly assumed that the score is between 0 and 1. However, in this paper, we generalize it so that it can take any real value (see next section).

Let binary variable $x_{ij} = 1$, if Record i in D_A is found to be a match with Record j in D_B and $x_{ij} = 0$, otherwise. The AP is to find $\mathbf{X} = \{x_{ij}, \forall i \in D_A, \forall j \in D_B\}$ that satisfies the one-to-one matching constraints: $\sum_{\forall i \in D_A} x_{ij} = 1, \forall j \in D_B$ and $\sum_{\forall j \in D_B} x_{ij} = 1, \forall j \in D_A$. Formally, AP is defined as:

$$\max_{\mathbf{X}} f_1 = \sum_{\forall i,j} s_{ij} x_{ij}$$

$$s.t. \sum_{\forall i \in D_A} x_{ij} = 1, \quad \forall j \in D_B$$

$$\sum_{\forall j \in D_B} x_{ij} = 1, \quad \forall i \in D_A$$

3 Experiments

In this short paper, we conduct an experiment on a network of 1000 individuals. Because no classification nor threshold is used, the size of the problem is $n \times n$. This full-size problem allows us to thoroughly examine the impact of error rate when it varies from 0.05 to 0.95. Also, because we use a synthetic network, we can designate the true matching individuals to evaluate the matching results. In particular, we set all diagonal elements to correspond to the same individuals.

We first generate the $n \times n$ similarity scores, $\mathbf{S} = \{s_{ij}, \forall i \in D_A, \forall j \in D_B\}$ according to a standard normal distribution. If these scores represent a real system, the similarity score of the pair of records pertaining to the same individual should have the largest score. Hence, for each individual i, we swap s_{ii} with the maximum score among all $s_{ij}, j = 1,\ldots, n$. Therefore, the true matches are all diagonal elements in the nxn matrix.

The next step is to make some errors. Each score has a chance of p_e to be replaced by a new value following a standard normal distribution independent of the previous value. We vary p_e from 0.05 to 0.95 at an increment of 0.05. For each p_e, five replications were made. This error generation process creates two types of errors: (1) changing s_{ii} to be smaller than at least one other score, s_{ij} and (2) changing at least one s_{ij} to be larger than s_{ii}. It is also possible that s_{ii} is first reduced to be less than only one s_{ij} and then s_{ij} is changed to be smaller than the new s_{ii}. In such a case, no effective error is made.

After the error generation process, a synthetic set of nxn similarity scores with errors is obtained. If one uses this set of similarity scores to determine the matching

Fig. 1 Performance (precision or recall) of GM and AP in changes of error rate

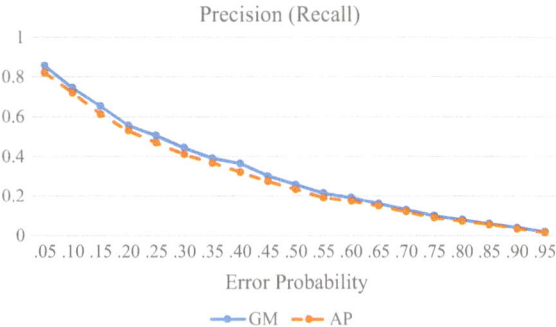

(i.e., based on the maximum scores), the potential error in the matching is the fraction of non-maximum scores within the diagonal line.

Two standard accuracy measures are precision and recall. Precision is the percentage of correct matches within all "matches" (correct and wrong) found by the algorithm, that is, correct matches/(correct matches + wrong matches). Recall is the percentage of correct matches within all actual matches in the data, that is, correct matches/(correct matches + false actual matches). Because there are exactly n individuals in this study, the denominator of both precision and recall are the same. That is, wrong matches (wrongly classified as matches) must equal false actual matches (true matches wrongly classified as un-matches). Therefore, in the case of one-to-one matching, precision is identical to recall.

Figure 1 presents the results of precision (same as recall). It is clear that GM outperforms AP regardless of the error rate (the superiority is minor especially at high error rate). This finding is consistent with the results in [6]. Also, as the error rate increases, the performance of both GM and AP degrade as expected.

Next, we extend the analysis by breaking down the matching results into four categories: (1) Match to Match (MM), (2) Match to Un-Match (MU), (3) Un-Match to Match (UM), and (4) Un-Match to Un-Match (UU). The diagonal element (i.e., s_{ii}) i s called MM if no error occurs in s_{ii} (that is, the similarity score still represents the right match) (Match) and then after the matching algorithm (either GM or AP), it was still classified as the same person (i.e., $x_{ii} = 1$) (Match). The diagonal element is called MU if no error occurs (Match) but after the matching algorithm, it was classified as different persons (i.e., $x_{ii} = 0$) (Un-Match). Similarly, UM means that an error occurs in s_{ii} (Un-Match) but was correctly classified (i.e., fixed) as the same person, and UU represents that an error occurs in s_{ii} (Un-Math) and the algorithm was unable to fix the error by classifying it as a match (Un-Match).

These four performance measures are shown in Fig. 2a–d. Figure 2a, b is consistent with Fig. 1a in that, GM obtains higher accuracy. Figure 2c further supports this result. It shows that AP can turn more originally correct matches into incorrect matches. The reason for this is that when AP tries to maximize the total similarity score, it could scarify a maximum similarity score to save several non-maximum scores, so the total score is maximized.

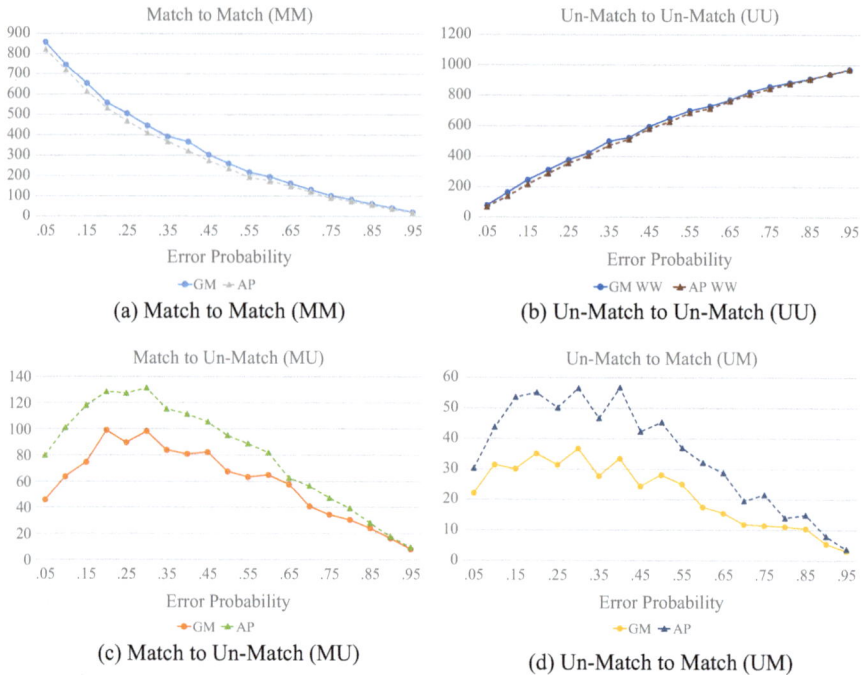

Fig. 2 Four performance measures

Figure 2d, however, reveals a situation where AP is better than GM: When examining UM, we found that AP, also due to its maximization objective, has a better chance to fix some erroneous un-matches.

Because of this special situation and to be fair, we define another performance measure: the "fixed to messed up ratio", which equals to UM/MU—the number of errors a method fixed divided by the number of errors it created. This measure for the two methods is shown in Fig. 3. It shows that the "marginal benefit" (number of errors fixed per unit of errors created) of AP is higher than that of GM in most

Fig. 3 Fixed to messed up ratio: UM/MU

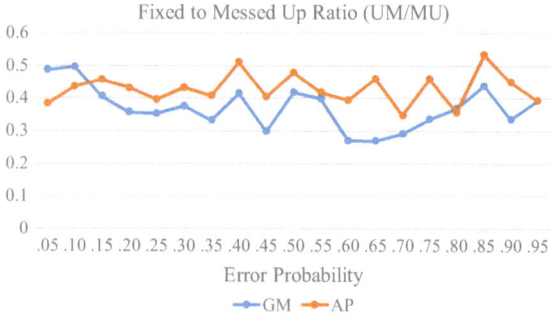

cases, in particular, when the error rate is higher than 10%. In privacy-preserving data matching, because identification information is removed, the error rate is likely to be high in computing the similarity scores. As such, AP may still be a better choice if the goal is to fix more errors.

4 Conclusion and Future Work

Although studies in the literature found that the simple greedy method outperforms AP, we show that when a more thorough comparison is conducted, AP may still be better under the measure of error correction (i.e., UM). Due to space limitation, some of the results were not included. There are many possible directions for future work. First, real data should be used to examine the four performance measures. Second, larger networks should be used to see if the results change in the size of the network. Third, the four performance measures should be analyzed theoretically rather than just experimentally. Last but not least, the reason for AP to perform better GM under UM should be analyzed analytically.

Acknowledgements This paper was partially funded by Shenzhen Municipal Development and Reform Commission, Shenzhen Environmental Science and New Energy Technology Engineering Laboratory, Grant Number: SDRC [2016]172.

References

1. R.K. Ahuja, T.L. Magnanti, J.B. Orlin, *Network Flows: Theory, Algorithms, and Applications* (Prentice-Hall, Inc, 1993)
2. P. Christen, in *Data Matching—Concepts and Techniques for Record Linkage, Entity Resolution, and Duplicate Detection*, ed. by M.J. Carey, S. Ceri (Berlin, Springer, 2012a)
3. P. Christen, A survey of indexing techniques for scalable record linkage and deduplication. IEEE Trans. Knowl. Data Eng. **24**(9), 1537–1555 (2012b)
4. I.P. Fellegi, A.B. Sunter, A theory for record linkage. J. Am. Stat. Assoc. **64**(328), 1183–1210 (1969)
5. J. Fisher, P. Christen, Q. Wang, E. Rahm, A clustering-based framework to control block sizes for entity resolution, in *Proceedings of the 21th ACM SIGKDD International Conference on Knowledge Discovery and Data Mining*, 2783396 (ACM, 2015), pp. 279–288
6. M. Franke, Z. Sehili, M. Gladbach, E. Rahm, Post-processing methods for high quality privacy-preserving record linkage (Cham, 2018)
7. Z. Fu, P. Christen, M. Boot, Automatic cleaning and linking of historical census data using household information, in *2011 IEEE 11th International Conference on Data Mining Workshops*, December 2011 (2011), pp. 11–11
8. Z. Fu, P. Christen, J. Zhou, A graph matching method for historical census household linkage (Cham, 2014)
9. C. Hsu, *Service Science: Design for Scaling and Transformation* (World Scientific and Imperial College Press, Singapore, 2009)

10. A. Karakasidis, V.S. Verykios, Advances in privacy preserving record linkage, in *E-Activity and Innovative Technology, Advances in Applied Intelligence Technologies Book Series*, Igi Global (2010)
11. J. Kilian, *Uses of Randomness in Algorithms and Protocols* (MIT Press, Cambridge, MA, 1990)
12. H. Kum, A. Krishnamurthy, A. Machanavajjhala, S.C. Ahalt, Social genome: putting Big Data to work for population informatics. Computer **47**(1), 56–63 (2014)
13. P. Maglio, C. Kieliszewski, J. Spohrer, *Handbook of Service Science* (Springer, New York, NY, 2010)
14. K. McCormack, M. Smyth, Privacy protection for Big Data linking using the identity correlation approach. 统计科学与应用: 英文版 (3), 81–90 (2017)
15. B. Schneier, *Applied Cryptography: Protocols, Algorithms, and Source Code in C*, 2nd edn. (Wiley, New York, 1996)
16. J. Spohrer, P.P. Maglio, The emergence of service science: toward systematic service innovations to accelerate co-creation of value. Prod. Oper. Manag. **17**(3), 238–246 (2008)
17. S.L. Vargo, M.A. Akaka, Value cocreation and service systems (re)formation: a service ecosystems view. Serv. Sci. **4**(3), 207–217 (2012)
18. D. Vatsalan, P. Christen, V.S. Verykios, A taxonomy of privacy-preserving record linkage techniques. Inf. Syst. **38**(6), 946–969 (2013)

Developing a Production Structure Model Using Service-Dominant Logic—A Hypergraph-Based Modeling Approach

Mahei Manhai Li, Christoph Peters and Jan Marco Leimeister

Abstract To make a fundamental shift toward value orientation, manufacturing companies strategically move to integrate services into their portfolio. While manufacturing firms rely on production information systems as the backbone of their operations, these systems are based on product structure models (e.g., bill of materials). This poses a problem because services do not adhere to the goods-dominant perspective of product structures. To solve this divide, this paper proposes an integrative mathematical model for both production systems and service systems. The model draws upon concepts of service-dominant logic and is based on hypergraph theory. To illustrate that the production structure model includes both product structures and process structures, we further demonstrate that the production structure model can be transformed into either. Therefore, our theoretical contribution lies in introducing a structural model into production systems that is compatible with structures of a service system model. For practice, the model enables the development of production information systems that can plan and control products, services, and hybrids.

Keywords Production service system · Service systems · Production structure model · Production graph · Product structure · Process production structure · Hypergraph model of production structure

M. M. Li (✉)
Information Systems, Research Centre for IS Design (ITeG), University of Kassel, 34121 Kassel, Germany
e-mail: mahei.li@uni-kassel.de

C. Peters · J. M. Leimeister
Institute for Information Management, University of St. Gallen, 9000 St. Gallen, Switzerland

Information Systems, Research Centre for IS Design (ITeG), University of Kassel, 34121 Kassel, Germany

© Springer Nature Switzerland AG 2020

169

H. Yang et al. (eds.), *Smart Service Systems, Operations Management, and Analytics*, Springer Proceedings in Business and Economics, https://doi.org/10.1007/978-3-030-30967-1_16

1 Introduction

Since the twenty-first century, services have steadily gained importance, playing an increasingly larger role for the world economy [1]. As of 2017, approximately 63% of the worldwide GDP originated from the service sector, while only 30% stemmed from the industry sector. Consequently, the widely accepted idea of service-dominant logic (SDL) emerged [2]. SDL considers production-related issues as relevant and the traditional characteristics of being immaterial or not as a distinguishing factor for services by introducing the concept of operand and operant resources. Hence, service-dominant logic treats products as goods that transport value for customers. SDL covers the entire economic spectrum, including both the service sector as well as the industry sector.

Services are primarily non-goods-reliant, such as digital services or knowledge-intensive services, and exclude issues of production [3]. Although there are product service systems (PSS) that rely on products or on how to incorporate services into an existing product portfolio as means of an added service, they look at service as a tool to identify business model innovations [4]. It is primarily servitization research that looks at traditional goods-dominant production to make inferences on how to transform goods-dominant organizations (e.g., manufacturing) into a service-dominant organization in hopes of improving their business model [5, 6]. Despite the claim that service-dominant logic is applicable to goods-dominant logic issues, it remains open how service logic can be integrated and mixed into traditional production structures and vice versa. In this paper, we aim to move toward an SDL perspective on production by introducing a production system structure that is compatible with the structure of service systems.

Following the service-dominant logic perspective, a mathematical model for service systems has recently been introduced [7]. Its hypergraph-based service system graph (SSG) is a step toward operationalizing service-dominant logic [2] and particularly its views on service systems [3]. Service and production are both transformation processes of resources [8]. The former is done with operant resources as acts and the latter is done with supplement resources, such as operation equipment and assets. A significant difference between service and production is the resource characteristic. For service, research often focuses on the immateriality of its resources, whereas in production the focus is on material resources. Furthermore, existing process models and product structure models are the foundation for modern production information systems (e.g., ERP systems). Yet, it is apparent that product structures (e.g., bill of materials) are not suitable for modeling service systems. For therein lies our overarching research goal: to find a common mathematical form to model both service systems and production systems. To move toward our research goal, in this paper, we apply its formalization approach, its underlying mathematical logic of using hypergraph theory, and mappings to develop a production graph to model production systems [9] and production structures in particular.

By using one integrative model, future engineers are enabled to model systems that include both production and services, more precisely and clearer, paving the way for a quantifiable service-dominant logic perspective for service systems and production. Additionally, for production systems, product models are essential because of their important role across the functional areas of product life cycles [10, 11], including but not limited to design, purchasing, inventory management, planning and scheduling, and logistics [12–14]. Due to the wide acceptance of product models and their importance for production and, more importantly, to demonstrate our production structure's wide applicability, this paper will conclude by deriving product structures and process structures from the introduced production structure. This intends to demonstrate that the production graph/production structure is compatible with existing product models (e.g., bill of materials/BOMs).

This paper thus seeks to answer the question of *how to represent production structures as foundation models for application information systems so that they are compatible with service system structures*. To explore this question, the paper starts with a brief related work section, which outlines characteristics of the service-dominant logic perspective and the mathematical service system graph as base models for service systems. Then, we introduce contemporary product structures for manufacturing systems. The next section introduces a mathematical model of production structures, which describe the production's factors and their relations, including raw materials, operating equipment, other assets, and processes. This model differs from previously introduced product structures, which describe the relations between a product and its composition. The next chapter introduces the application of the production structure, which shows that traditional product structures can be induced as subgraphs from production structure graphs and it also shows how the mathematical model of a production structure can additionally be an induced subgraph for a process flow model. Additionally, we explain the versatile role of functions within the production structure application. The paper concludes with a discussion and future work section on the three major contributions of our production graph: (1) conceptually integrating production into service, further diminishing its divide, (2) showing the usefulness and versatile application possibilities of a production structure regarding product models, and (3) providing a common mathematical model for both production structures and service system structures.

2   Related Work

In this section, we briefly review our underlying perspective of service-dominant logic [2] and corresponding service systems, a recent mathematical model for service systems represented as graphs. Second, in order to explore and model production systems, we first introduce how the structure of products is typically represented in industrial engineering and manufacturing [12, 14]. Then the process models will be described briefly because both of them are used as items compared to the production structure to show that they are contained within it.

2.1 Service-Dominant Logic and Service System Model

Service-dominant logic exists in relation to goods-dominant logic [2]. In the traditional goods-dominant logic, a typical manufacturing process consists of resources that are transformed, such as raw materials, plus factors that are needed for the transformation but are not transformed themselves, such as equipment, tools, and workers. Depending on what is to be transformed, the resources that are transformed could be either physical attributes of the resource (e.g., manufacturing) or are transformed by moving the resources to other locations (e.g., logistics). Furthermore, a transformation of human resources can occur through physical changes (e.g., health care) and also through changes of the mind (e.g., mobile learning and entertainment services) [15]. In each example, the conditions of certain resources were transformed while some resources were not affected by the transformation [8]. We acknowledge the operand and operant resource perspective and integrate both types of resources into our understanding of service as transformation of all input factors into the output factors [15], also known as value proposition [16].

Complimentary to our service as a transformation process perspective, research on service science regards the service system as its basic unit of analysis [3, 17, 18]. All constituent factors of a service system and its interrelations make up a service system's structure, which is defined mathematically using hypergraph theory [19]. The service system structure is also known as service system model (SSM) and its graph is known as service system graph (SSG). The equivalent of a service system structure would be a production system structure that encompasses both process structure and product structures. The production system structure is a synonym for production structure and is the key contribution of this paper. Furthermore, it is exactly its mathematical underpinning that enables the compatibility with service system structures.

More precisely, constituent factors of a service system are **actors** utilizing operand and operant **resources** [3], whose **activities** describe the "transformation process". Actors can be individuals, teams, organizations, cross-organizational business units, or even systems, as long as they mobilize the required resources. This mobilization includes conceptual actors that describe any additional restriction in the resource configuration. A formal model of the structure and composition of service systems has been introduced, which can model the relationships of all resources and activities of services and corresponding service systems [7, 19]. However, their model is not applicable to product models. Therefore, we define a production graph for modeling product structures and its accompanying process structure.

2.2 Structures of Products and Process Models

The goal of this paper is to introduce a production graph as a structure model that covers both product structures and process structures. To understand how a mathematical model can replace product and process structure models, this subchapter briefly introduces what a unifying production structure would include.

To model structures of products, it has been generally accepted to use BOMs (bill of materials) for the composition of products. They list the components that are required to produce a specific product, such as parts, subparts/assemblies and products, variants, and alternatives and its relationships [20]. The typical bill of materials is based on a tree structure, and a collection of BOMs define the product structure. Furthermore, product structures are used in conjunction with process models that explain how specific product structures are made, by detailing each step that is required and capture the required, sequence of activities while considering the information of the product structure. Therefore, BOMs are fundamental for planning and scheduling production and contain information on the quantity of each required part, which is used to assess further planning and scheduling tasks [21, 22]. The product model is therefore used across the entire product life cycle management, spanning from product design [23] production, [21, 22], inventory [24] and supply chain and logistics [25]. Any viable production must, therefore, be able to have the required information of a given product structure.

The process model of the production process corresponds to the product structure and must be established simultaneously. Based on diverse requirements, such as time, cost, and quality, both models are required to collect the necessary data to find possible production pathways, such as optional or alternative sequences and material requirements for a product. To model the activities into production processes, different modeling approaches and tools are available, with BPMN and Petri-nets or variants thereof being widespread [26, 27].

3 Production Structure

The purpose of this section is to model production structures, which describe the relations and the combinations of various resources, such as input material, operating resources, such as equipment, assets, and other necessary things, and the technology process (plans, know-how) to make something for consumption (as output). Usually the production structure model is built by two models separately and work together for application; one of it is the product structure model, often used to plan materials, and another is the process model, usually for planning the use of equipment. In our research, an integrated model for the production structure is developed to consider the more efficient application optimization, which integrates both product model and process model.

3.1   Resources, Actors, Activities

Resources in the service-dominant logic represent the required production factors [28, 29]. Some resources are soft and intangible things, such as skills and knowledge, whereas other resources include both materials and manufacturing machines. They can consist of single materials, a set of materials or parts, modules, and collection of resources. Yet, as resources can be applied to other resources in order to create value [29], it is beneficial to distinguish between **subject resources** and **supplementary resources**.

Subject resources are needed for creating value and are subject to value creation itself, changing its own form in the process [28]. For example, during production, raw materials are changed into refined materials, adding to its value. This entails changing the very form and nature of the resources, the same holds true for parts that get assembled into products, forming a new entity. However, supplementary resources are also necessary for value creation, yet are not subject to the transformational process [8]. A classic example of supplementary resources is operating equipment, which is required for many manufacturing processes but does not get changed during manufacturing. Other factors often include capital intensive assets such as land, which are non-depreciated after its use, and deprecated assets, such as machinery, plants, building, factories, etc., all of which have in common that they do not transform their own characteristics during production. Operand resources are therefore usually subject resources (e.g., material), whereas supplementary resources can include operand (e.g., machinery) and operant resources (skills and knowledge).

An actor is used to describe the executor of an action, that is to say, an actor is an entity that makes use of relevant resources to complete an activity. In the context of production, production equipment, including factory, business premises, and so on can be regarded as actors. Activity is a general term describing the process of resource transformation.

3.2   Production Graph

The previously introduced elements of interests can be modeled as either actor, resource, or activity using hypergraph theory because, in accordance with [30], hypergraphs can be ascribed three meanings: vertices, hyperedges, and the entire hypergraph. As we introduced, there are two types of input variables that are relevant for operations: subject resources and supplementary resources. Additionally, following SDL, actors, activities, and objects need to be defined in a rigorous and formal matter. This chapter will model a production graph using hypergraph theory to define the interrelations of all five terms and represent the value creation structure for production.

In mathematics, the generalization of a graph is called a hypergraph. A hypergraph $G = (V, E)$ exists as a pair of edges E and a set of vertices V, where the edges $e \in E$

do not only connect two but any number of vertices v ∈ V, thus calling E a set of hyperedges. A hyperedge e ∈ E is, therefore, a subset of all vertices V, which are connected by it, e ⊆ V. Additionally, E is a subset of P(V)\Ø, where P(V) is the power set of V.

From the input–output perspective, operating consumables can either be parts that get assembled or consumables that get "used up". Either way, after operations, at least one part with increased value gets produced (sometimes excess waste as well, but usually it is not part of planning). Both parts and consumables are modeled as hypergraph vertices. Thus, we replace v ∈ V with resources r ∈ R. We define actors a ∈ A as the hyperedge of R, replacing e ∈ E. As mentioned, assets and a set of resources represent a form of unit that shows a logical high cohesion. We name the unit of actors and resources as product objects. Hence, product objects represent all required input factors for operations. We define product objects as follows:

Definition 1 A finite non-empty set O_P is called product object denoted by $O_P =$ (R; A = $(a_i)_{i \in I}$) on a finite set R of resource is a family $(a_i)_{i \in I}$, (I is a finite set of indexes) of subsets of A called actor which $a_i \neq \emptyset$ and R = $U_{i=i \in I} a_i$.

This definition shows that a product object is a hypergraph. Set A = (a_i) describes all combinations of resources in production with subset a_i representing a logical relation describing the cohesion of a group of materials for a specific product or part. This is very much in alignment with the general principles of object-orientation in which objects reflect a functional cohesion [31]. Apart from the required resource, to create value, we still need to define production functions that incorporate the transformational process. This approach applies to hypergraph theory, similar to how [7] the mathematical approach is applied to service systems. They provide the value addition for resources while using supplementary resources.

Definition 2 Given hypergraph $O_p = [R, A]$ as a set of product objects with resource set R ≠ Ø; actor set A ≠ Ø and mapping Ψ(Ψ−, Ψ+): $O_p \to O_p$, where Ψ−(O_p) ∩ Ψ+(O_p) ≠ Ø and ψ−(o_p) ∩ ψ+(o_p) = Ø for ∃ o_p ∈ O_P and ψ ∈ Ψ, then the mapping Ψ is called operating function or production function, for which tuples (Ψ−, Ψ+) represent the input and output function; Ψ−(Op) input object and Ψ+(O_p) output object; the tuple $G_p(O_p, Ψ)$ is called **production graph** and represents the *production structure*.

The production graph is a directed graph and function Ψ describes the transformation between input resources and output resources as performed by the actors. The set of functions Ψ−(O_P) define which product objects are required as input factors and function Ψ + (O_P) defines the output product objects. Input and output factors for manufacturing operations are therefore product objects. Graphically, the arrow directions illustrate which product objects are considered input and which as the output of a function. In addition to describing the relationship of the function, the most important part is that the relationship can have different functions, each representing a business logic, such as determining product and component amount required, required time, or even functions that are used for calculating costs. This makes the production structure highly adaptable to different functional purposes. Furthermore,

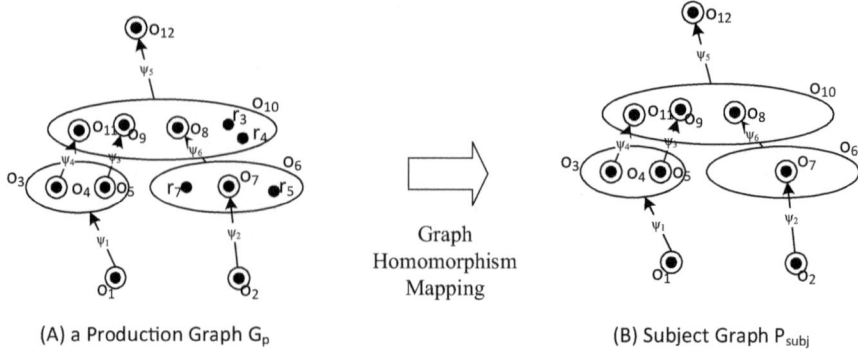

(A) a Production Graph G_p Graph Homomorphism Mapping   (B) Subject Graph P_{subj}

Fig. 1 Graph homomorphism

due to the nature of these functions, their values are determined dynamically to its input parameters and therefore are not restricted to linear functions but can, for example, include discrete functions.

Since the production graph is a graph, it has a graphical representation. We adhere to the graphical representation of hypergraphs and mappings of similar models, in which hypergraphs are illustrated as a combination of black dots (resources) and circles or ellipses (hyperedges) and mappings as arrows [7]. An example production graph can be seen in Fig. 1.

In this chapter, we have modeled a holistic model for the entire production structure. Due to G_P being a graph consisting of objects O_P and connecting activities Ψ and because we follow SDL, using G_P means that there is an underlying service object-oriented perspective on production structures. However, to make use of the production graph, it requires additional transformations based on its application purpose.

3.3 Operations and Production Graph Projections

For operations, management must address two major issues, requirements planning and scheduling (RPS) and resource capacity scheduling (RCS). During RPS, managers need to assess whether an order can be manufactured and delivered in time. Resource capacity scheduling assesses how to optimally schedule manufacturing equipment efficiently. In this chapter, we will show how production graphs can be applied to address both tasks by means of mathematical projections.

Both RPS and RCS require different perspectives on the production graph, where the former focuses on subject resources R_{subj} and the latter on supplementary resources R_{sup}. Therefore, $R = R_{Subj} \cup R_{Sup}$. Since the *production structure* $G_P = [O_P, \Psi]$ and we know that O_P is a hypergraph with R as vertices, we require a projection $Proj_{Subj}$ that projects a new production graph $G_{P.Subj} = [O_{P.Subj}, \Psi]$, which consists of all subjective resources, while keeping the original graph structure (activities). Conversely, for RPS, we also require a projection $Proj_{Sup}$ to derive graph $G_{P.Sup}$ = $[O_{P.Sup}, \Psi]$. In other words, the projection that is required is a homomorphism. The homomorphism is a mapping from one graph to another graph that maps adjacent objects from the source graph to adjacent objects of the target graph. This means that the structure of both graphs is the same, as illustrated by Fig. 1.

For applying production graphs, different production graphs need to be projected while selecting different vertices and retaining its production structure. Therefore, to find a homomorphism that matches our application conditions, we require a homomorphism f. f reflects the selection criteria, which in our case can be choosing only the subset for all subjective resources.

Theorem 1 *Homomorphism between production graphs* $G_P = [O_P, \Psi]$ *and projected production graphs* $G_{P.Subj} = [O_{P.Subj}, \Psi]$ *and* $G_{P.Sup} = [O_{P.Sup}, \Psi]$ *retain their production structure.*

Proof For $G = [O, \Psi]$, we search $f: G \to H:G \mapsto [O_f, \Psi_f]$, with $\{G = [O, \Psi] \mid o \in O, o_f \in O_f, \psi \in \Psi, \psi_f \in \Psi_f$ and $O_f \subseteq O, \Psi_f \subseteq \Psi\}$. $f(G) = [O_f, \Psi_f] \therefore f(G) = [f(O), f(\Psi)] \because o_f \in O_f$ and $O_f \subseteq O \therefore o_f \in O. \because \psi_f = \Psi_f$ and $\Psi_f \subseteq \Psi \therefore \psi_f \in \Psi.$

Before order production on the shop floor can begin, both marketing and production need a requirement plan, in which usually the bill of materials lists what parts or materials are required. In other words, they require a detailed list of subject resources R_{Subj}. All the required information is also included in the base production graph G_P. However, in G_P, both subject and supplementary resources are included, while the base structure tells what subpart or material is needed. Therefore, we make use of the above-mentioned homomorphism of a new target graph $G_{P.Subj}$ that only includes the subject resources and retains the object structure. Based on $G_{P.Subj}$, a detailed requirements plan can be derived. A type of input resource, coupled with the retained structure, which is in essence formalized as an induced graph, is needed to make a time schedule. The time schedule contains the information on how long an order needs, which is often essential for marketing decision-makers to reduce lead time. Additionally, for a more accurate assessment of time-to-delivery, information of the capacity schedule is needed. Therefore, capacity scheduling ideally should happen simultaneously, if possible. Therefore, another induced graph $G_{P.Sup}$ with R_{Sup} as the product object's element set is required. $G_{P.Sup}$ and $G_{p.sibj}$ are both homomorphic and therefore the same in terms of structure yet differ in terms of resources. For capacity scheduling, supplementary resources are required.

4 Application Example

This chapter demonstrates the production structure's application potential by introducing how a production structure P_{Subj} can be transformed into a product structure and process flow model. However, its application is not limited to product structures and process flows. Depending on the chosen resources and especially the functional purpose, as defined by production functions, its application changes accordingly. However, we chose product models and their corresponding process flows for their central role in production [14, 21, 22, 24, 25]. Therefore, we specifically explore how functions are employed for different functional purposes, such as in determining quantity, time, or costs of the production (structure).

Production graphs G_p are graphs that are based on hypergraphs and therefore are able to express a multitude of complex structures, which this paper is not able to cover. However, by choosing the above-mentioned subject graph P_{Subj} as an example, using mathematical operations, we can derive an induced subgraph that can represent a product structure. This induced subgraph takes on the form of a simple graph, as defined by Berge [30], more precisely, it is a tree structure with the product as its root node. Therefore, consider P_{Subj} a production structure and homomorphism of production structure P_G, only consisting of subject resources, devoid of supplementary resources for the purpose of this application example. Therefore, the original production structure G_P includes both subject and supplementary resources, such as materials and parts, as well as manufacturing equipment and utilities. The graph P_{Subj} includes all the input–output subject resources, relations, and the necessary activities before a product O_{12} can be produced, including product variants. Since production relies heavily on product models, usually represented by BOMs [14], we transform the model into a tree, listing the products and its components (see Fig. 2D). Mathematically, the graph P_D is an induced subgraph of P_{Subj}, which sometimes requires additional mathematical operations depending on the graph structure, as Fig. 2 shows. The graph P_D includes the necessary information required for a useful product model. However, there is a subtle difference at O_{11}, where some information has been lost. As Fig. 2 shows, r_1 is not required as input for ψ_5, yet it is a byproduct of ψ_4. Therefore, P_D does not include r_1, since a BOM only includes the required materials and parts for a given product. Although this information is not relevant for product models, it shows the informational advantages of our production model toward conventional product models.

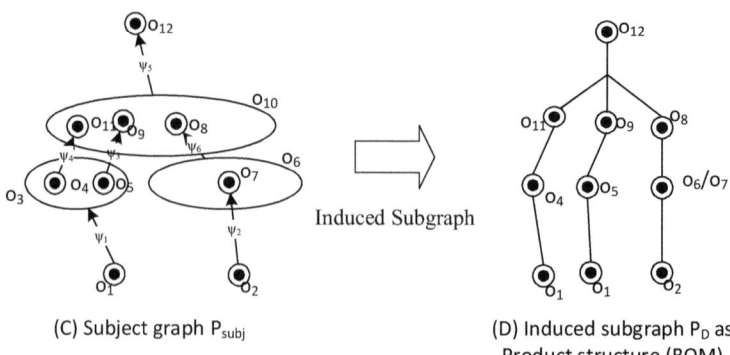

(C) Subject graph P_{subj}

(D) Induced subgraph P_D as Product structure (BOM)

Fig. 2 Product model

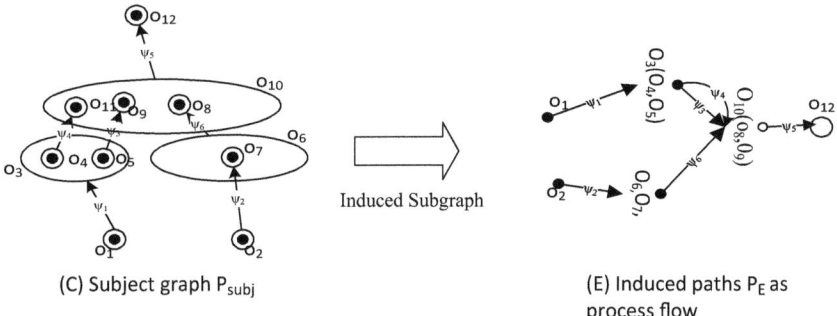

Fig. 3 Process flow model

Product models are used in close conjunction with process models for different purposes, most notably for planning and scheduling [32]. Since our production model also includes information on production sequences, we can select the induced paths of P_{Subj} to model process flows, analogous to the induced subgraph for creating a product model. In Fig. 3, P_D shows how our production model includes alternative production paths, as illustrated by the simultaneous existence of ψ_3 and ψ_4.

After showing how a production graph can be used to derive conventional simple graph structures that are prevalent in production, we move on to touch upon the topic of functions. As briefly explained in Sect. 3, activities are mappings between product objects. Its application potential lies in the chosen functions those mappings represent. As activities are transformational by nature [8], we did not specify precisely how they transform the subject resources of the input product objects. This is reliant on context. For example, to determine the lead time, at least the entire production time needs to be determined. This can be done by first determining the total quantity of products and components that need to be produced. This is achieved by analyzing a product structure, in our case P_D (see Fig. 2). In P_D, the activities represent the relationship of what elements are required and each function can determine the quantity of each required subject resource. After determining the specific quantity of each required part, the parameters can be taken as input for time-determining functions of a process flow model P_E (Fig. 3). Since functions can be freely defined in software systems, a common production structure would, therefore, allow a combined function, determining both quantity and time simultaneously. Therefore, our production structure can be applied to different application scenarios. Quantity and time are just two examples. If quantity and time parameters can be determined, cost structures follow the same logic.

5   Discussion and Conclusion

Our production structure model is a step toward providing a unified model based on the service-dominant logic perspective that can be used for both services and production. We argue that our production structure model is compatible with service systems and can integrate both production and service activities into one structure. In this paper, we show that the production structure model can be used to induce both product models and process models, both being fundamental for production information systems.

Thus, our first contribution lies in the mathematical model of a novel production structure model using SDL as an underlying perspective. The production graph integrates (a) an integrative model of both product structure and its input–output relations and (b) a production structure model that introduces functions so that input functions and output functions can be applied within a production structure's application system to calculate the value. This was made possible by applying a service-dominant logic perspective and using the generalization of simple graphs into hypergraphs and its highly expressive power [30], which enables us to model more complex scenarios than conventional models that rely on simple graphs.

Our second main contribution lies in demonstrating the usefulness and versatility of our proposed production structure model. By exploiting the expressiveness of hypergraphs and applying the concept of homomorphisms and induced subgraphs, we demonstrate that the introduced production structure model is a viable alternative to both product models and process models. In other words, this paper (a) introduced the production structure model and (b) demonstrated that product models and process models are included in the introduced production structure.

For practice, in line with our research question, we explored a modeling approach for production that is compatible with service system graphs [7] yet addresses production. Future research should consider resource planning systems that are based on joint production graphs and service system graphs because such a system would have three main advantages to conventional production systems: (1) A mathematical model would simplify an application system's interface for databases, is a basis for simulations and can readily be modeled using conventional graph tools. (2) A production structure model that encompasses product and process structures enables innovative algorithms and heuristics for simultaneously planning for production time and requirements. (3) The hypergraph-based graph enables modeling previously hard to model, complex production endeavors such as chemical production and cyclical manufacturing. All three practical implications would require additional research. However, we call upon future researchers and practitioners to further explore the advantages of the production structure.

References

1. Central Intelligence Agency, *The CIA World Factbook 2017* (Skyhorse Publishing, New York, 2016)
2. S.L. Vargo, R.F. Lusch, Service-dominant logic: continuing the evolution. J. Acad. Mark. Sci. **36**, 1–10 (2008). https://doi.org/10.1007/s11747-007-0069-6
3. P.P. Maglio, J. Spohrer, Fundamentals of service science. J. Acad. Mark. Sci. **36**, 18–20 (2008). https://doi.org/10.1007/s11747-007-0058-9
4. M. Berkovich, J.M. Leimeister, A. Hoffmann, H. Krcmar, A requirements data model for product service systems. Requir. Eng. **19**, 161–186 (2014). https://doi.org/10.1007/s00766-012-0164-1
5. H. Lightfoot, T. Baines, P. Smart, The servitization of manufacturing: A systematic literature review of interdependent trends. Int. J. Oper. Prod. Manag. **33**, 1408–1434 (2013). https://doi.org/10.1108/IJOPM-07-2010-0196
6. C. Peters, I. Blohm, J.M. Leimeister, Anatomy of successful business models for complex services: insights from the telemedicine field. J. Manag. Inf. Syst. **32**, 75–104 (2015). https://doi.org/10.1080/07421222.2015.1095034
7. M.M. Li, C. Peters, J.M. Leimeister, A hypergraph-based modeling approach for service systems, in *INFORMS International Conference on Service Science* (2018)
8. T.P. Hill, On goods and services. Rev. Income Wealth **23**, 315–338 (1977). https://doi.org/10.1111/j.1475-4991.1977.tb00021.x
9. S.B. Gershwin, *Manufacturing systems engineering* (Prentice Hall, Englewood Cliffs, 1994)
10. J.R. Jupp, L. Rivest, D. Forgues, C. Boton, Comparison of shipbuilding and construction industries from the product structure standpoint. IJPLM **11**, 191 (2018). https://doi.org/10.1504/IJPLM.2018.10015944
11. G. Schuh, H. Rozenfeld, D. Assmus, E. Zancul, Process oriented framework to support PLM implementation. Comput. Ind. **59**, 210–218 (2008). https://doi.org/10.1016/j.compind.2007.06.015
12. A. Okamoto, M. Gen, M. Sugawara, Integrated data structure and scheduling approach for manufacturing and transportation using hybrid genetic algorithm. J. Intell. Manuf. **17**, 411–421 (2006). https://doi.org/10.1007/s10845-005-0014-9
13. H. Stadtler, Supply chain management and advanced planning—basics, overview and challenges. Eur. J. Oper. Res. **163**, 575–588 (2005). https://doi.org/10.1016/j.ejor.2004.03.001
14. L. Hvam, A procedure for building product models. Robot. Comput.-Integr. Manuf. **15**, 77–87 (1999). https://doi.org/10.1016/S0736-5845(98)00030-1
15. H. Fromm, J. Cardoso, Foundations, in *Fundamentals of Service Systems*, 1st edn., ed. by J. Cardoso, H. Fromm, S. Nickel, G. Satzger, R. Studer, C. Weinhardt (Springer International Publishing, Cham, 2015), pp. 1–32
16. J.D. Chandler, R.F. Lusch, Service systems: a broadened framework and research agenda on value propositions. Engag. Serv. Exp. J. Serv. Res. **18**, 6–22 (2015). https://doi.org/10.1177/1094670514537709
17. T. Böhmann, J.M. Leimeister, K. Möslein, Service systems engineering. Bus. Inf. Syst. Eng. **6**, 73–79 (2014). https://doi.org/10.1007/s12599-014-0314-8
18. J.M. Leimeister, *Dienstleistungsengineering und -management* (Springer, Berlin, Heidelberg, 2012)
19. M.M. Li, C. Peters, Reconceptualizing service systems—introducing service system graphs, in *International Conference on Information Systems (ICIS)* (2018)
20. M.H. Jansen-Vullers, C.A. van Dorp, A.J.M. Beulens, Managing traceability information in manufacture. Int. J. Inf. Manage. **23**, 395–413 (2003). https://doi.org/10.1016/S0268-4012(03)00066-5
21. H.A. Reijers, S. Limam, W.M.P. van der Aalst, Product-based workflow design. J. Manag. Inf. Syst. 20, 229–262 (2003). https://doi.org/10.1080/07421222.2003.11045753
22. A. Balakrishnan, J. Geunes, Requirements planning with substitutions: exploiting bill-of-materials flexibility in production planning. Manuf. Serv. Oper. Manag. **2**, 166–185 (2000)

23. F. Bernstein, A.G. Kök, L. Xie, The role of component commonality in product assortment decisions. M&SOM **13**, 261–270 (2011). https://doi.org/10.1287/msom.1100.0317
24. M.K. Doğru, M.I. Reiman, Q. Wang, Assemble-to-order inventory management via stochastic programming: chained BOMs and the M-system. Prod. Oper. Manag. **26**, 446–468 (2017). https://doi.org/10.1111/poms.12658
25. G.Q. Huang, X.Y. Zhang, L. Liang, Towards integrated optimal configuration of platform products, manufacturing processes, and supply chains. J. Oper. Manag. **23**, 267–290 (2005). https://doi.org/10.1016/j.jom.2004.10.014
26. K. Salimifard, M. Wright, Petri net-based modelling of workflow systems: an overview. Eur. J. Oper. Res. **134**, 664–676 (2001). https://doi.org/10.1016/S0377-2217(00)00292-7
27. OMG, Business Process Model And Notation 2.0 (2011)
28. E. Gutenberg, *Grundlagen der Betriebswirtschaft Bd1: Die Produktion*, 23rd edn. (s.n, Berlin, 1979)
29. J.A. Constantin, R.F. Lusch, *Understanding Resource Management: How to Deploy Your People, Products, and Processes for Maximum Productivity* (Planning Forum, Oxford, Ohio, 1994)
30. C. Berge (Ed.), *Hypergraph: Combinatorics of Finite Sets* (Elsevier, 1989)
31. I. Jacobson, *Object-Oriented Software Engineering: A Use Case Driven Approach* (Addison-Wesley, Harlow, 1998)
32. H.T. Papadopoulos, C. Heavey, J. Browne, *Queueing Theory in Manufacturing Systems Analysis and Design*, 1st edn. (Chapman & Hall, London, 1993)

Airworthiness Evaluation Model Based on Fuzzy Neural Network

Jie-Ru Jin, Peng Wang, Yang Shen and Kai-Xi Zhang

Abstract Based on fuzzy neural network, this study explores the quantifying calculation problem of airworthiness of specific rescue operations. Aiming at the problem of quantification assessment of flightworthiness for rescue flight operation, this study starts from the perspective of the mechanical properties of the aircraft itself, and use a fuzzy neural network model for rescue operations of airworthiness evaluation modeling. The rescue historical data of the EC-135 helicopter model is used for model training, in order to form a quantitative model of airworthiness assessment for flight operation. On the one hand, the quantitative output results provide qualitative guidance for aircraft's competency, and on the other hand, it provides a basis for comparing the advantages and disadvantages of multitask allocation schemes. Optimization of the whole system task allocation effect is formed through the best way of individual utility. Aiming at the heterogeneous problem for rescue system, this paper introduces the concept of "task ability vector", quantitative representing the ability of heterogeneous aircrafts and requirements of mission. The comprehensive ability of quantitative calculation of multi-aircrafts alliance is discussed, as well as the single and multi-aircrafts cooperative task ability.

Keywords Rescue task allocation · Fuzzy neural network · Aviation emergency rescue

1 Research Background and Significance

Assessing the airworthiness of a task assignment is a prerequisite for large-scale task assignment. Its significance is to judge whether a pair of task assignments is feasible and the rationality of the assignment. This chapter aims to design an airworthiness

J.-R. Jin · Y. Shen (✉) · K.-X. Zhang
College of Economics and Management, Nanjing University of Aeronautics and Astronautics, Nanjing 211016, Jiangsu, China
e-mail: shen.y@nuaa.edu.cn

P. Wang
Haifeng General Aviation Technology Co., Ltd, 100070 Beijing, China

© Springer Nature Switzerland AG 2020
H. Yang et al. (eds.), *Smart Service Systems, Operations Management, and Analytics*, Springer Proceedings in Business and Economics, https://doi.org/10.1007/978-3-030-30967-1_17

assessment model from the perspective of aircraft mechanical performance and to have the ability to quantitatively assess operational airworthiness through fuzzy neural network learning.

This paper uses the fuzzy neural network to model the airworthiness assigned to the navigation rescue operation [1]. Taking the EC-135 rescue helicopter as an example, the network training is carried out in combination with the flight operation history data of the aircraft to complete the aircraft's different rescue operations and to evaluate the calculation of airworthiness. In this paper, the results of the airworthiness assessment assigned by a rescue operation are represented by real numbers on the interval [0, 1], and the larger the value, the higher the airworthiness. Intuitively, if the airworthiness of dispatching an aircraft to perform a rescue mission is small, which means that the aircraft is less suitable to perform the task, the assignment may let the mission cannot be completed successfully, or the risk of the flight operation is high or even it is impossible to be actually carried out.

2   Airworthiness Evaluation Model Based on Fuzzy Neural Network

2.1   Airworthiness of Neural Network Based on the Idea of Assessment

To evaluate the ideal degree of a rescue mission assignment result in the rescue scenario, it is necessary to consider two aspects: the tolerance of the aircraft to the load in actual operation, and the adaptability of the aircraft performance to the mission environment. The above problems can be summarized as the airworthiness assessment of rescue aircraft, which is ubiquitous in flight operation evaluation. Because it is a complex nonlinear problem that relies on professional knowledge and rescue and rescue experience, this study uses fuzzy neural network model to model and learn the competency evaluation to obtain quantitative results. The idea of this study are the following:

(1) Analyze the impact mechanism of airworthiness of rescue flight operations, analyze aircraft performance, mission requirements and influencing factors related to rescue scenarios and prepare for FNN's input layer node construction;
(2) Model with fuzzy neural network as a tool and design a reasonable network structure based on problem analysis;
(3) Use the membership function to represent the domain knowledge and design the membership level node based on the knowledge of airworthiness evaluation,;
(4) Perform network training using a gradient-based backpropagation learning algorithm;

(5) On the SPSS Modeler platform, calculate the airworthiness experiment through the BP learning experiment and the training set of the test set, and compare the airworthiness calculation result of the model with the traditional ANN model. Thus, the feasibility and airworthiness evaluation effect of the FNN model are tested.

2.2 FNN's Principle of Airworthiness Calculation

The construction of dynamic airworthiness measurement evaluation model has the following difficulties: First, the sudden and accidental nature of emergency rescue, it means that there is no established daily route or fully predictable task content. The evaluation problem is quite different and the dynamic airworthiness evaluation method is needed. Second, the mapping relationship between the requirements of the mission and the aircraft performance parameters is relatively vague. In actual rescue, it usually relies on expert knowledge to make empirical judgments. It is difficult to perform accurate mathematical modeling and calculation. Third, the decision variables has the ambiguity, because the concept definition itself has ambiguity (such as "plateau") or it cannot be known exactly when the aircraft selection decision (such as "job duration"), It is difficult to reasonably characterize and calculate, but such factors have an important impact on the judgment of aircraft selection.

In this system, FNN is still a neural network in essence, retaining the basic properties and structure of the neural network, only using fuzzy neurons, starting from improving the heuristic, transparency and robustness of the neural network, fuzzy concepts, and fuzzy inference rules of the neural network of neurons, connection weights, and the network learning. The structure and basic properties of neural networks are preserved in fuzzy neural networks [2]. "Domain knowledge" is expressed in a fuzzy set to improve the transparency of the network and the ability of the network to interpret and to enhance the robustness of the network and the controllability of the learning algorithm.

2.3 How FNN Works

Due to the particularity of the task assignment problem, the fuzzy neuron construction method is combined with the design of the neuron node to integrate the expert knowledge in the field of aviation rescue. The fuzzy node design and calculation method are as follows.

Design of fuzzy nodes

The calculation of fuzzy nodes is used to determine the mapping of tasks to aircraft. The establishment of mapping rules involves the expertise of aviation rescue mission assignments. Therefore, the fuzzy theory is combined in the design of the nodes of this problem to fuse expertise [3]. The construction process is mainly divided into three steps:

1. Construct a fuzzy knowledge base: According to the expert knowledge summary task mapping rules, disassemble the knowledge elements and form a fuzzy knowledge set. For example, the "material transport" task of "distance greater than 150 km" requires "fixed wing" "B series transport aircraft".
2. Determine the domain of each feature: define the domain for each dimension of the task vector. The definition process needs to combine the content of the data and the content of the fuzzy knowledge to keep the semantics consistent. For example, the terrain dimension of the rescue mission can be defined as rescue place Type x1: [city, plain, ocean, mountain, and plateau].
3. Construct membership function: Determine the membership degree of the corresponding aircraft by the membership function, and the membership function determines the division rules of the mission.

(2) FNN model structure

In this study, a multilayer feedforward neural network is used as the basic structure, and each layer is numerically calculated by means of full connection. Its various layers are constructed as follows:

1. Input layer: It is used for the input of task features; each feature of the task to be evaluated corresponds to a node in the input layer. The input layer is the interface of the task feature data and the context data input network for data transfer.
2. Membership function determination layer: Determine the domain of each feature in the rescue mission vector, and then combine the expert knowledge of navigation rescue to design the membership function of each task feature, and determine the membership degree of the feature factor set according to the input feature value.
3. Fuzzy inference layer: realizes the image between the fuzzy input vector X and the fuzzy output vectors.
4. Output layer: The last calculation in the neural network outputs the value of the airworthiness.

(3) Calculation method of neurons

At the neuron level, each node linearly weights multiple input values passed by the nodes in the previous layer and then transforms the linear weighted results through the activation function. At the network level, the neural network obtains multiple nonlinear transformations through multiple levels of computation and finally obtains nonlinear output results from the output layer.

(4) Network connection and calculation method

The multilayer feedforward network connects all the nodes in a fully connected manner and performs layer-by-layer numerical transfer and calculation from the input layer to the output layer. Linear calculation and a nonlinear calculation are performed at each neuron node to obtain the calculation result of the node, and the calculation result is transmitted to the next node according to the network connection direction (exception: the input layer node is not calculated, only numerical transfer) [4].

3 FNN-Based Operational Airworthiness Calculation Modeling

3.1 Airworthiness Feature Selection and Design of the Input Layer Nodes

Because there are many factors affecting the operational airworthiness, in which the interaction between various factors is complex, some factors are not easy to obtain and quantify. Based on the mechanism of the impact of rescue mission on aircraft demand, this study screens the model decision indicators according to the principles of feasibility, comprehensiveness, stability, and quantifiability to ensure the validity and simplicity of the model. The factors are shown in Table 1.

Among the 10 fields in the above table, the first 8 items are variables that affect the size of airworthiness, while the latter 2 "Task Results" and "Service Record" are not influencing factors but task execution results. The reason why these two fields are used as decision variables here is to facilitate the subsequent supervised learning of the model, which is equivalent to the sample label, that is, the model can adjust

Table 1 List of decision variables for airworthiness assessment

Name	Type	Unit	Ranges	Statement
Altitude	int	m	[100,1000]	The altitude of the place
Condition	Enumeration	–	{0,1}	0:without-airport; 1:airport
Distance	int	km	[0, 1000]	Total flight from the start to the end of the rescue
Time	int	min	[0, 180]	Effective working time of the aircraft during the rescue
Manned number	int	–	[0, 25]	Number of people on the rescue mission (excluding pilots)
Manned number	int	kg	[0, 5000]	Total business load of the mission (including passenger weight)
Rain	int	mm	[0, 60]	The rain of the place
Wind force	float	m/s	[0, 20]	The wind force of the place
Result	Enumeration	–	{0, 1}	0:fail; 1:success
Record	Enumeration	–	{0, 1, 2}	0:no record;1:unusual; 2:normal

the network weight according to the value of the field to generate the desired output during training, thereby achieving the purpose of network learning.

3.2 FNN Network Structure Design

This study fully considers the characteristics of the rescue aircraft. After several attempts to combine the existing resources and data in China, the FNN model of this problem is constructed by adding the fuzzy layer on the traditional BP neural network. This model has the following main advantages: First, the use of aviation expertise to guide the design of nodes and network structures enhances the interpretability of neural network models, making the model easier to understand and use. Second, because the same performance attribute of different models often exhibits an order of magnitude difference in numerical values, the design of membership degree through the fuzzy layer can better stratify the characteristics of large span values. Third, compared to the fuzzy inference network that relies entirely on fuzzy knowledge construction, this model still maintains the structure and function of the traditional BP neural network in reasoning, which makes the model better use historical rescue data in training. (Rather than rely entirely on fuzzy knowledge bases). The model structure is shown (Fig. 1).

Input layer: The input layer is the entry point for decision information in this FNN model. Each node in the layer represents a decision variable, which is calculated as X_i (i = 1, 2, ..., n), where n is the total number of decision variables and is the number of nodes in the input layer. The input layer nodes only perform numerical values and do not perform numerical calculations.

Fig. 1 Fuzzy neural network model assigned by emergency rescue aircraft

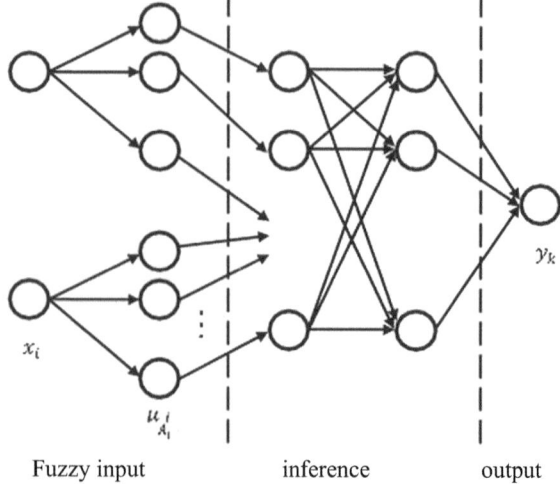

The fuzzy layer: The fuzzy layer fuzziest the variables passed in the input layer, and the fuzzy process is completed by the calculation of the membership function of each node. In the fuzzy layer, the input variable X_i, will have m_i fuzzy nodes corresponding to it, and the total number of fuzzy nodes of the layer is 为 $\sum_{i=1}^{n} m_i$, which will input the jth node corresponding to the variable X_i. The fuzzy set is denoted as $A_i^j (j = 1, 2 \ldots, m_i)$, and the membership function corresponding to the fuzzy set A_i^j is denoted as $\mu_{A_i}^j (X_i)$. This function nonlinearly transforms X_i and maps it into the interval $[0, 1]$ to characterize the degree of membership of the input variable X_i to the set A_i^j. The construction of the degree of membership will be described in detail below.

Output layer: The output layer is used to obtain the output of the membership calculation structure, which is calculated as y_k. In this study, the range of airworthiness is $[0, 1]$.

3.3 Membership Function Design

Because the mechanical properties of different models are different, this paper uses the more common model EC-135 in China's navigation rescue as an example to construct the model and design the membership function according to the aero machine knowledge related to the model.

Aviation emergency rescue has strong practicality and operability. The influencing factors involved in the model selection are often difficult to use accurate mathematical to characterize. The decision-related rescue mission requirements, actual aircraft performance, and other factors all show boundary blur. In this study, fuzzy mathematics is used to characterize fuzzy decision variables, and the membership function $\mu A (x)$ is constructed to describe the degree of membership of a decision variable x to a fuzzy set A, whose value is $[0, 1]$. When the value of the membership degree $\mu A (x)$ is closer to 1, the degree that the variable x belongs to the set A is larger, and if it is close to 0, the degree of membership is smaller. The membership function $\mu A (x)$ has various structural forms. The commonly used types include trigonometric functions, Gaussian functions, sigmoid functions, etc. The selection of functional forms is related to the self-property of set A and the membership characteristics of variable x.

3.4 Activation Function Selection and BP Learning

(1) Selection of activation functions on network nodes

In this problem, the influence of mission characteristics on the aircraft can be divided into strong constraints and weak constraints. For example, the choice of aircraft is strict and uncoordinated for the selection of mission type and rescue terrain. This part of the feature has strong constraints on task assignment. Therefore, it is proposed to use the threshold function as the activation function of the node. Let the node k be linear. The output is v_k and the activation function is $\varphi(v_k)$. The activation function of the node is of the form

$$\varphi(v_k) = \begin{cases} 1, v \geq 0 \\ 0, v \leq 0 \end{cases}$$

In which

$$v_k = \sum_{j=1}^{m} w_{kj} x_j + b_k$$

For constrained features in the task vector that are easy to calculate and compare, such as the distance from the aircraft to the rescue point, available load, and climate, such nodes are intended to use the sigmoid function as the activation function, which can be tried with a typical logistic function:

$$\varphi(v_k) = \frac{1}{1 + e^{-av}}$$

Through a large number of sample learning, the accuracy of the task matching of the neural network will be continuously improved. When the classification accuracy rate converges a certain threshold, the learning process can be considered complete. The trained algorithm will have the ability to complete the assignment of aviation emergency rescue tasks. The knowledge acquired through the training will be stored in the model in the form of network weights, so that when new samples enter the network, the classification results can be obtained through multilayer calculation.

(2) Based on gradient descent learning BP

There are three basic learning methods for fuzzy neural networks: learning algorithms based on gradient descent, learning algorithms based on recursive least squares and clustering methods. In this paper, the BP learning algorithm based on gradient descent is used for network learning. The purpose of the learning experiment is as follows: 1. Obtain a neural network model that can be used in the assignment of aviation rescue missions. Through the learning of the task sample data, the connection weight of the neural network is continuously modified and optimized and the distribution knowledge can be stored in the neural network in the form of weights. 2. Explore the appropriate activation function form and reasonable data form, optimize the network structure and parameters and complete the convergence of network training. 3. According to the analysis of the experimental results, the importance of each mission characteristic of the aircraft is determined, and suggestions for task assignment work are provided.

The specific steps for conducting a neural network learning experiment are as follows:

1. Selection of training samples

After obtaining the training samples matching the neural network structure through the data preprocessing, the appropriate sample set is selected as the training set for network training, and part of the samples are reserved as the test set for subsequent training effect test.

2. network weight initialization

Initializing network weight is the starting point of network training. In this study, it is proposed to use uniform distribution for weight initialization. Under the set of initial network weights, the initial loss value of the corresponding loss function will be used as the starting point of the gradient descent when the network is training.

3. given input and output

Given input $x_i (i = 1, 2, \ldots, n)$ and output $y_k (k = 1, 2)$

4. Calculate the forward propagation value of the input

That is, the calculated value of the input information through the hidden layer can be calculated by the following formula:

$$z_j = f\left(\sum_{i=0}^{n} w_{ij} x_i\right), j = 1, 2, \ldots, m$$

The final output of the hidden layer is

$$y_k = f\left(\sum_{j=0}^{m} w_{kj} z_j\right), k = 1, 2$$

5. Reverse correction weight

The learning process of the backpropagation algorithm can be summarized as forward transfer $\rightarrow$ loss calculation $\rightarrow$ gradient calculation $\rightarrow$ weight update, that is, the training sample is input into the network through the input layer of the network; the output result is obtained through multilayer calculation of the neural network; Compare the label of the result and the training sample to calculate loss, then calculates the gradient of the loss function at the current point relative to each parameter, and updates the network weight by the gradient descent method.

6. Judgment convergence: If the error accuracy requirement is met, the training is completed, otherwise it returns to (3) retraining.

This study intends to use BP learning algorithm to learn the network. The principle is to reduce the error between the sample and the calculation result by gradient descent to feedback the learning result so that the weight can be optimized. It is essentially a mathematical model of neural network learning, which can correct the weight coefficients of each layer in the network and is suitable for the learning of multilayer networks. The topology of the neural network is a multilayer forward network without feedback.

4 Network BP Learning Experiment

4.1 Airworthiness Learning Experiment Using Traditional ANN Model

The model tends to converge within 200 trainings and can no longer reduce the error. It is regarded as training completion. At this time, the accuracy of airworthiness based on the training set can reach 78.5%.

4.2 Airworthiness Learning Experiment Using FNN Model

The design of the membership degree layer enables the membership function to act on the corresponding nodes, so as to play the navigational related professional knowledge to model training. Guide the role to improve the learning effect of the training set. From the perspective of training effect, based on the FNN model, the accuracy of the training set after BP learning convergence reached 98.6%.

4.3 Test and Comparison of Airworthiness Evaluation Effects of ANN and FNN

Using the experimentally trained ANN model and the experimental two-trained FNN model, the airworthiness of the same test set sample (70) is predicted separately. The test result of the test set sample is shown in the scatter plot. The coordinates are the actual airworthiness of the test sample, and the abscissa is the airworthiness calculation result of the model output. From the perspective of scatter plot distribution, the correlation between the output of the FNN model and the actual value of the test sample is significantly stronger than that of the ANN model, which indicates that the FNN model is more effective than the ANN model for airworthiness prediction. At the same time, the scatter diagram in Fig. 2b is more concentrated, which means that the variance of the error is also smaller.

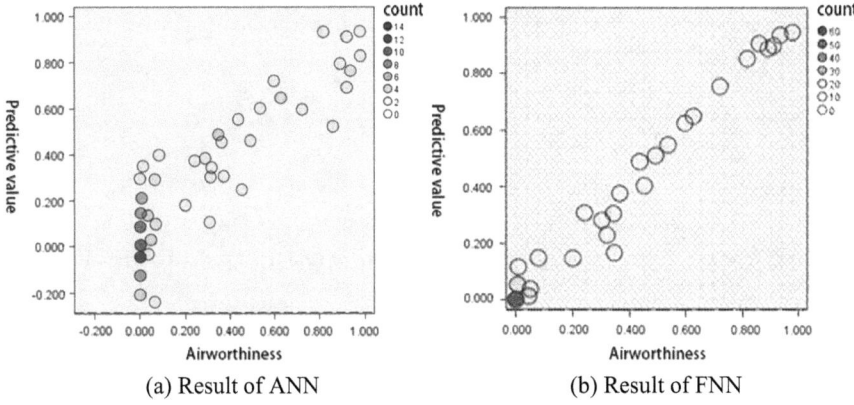

(a) Result of ANN   (b) Result of FNN

Fig. 2 Comparison of the scatter plots of the airworthiness calculation results of the ANN model based on the test set and the FNN model

Table 2 Comparison of network learning results based on test sets	–	BP neural networks	Fuzzy neural networks
	Average error	0.114	0.027
	Qualitative accuracy	82.86	91.43%

In this study, the value of airworthiness is defined as a real number in the range [0, 1]. From the error of the prediction results of the test set, the average error of the model's airworthiness calculation is 0.027, which is significantly lower than the average error of the ANN model by 0.114 (Fig. 2 and Table 2).

5 Summaries

In summary, the FNN model proposed in this paper can obtain a more reliable evaluation result for the flightworthiness of the flight operation of the EC-135 model. At the same time, compared to the ordinary ANN model without using the fuzzy method, the model has the following advantages: in the network structure, after constructing the fuzzy layer, the model can reduce the number of hidden layer nodes and achieve better prediction effect, so that the network structure is easier to understand and explain than which using the ordinary ANN model; in terms of training effect, the introduction of domain knowledge to guide network training enables the model to converge faster and achieve higher accuracy.

Acknowledgements Supported by the National Key Research and Development Program of China(Grant No.2016YFC08022603).

References

1. X. Deng, X. Wang, Incremental learning of dynamic fuzzy neural networks for accurate system modeling. Fuzzy Sets Syst. **160**(7), 972–987 (2009)
2. A.F. Gobi, W. Pedrycz, The potential of fuzzy neural networks in the realization of approximate reasoning engines. Fuzzy Sets Syst. **8**(22), 2954–2973 (2006)
3. C.F. Juang, T.M. Chen, Birdsong recognition using prediction-based recurrent neural fuzzy networks. Neurocomputing **71**(1–3), 121–130 (2007). C.F. Juang, L.T. Chen, Moving object recognition by a shape-based neural fuzzy network. Neurocomputing **71**(13), 2937–2949 (2008)
4. M. Singh, S. Srivastava, M. Hanmandlu et al., Type-2 fuzzy wavelet networks (T2FWN) for system identification using fuzzy differential and Lyapunov stability algorithm. Appl. Soft Comput. **9**(3), 977–989 (2009)

Two-Level Trip Selection and Price Incentive Scheduling in Electric Vehicle-Sharing System

Zihao Jiao, Xin Liu, Lun Ran and Yuli Zhang

Abstract The rebalance operations have been an essential problem in car-sharing service. In this paper, a two-level price incentive trip selection process is proposed to mitigate the imbalance issue in an electric vehicle-sharing (EVS) system. Specifically, at the perspective of customers, a trip price plan is made based on the adoption rate incorporating stochastic utility function in the first-level trip selection. The second-level selection adopts part of customers kept in the first-level selection, which brings less reposition cost happened in the scheduling operations in the EVS service. In the two-level trip selection process, the uncertain parameters, i.e., customers' price expectation, potential travel demand, are assumed as random variables with known statistical measures, e.g., marginal moments, obtained from the real world. And the corresponding worst-case chance constraints combined with these random variables are further approximated as the convex optimization. In a real-world case study, the computational results demonstrate several economic and environmental benefits of our two-level selection program in the EVS system.

Keywords Electric vehicles · Car-sharing service · Robust scheduling · Price incentive policy · Repositioning scheduling

1 Introduction

Achieving short-term and convenient rent, electric vehicle-sharing (EVS) service allows customers to rent a car by a mobile terminal, which becomes an effective alternative to vehicle ownership and reduces the usage rate of personal vehicles. To some extent, EVS improves vehicles' utilization and maintain the total number of traffic vehicles with easing air pollution and other environmental issues. Moreover,

Z. Jiao · X. Liu · L. Ran (✉) · Y. Zhang
School of Management and Economics, Beijing Institute of Technology, Beijing 100081, People's Republic of China
e-mail: ranlun@bit.edu.cn

Sustainable Development Research Institute for Economy and Society of Beijing, Beijing 100081, People's Republic of China

195
H. Yang et al. (eds.), *Smart Service Systems, Operations Management, and Analytics*, Springer Proceedings in Business and Economics, https://doi.org/10.1007/978-3-030-30967-1_18

studies have shown that EVS can not only reduce the number of private cars [1–4], but also decrease the vehicle kilometers traveled (VKT) [5], reduce emissions of CO_2 and other harmful gases (e.g., CO, NO_x, PM, and NMVOC) [2–5], and cut down energy consumption [4], etc.

Behaving many eco-friendly attributes and Internet access, electric vehicles (EVs) have become an important component of EVS, which also perform conveniently for the customers to rent and return the vehicle anywhere in the service regions. However, the EVS operations still exist lots of problems mainly in two aspects. The first is imbalance problem: Due to the random touring habits of customers and traffic condition [6], frequent reposition operations, which are essential to balance vehicles between different regions, bring the extra cost for EVS operators. The second is unstable external factors: Uncertain travel demand [7] is the key problem, which increases the imbalance between demand and supply and hinders ordinary EVS operation in EVS. In this regard, there are many relevant studies focusing on mitigating imbalance problem by proposing efficient reposition plans [6–8]. Moreover, robust optimization [7] and stochastic programming [9] approaches are applied to put forward a robust scheduling plan in managing travel demand.

In practice, general transportation operators like DiDi and Lyft, apply the surging price to incentive drivers to provide extra service. It motivated us to adopt "Price Incentive" policy to improve operational efficiency in EVS systems. Specifically, in real practice, potential customers face several transportation choices, e.g., taxi, subway, and bus. Thus, we assume that rational potential customers are willing to adopt EVS only if "the price is right". Further, operators select the most "economic" travel demand proposed by these customers to meet. Based on these realities, two-level selection process is incorporated into our price incentive EVS reposition framework in this paper. The contributions of this paper are listed as follows:

1. A two-level trip selection is incorporated into EVS service to improve the quality of service and operational efficiency.
2. Customers' price expectation and uncertain rental demand are considered under a distributed robust framework.
3. Price incentive policy is applied in our reposition plan in EVS.
4. A real-world case study in CAR2GO Amsterdam is conducted to identify the economic efficiency of our reposition plan.

The remainder of this paper is organized as follows. The two-level trip selection process in EVS and the corresponding convex approximation are demonstrated in Sect. 2. Section 3 conducts a case study incorporating real-world data, where we propose some management insights for improving the efficiency of EVS operations under the price incentive framework. Finally, a conclusion and future research directions are given in Sect. 4.

Table 1 List of notations

Sets	
I	Set of candidate regions
T	Set of time periods
Decision variables	
x_{ijt}	Number of reposition EVs from region i to j during operation period t, $\forall i \neq j \in I$
y_{ijt}	Actual adopted customers by EVS operators from region i to j during operation period t
$z_{it}, \hat{z}_{it}$	Number of available\unavailable EVs in region i during operation period t
p_{ijt}	The price paid for each trip from i to j
Parameters	
θ_{ij}	The distance between region i and region j
ϕ_{ijt}	The travel time between region i and region j at operation period t
η_{ijt}	The charge duration for EV from region i to j at operation period t
λ_{ijt}	The number of customers arrived at stations what want to trip from region i to j during operation period t
a_i	Standard capacity, i.e., parking spots, in region i
C_{ijt}^{R}	Reposition cost from region i to j at operation period t
C_i^{A}	Initial allocation cost in each region i
$\hat{p}_{ijt}$	Customer expected prices for per trip from region i to j during period t
d_{ijt}	Potential travel demand from region i to j during period t

2 Model

2.1 Parameters Setting

The notations applied in our model are listed in Table 1.

2.2 The Two-Level Trip Selection Process in EVS Service

We consider a two-level trip selection process in EVS service systems, where the first step is for customers to choose whether to adopt EVS trip service by comparing provided prices and their excepted prices, and the second step is for operators to further select "valuable" trips based on the adopted travel demands.

2.2.1 Travel Price Formulations

In practice, the price of each trip is mainly depended on the travel distance and travel time. Thus, we denote the actual price $p_{ijt}(p^{\mathrm{T}}, p^{\mathrm{D}}, \phi)$, $\forall i \neq j \in I, t \in T$ consists of two parts, including the fees on time p^{T}, and the extra fees on distance p^{D} for customers who drive more than the max-free limitation Θ specified by the operators. Thereby, as stated, the following formulation holds:

$$p_{ijt}(p^{\mathrm{T}}, p^{\mathrm{D}}, \phi) = p^{\mathrm{T}}\phi_{ijt} + p^{\mathrm{D}}(\theta_{ij} - \Theta)^{+} \tag{1}$$

We also note that two main prices p^{T} and p^{D} are integrated into a total price based on trips, i.e., p_{ijt}, $\forall i \neq j \in I, t \in T$.

2.2.2 The Upper Level Selection: Potential Customers Adoption

In the first "selection" level, customers have the priority to decide whether to adopt the EVS service, which depends on whether they will acquire the corresponding positive utilities, i.e., $U(\hat{p}_{ijt}) \geq 0$, where $U(\hat{p}_{ijt}) := \hat{p}_{ijt} - p_{ijt}$, $\hat{p}_{ijt}$ represents the random expected price requested by customers. Assuming d_{ijt} is the potential travel demand[1], the number of actual customers who are willing to adopt EVS service is $\lambda_{ijt}, \lambda_{ijt} := \mathbb{P}_{\hat{p} \sim \mathcal{P}}\{U(\hat{p}_{ijt}) \geq 0\} \cdot d_{ijt}$, $\forall i \neq j \in I, t \in T$, where we denote $\hat{p} := ((\hat{p}_{ij, \forall i \neq j \in I})_t, \forall t \in T)$. According to the above statements, we give the following definitions:

Definition 1 The optimal trip price after the first-stage selection is denoted by $p_{ijt}^{*} := \mathrm{argmax}_{p, d \in \mathcal{D}}\left\{\sum_{i \in I}\sum_{j \in I/\{i\}}\sum_{t \in T} d_{ijt}\mathbb{P}_{\hat{p} \sim \mathcal{P}}\{\hat{p}_{ijt} - p_{ijt} \geq 0\}\right\}$, where $\mathcal{D}$ denotes the total uncertain set of potential customers demand, and the number of corresponding optimal adopted customers in the first-level selection is λ^{*}.

EVS operators design price plan for maximizing adopted customers, whose adoption decisions are assumed mainly depend on price. Potential customers are defined as the people who are willing to travel from a region i to j. There are many substituted travel transportations, e.g., Taxi, Bus, Subway for them. They would adopt EVS service only if "Price is right" as mentioned. Thus, the first-level selection is made by managing price. According to Definition 1, operators should decide p^{*} to improve service adoption, which relies on the random expectation $\hat{p}$. In practice, it is hard to collect any valid data to estimate accurate expectation for all kinds of people. In this regards, the other transportation fees like taxi or ride-sharing offer reference values, by which the marginal moment information, e.g., mean and variance can be obtained.

---

[1] Here we assume travel demand denote the potential customers who need a transportation.

2.2.3 Lower Level Selection: Operator Trip Selection

In the second-level selection, operators just adopt part of potential customers from the first-level trip selection. In this way, overload or imbalance problem in specific trips can be mitigated by refusing other adopted customers. Hence, the number of actual trips y_{ijt} in EVS operations must satisfy the following condition:

$$y_{ijt} \leq \lambda_{ijt}, \forall i \neq j \in I, t \in T \tag{2}$$

In the above constraint, λ_{ijt} is defined as potential customers willing to adopt EVS trips, as mentioned, $\lambda_{ijt}(\boldsymbol{p}, \boldsymbol{d}) := \mathbb{P}_{\hat{p} \sim \mathcal{P}}\{\hat{p}_{ijt} - p_{ijt} \geq 0\}d_{ijt}, d_{ijt} \in \mathcal{D}$, where $\boldsymbol{d}$ denotes decision vector $\left((d_{ij, \forall i \neq j \in I})_t, t \in T\right)$.

Above all, the maximum trip price limitation is considered into a probability formation incorporating partial information of customers' expectation $\hat{p}$. Then we consider actual transportation demand in a robust way [10], because it is hard to predict the changing demand in practice. In Proposition prop: RO1, formulation 3 holds strong duality property. Then we get a nonlinear robust counterpart considering budget uncertain set.

Proposition 1 Assume that potential travel demands vary in the uncertain polyhedron set $\mathcal{D} := \left\{d_{ijt} : d_{ijt}^{\mathrm{L}} \leq d_{ijt} \leq d_{ijt}^{\mathrm{U}}, \sum_{t \in T} d_{ijt} \leq D_{ij}, \forall i \neq j \in I, t \in T\right\}$, where $d_{ijt}^{\mathrm{U}}, d_{ijt}^{\mathrm{L}}, D_{ij}$ denote upper bound, lower bound and maximum trip numbers in each trip from i to j, $\forall i \neq j \in I$, respectively, and $d_{ijt}^{\mathrm{U}} = \tilde{d}_{ijt} + \epsilon \hat{d}_{ijt}$, $d_{ijt}^{\mathrm{L}} = \tilde{d}_{ijt} - \epsilon \hat{d}_{ijt}$. There is a robust counterpart formulation of the second-level selection constraints:

$$y_{ijt} \leq \alpha_{ijt}d_{ijt}^{\mathrm{U}} - \beta_{ijt}d_{ijt}^{\mathrm{L}} + \gamma_{ij}D_{ij}, \forall i \neq j \in I, t \in T, \tag{3}$$

$$\alpha_{ijt} - \beta_{ijt} + \gamma_{ij} \leq s(\boldsymbol{p}, \hat{\boldsymbol{p}}), \forall i \neq j \in I, t \in T, \tag{4}$$

$$\alpha_{ijt}, \beta_{ijt}, \gamma_{ij} \geq 0, \forall i \neq j \in I, t \in T. \tag{5}$$

where $\alpha_{ijt}, \beta_{ijt}, \gamma_{ijt}$ are the dual variables and $s(\boldsymbol{p}, \hat{\boldsymbol{p}})$ represents the adoption rate $\mathbb{P}_{\hat{p} \sim \mathcal{P}}\{\hat{\boldsymbol{p}} - \boldsymbol{p} \geq 0\}$.

As stated in Proposition 1, a linear robust counterpart is proposed by dual theory. We also adopt the marginal moment of customers' expectation $\hat{\boldsymbol{p}}$ to reformulate our model into a Second-Order Conic Programming (SOCP) form according to the following Theorem 1.

Theorem 1 We assume that the demand $\hat{p}_{ijt}$ follows an unknown probability distribution with mean μ_{ijt} and variance σ_{ijt}. Let the uncertain expected prices $\hat{p}_{ijt}$ depend on a random variable κ, i.e., $\hat{p}_{ijt} = \mu_{ijt} + \sigma_{ijt}\kappa$, where the mean and variance of κ are μ^{r} and σ^{r}, respectively. For notation simplification, we denote $\varGamma = \begin{bmatrix} \sigma^{\mathrm{r}} + (\mu^{\mathrm{r}})^2 & \mu^{\mathrm{r}} \\ \mu^{\mathrm{r}} & 1 \end{bmatrix}$

as second-order moment matrix of κ. The nonlinear constraint (4) is equivalent to the following SOCP formulations:

$$\left\|\left(2\sigma_{ijt}\sigma^{r\frac{1}{2}}, 1 - \delta_{ijt} - \Delta_{ijt} - \sigma_{ijt}^2\sigma^r\right)\right\|_2 \le 1 - \delta_{ijt} + \Delta_{ijt} + \sigma_{ijt}^2\sigma^r \qquad (6)$$

where $\delta_{ijt} := \alpha_{ijt} - \beta_{ijt} + \gamma_{ij}$, and $\Delta_{ijt} := \left(\mu_{ijt} + \sigma_{ijt}\mu^r\right)^2 + \iota_{ijt}p_{ijt}^2 - 2\iota_{ijt}p_{ijt}\left(\mu_{ijt} + \sigma_{ijt}\mu^r\right)$.

Proof. According to the literature [11, 12], the nonlinear constraint (4) is equivalent to the following SOCP formulations:

$$\left(1 - \delta_{ijt} + \left(\mu_{ijt} + \sigma_{ijt}\mu^r - p_{ijt}\right)^2 + \sigma_{ijt}^2\sigma^r\right)^2$$
$$\ge 4\sigma_{ijt}^2\sigma^r + \left(1 - \delta_{ijt} - \left(\mu_{ijt} + \sigma_{ijt}\mu^r - p_{ijt}\right)^2 - \sigma_{ijt}^2\sigma^r\right)^2$$

We also introduce new variables $\sum_{m\in M}\iota_{ijtm}p_{ijtm}$ to linearize continuous variable p_{ijt}, where $\iota_{ijtm} \in \left\{\iota_{ijtm} : \sum_{m\in M}\iota_{ijtm} = 1, \forall i \in I, j \in I/\{i\}, t \in T\right\}$, M denote binary variables and price policy set, respectively. Which is the same,

$$\left\|\left(2\sigma_{ijt}\sigma^{r\frac{1}{2}}, 1 - \delta_{ijt} - \Delta_{ijt} - \sigma_{ijt}^2\sigma^r\right)\right\|_2 \le 1 - \delta_{ijt} + \Delta_{ijt} + \sigma_{ijt}^2\sigma^r \qquad (7)$$

2.3 Model Formulation

Our objective is to minimize the opposite number of operator profits, which includes the repositioning costs, initial EVs allocation costs, and revenue from adopting customers. The objective function is given as follows:

$$\min_{p,x,y,z} \sum_{t\in T}\sum_{i\in I}\sum_{j\in I/\{i\}} \left(C_{ijt}^R x_{ijt} + C_i^A z_{i0} - p_{ijt}y_{ijt}\right) \qquad (8)$$

Auxiliary variables $z, \hat{z}$ are introduced to represent the number of available EVs and unavailable EVs, which also satisfy the following constraints according to the literature [13]:

$$z_{it+1} = z_{it} + \sum_{\tau=1}^{t}\sum_{j\in I/\{i\}:\tau+\phi_{ji\tau}+\eta_{jit}\in(t,t+1]} (x_{ji\tau} + y_{ji\tau}) - \sum_{j\in I/\{i\}} (x_{ijt} + y_{ijt}), \forall i \in I, t \in T,$$
$$(9)$$

$$\hat{z}_{it+1} = \sum_{\tau=1}^{t} \sum_{j \in I/\{i\}: \tau + \phi_{ji\tau} \in (t+1-\eta_{jit}, t+1]} (x_{ji\tau} + y_{ji\tau}), \ \forall i \in I, \ t \in T, \qquad (10)$$

According to the recursion Eqs. (9) and (10) for available and unavailable EVs in each operation region and period, constraints (11, 12) denote the parking spots and EVs quantities limitations for reposition and travel:

$$z_{it} + \hat{z}_{it} \le a_i, \ \forall i \in I, \ t \in T \qquad (11)$$

$$\sum_{j \in I/\{i\}} (x_{ijt} + y_{ijt}) \le z_{it}, \ \forall i \in I, \ t \in T \qquad (12)$$

2.4 Model Reformulation

In the objective (8), there is a nonlinear term $p_{ijt} y_{ijt}$. To this end, a corresponding SOCP equivalent reformulations can be obtained according to the following Theorem 2.

Theorem 2 Let the binary variable $\iota_m := (\iota_{ijt, \forall i \ne j \in I, t \in T})_m$, $\forall m \in M$, be the decision variables associated with several provided prices, where $\iota_m = 1$ means that operators choose the mth price as its trip price. Then, the equivalent objective is given as follows:

$$\min_{\iota, x, y, z, \omega \in \mathcal{G}} \sum_{t \in T} \sum_{i \in I} \sum_{j \in I/\{i\}} \left(C_{ijt}^{R} x_{ijt} + C_i^{A} z_{i0} - \sum_{m \in M} (\omega_{ijtm} p_{ijtm}) \right) \qquad (13)$$

We introduce a linear polyhedron set $\mathcal{G} := \Big\{ \omega_{ijtm} : \omega_{ijtm} \le \iota_{ijtm} y_{ijt}^{U}, \ \omega_{ijtm} \le y_{ijt}, \omega_{ijtm} \ge y_{ijt} - (1 - \iota_{ijtm}) y_{int}^{U}, \ \omega_{ijtm} \ge 0, \forall m \in M, i \ne j \in I \Big\}$ to denote new domain of ω_m, where y^{U} denotes the upper bound of selected trips in the second stage. Then the nonlinear terms in the objective (8) can be linearized as new variables ω_{ijt} and corresponding new constraints as shown in $\mathcal{G}$. We note that p_m here represents the price parameters obtained from real-world data. Then we get an integrated model as follows:

$$\min_{\iota, x, y, z, \omega \in \mathcal{G}} \sum_{t \in T} \sum_{i \in I} \sum_{j \in I/\{i\}} \left(C_{ijt}^{R} x_{ijt} + C_i^{A} z_{i0} - \sum_{m \in M} (\omega_{ijtm} p_{ijtm}) \right)$$

s.t., Constraints (3, 5, 6, 9, 10, 11, 12)

It is easy to prove $\left(\sum_{m \in M} \iota_{ijtm} p_{ijtm} \right)^2 \to \sum_{m \in M} \iota_{ijtm} p_{ijtm}^2$ in that the cross item of ι_m, $\forall m \in M$ must be zero.

3   Case Study

We conduct a case study and several computational experiments to explore the price incentive effect from the two-level trip selection robust plan obtained by our model. Specifically, this part is designed to (1) Evaluate the performance of the two-level selection process in rebalancing trip demands among different service regions; (2) Explore how the price plan impacts the numbers of adopted EVS trip service. In the future research, we also try to conduct several numerical studies to (3) Explore how customers' behaviors, for example, range anxiety, impact the reposition operations in EVS service; to (4) Illustrate how the robust policy mitigates the negative influence on the EVS operation caused by uncertain travel demands. We adopt a real-world case of CAR2GO in Amsterdam. CAR2GO, a one-way car-sharing service project of internationally renowned luxury car manufacturer Daimler, provide car-sharing trip service in many countries and provide special EVS service for customers in Amsterdam. The entire service area is divided into 12 different regions as shown in Fig. 1. We collect the real historical trip data with the time span of more than 2 months, from November 23, 2017 to January 31, 2018, and it includes 27714 active trips in these service regions. The unit operation period is set to 1 h.

The statistical features of these trips in the EVS system for 24 h are shown in Fig. 2, which shows the temporal imbalance of EVS service demand obviously. There are trip demand peaks with about 27 trips at 11–12 o'clock while there is a valley with only 3 trips at 21–22 o'clock. The difference should be noticeable in daily operations. In addition, Fig. 3 shows that customers' trip demands also have spatial imbalance. We can see that the trip demands between region 2 and region 5 are prominently high, nearly 30 respectively, while trips between region 1, 12, and other regions are few. So, region 2 and 5 are the most active areas with the highest vehicle using and returning demands.

Fig. 1 Service region in Amsterdam

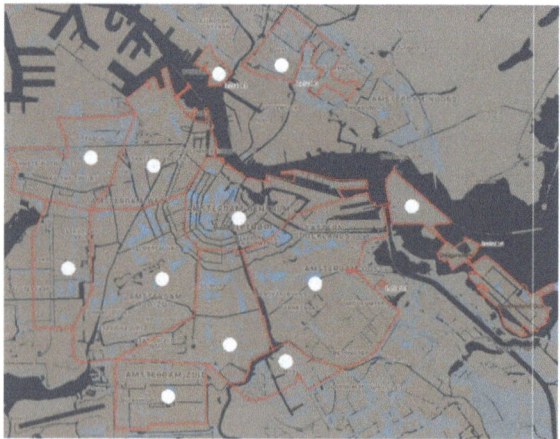

Fig. 2 Average demand in each hour

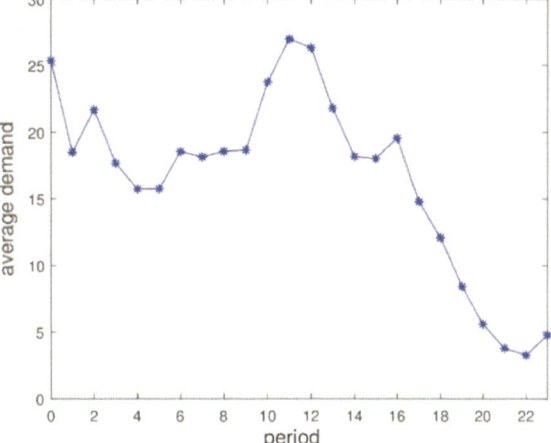

Fig. 3 Average demand between regions

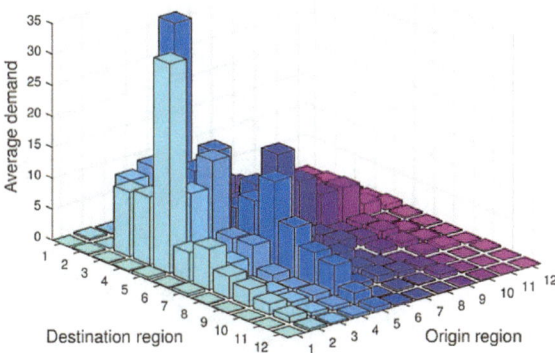

We calculate the mean and standard deviation of the trip numbers on each path from one region to another region in each hour from the real data and apply them in the model to meet the actual customer demand situation, as well as the mean and standard deviation of travel time. On the one hand, the travel time on each path in each hour is obtained by the mean plus the random multiple of standard deviation. On the other hand, since customers pay per minute 0.26–0.36 £ according to the CAR2GO official website in Amsterdam, the provided price vectors are built from the travel time multiply highest, middle and lowest prices. To ensure the quality of service and reduce customers' travel anxiety, the vehicles are supposed to begin charging once the trips are over. In the CAR2GO EVS system, the current vehicle model is smart for two ED and the charging duration is set to 1 h, according to the fact that the vehicles are fully charged in 45 min with 22 kW charger, and two and a half hours with 7 kW charger [14]. The reposition cost for each trip and the initial allocation cost in each region are assumed as 5 and 20, respectively. In the experiments, 1000 simulation trials are applied based on different situations to verify the several features of the model. All models are solved by Gurobi 8.0 on a macOS 10.14 machine with Intel(R) Core(TM) i7-6700HQ CPU 2.30 GHz and 16 GB memory specification.

Numerical result 1: Repositioning plan. We make EVS reposition operations in a 12-h period based on the solution of our proposed model as shown in Figs. 4, 5, 6, and 7, where the dark lines and near numbers represent travel trips with their numbers more than 5, the white lines and near numbers denote optimal reposition trips. The results confirm the fact that reposition operations are really essential and important for operators to balance vehicles in different regions and improve their profits, although they will cost a certain amount of money.

Numerical result 2: Performance Price Incentive Plan. We evaluate the recommendation schemes of price incentive in the trip sessions with variable service

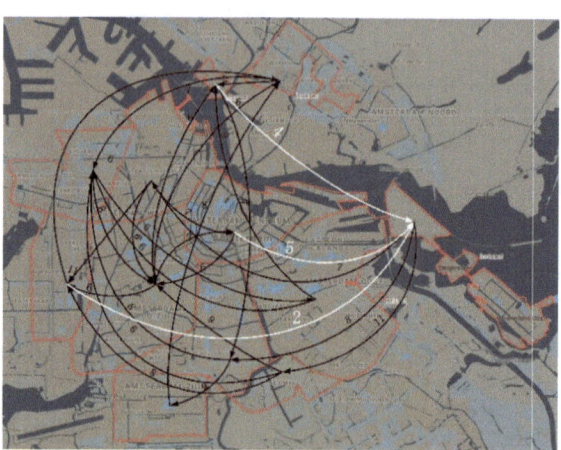

Fig. 4 Reposition operations in the second hour

Fig. 5 Reposition operations in the third hour

Fig. 6 Reposition
operations in the fourth hour

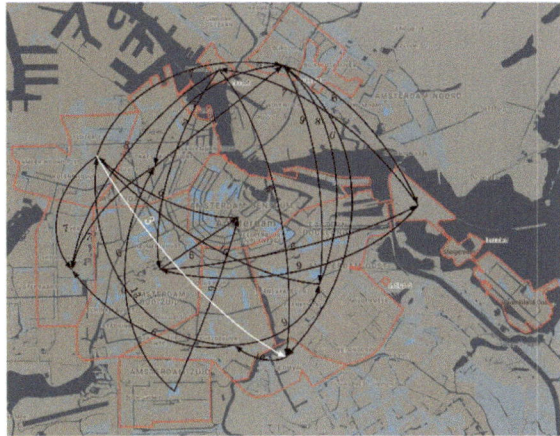

Fig. 7 Reposition
operations in the fifth hour

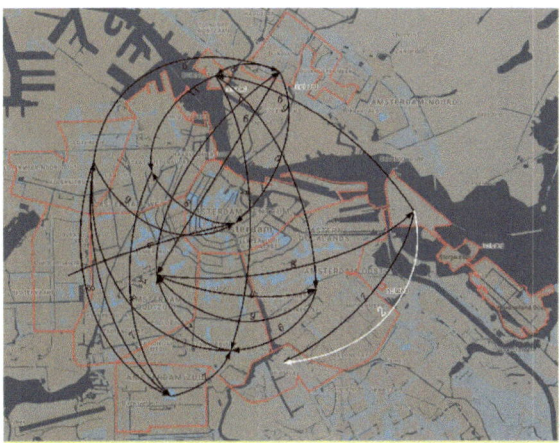

demands. Specifically, as shown in Fig. 8, in the most range of operation period (e.g., 1–7 and 10–12 period), the EVs flows increase with the decrease in the trip prices. It indicates that users have a clear preference for low-price service and low prices can motivate potential customers to adopt EVS service, which is consistent with the normal price–demand law and our result expectations. The results show that our price plans in the two-level process are effective to balance customer demands. Hence, operators are advised to implement a price incentive strategy so that they can reduce lost demand and increase profits.

As a result, the presented selection process is shown in line with the actual situation, which provides a perspective on considering operation problems for EVS operators. They need to understand and mine the value in the process to maximize their profit. For example, they can lower the price in the early stage of sharing service promotion to attract more users and increase the number of potential users. Then they

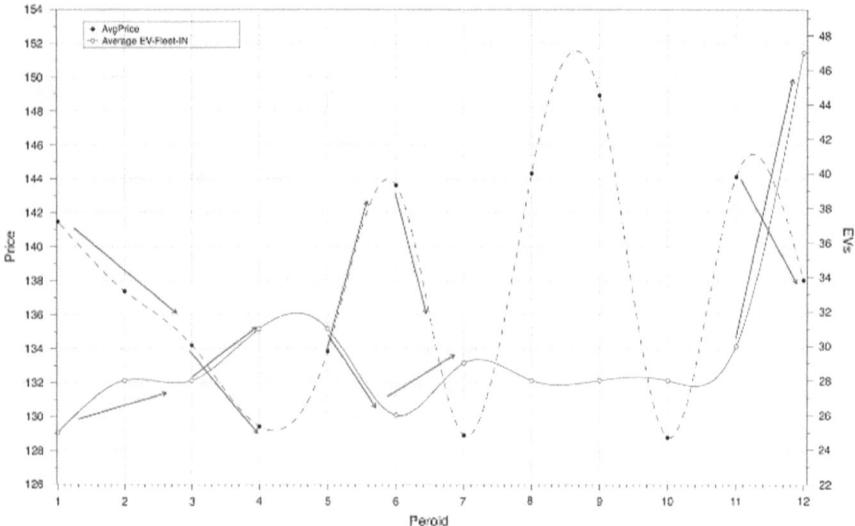

Fig. 8 Price incentive EVs flows

can select service trips according to their experience, users' trip history and specific itinerary information, including trip time, starting point, and ending point. Last but not least, the number of reposition vehicles need to be decided to avoid service region overstocked and vacant. Understanding the importance of the selection process can lead to a clearer analysis of how to increase profit through methods mentioned, including price incentive, trip selection, and reposition.

4   Conclusion

This paper develops a two-level trip selection process to optimize reposition operations in EVS sharing systems. In the scheduling model, the uncertain parameters, i.e., customers' expected price, and travel demand, are assumed as random variables with known statistic information or ambiguous set, e.g., marginal moments, which are obtained from the real world. And the corresponding worst-case adoption probabilities combined with these random variables are further approximated as SOCP. In a real-world case study in Amsterdam, the numerical results demonstrate the advantages of our two-level selection price incentive program in the EVS operations.

Acknowledgements This paper is supported by the China Scholarship Council, the Key Project of the Major Research Plan of the National Natural Science Foundation of China [Grants 91746210], the National Natural Science Foundation of China [Grants 71871023], the Beijing Natural Science Foundation [Grants 9172016], Special Fund for Joint Development Program of Beijing Municipal Commission of Education, the Beijing Institute of Technology Research Fund Program for Young Scholars, and Graduate Technological Innovation Project of Beijing Institute of Technology [Grants 2018CX20016].

References

1. E. Martin, S. Shaheen, J. Lidicker, Carsharing's impact on household vehicle holdings: results from a North American shared-use vehicle survey. in *Institute of Transportation Studies Working Paper*, vol. 46, no. 2143 (2010), pp. 150–158
2. J. Firnkorn, M. Müller, What will be the environmental effects of new free-floating car-sharing systems? The case of car2go in Ulm. Ecol. Econ. **70**(8), 1519–1528 (2011)
3. A. Musso, M.V Corazza, M. Tozzi, Car sharing in rome: a case study to support sustainable mobility. Procedia Soc. Behav. Sci. **48**, 3482–3491 (2012)
4. P. Baptista, S. Melo, C. Rolim, Energy, environmental and mobility impacts of car-sharing systems. empirical results from Lisbon, Portugal. Procedia Soc. Behav. Sci. **111**, 28–37 (2014)
5. E.W. Martin, S. Shaheen, Greenhouse gas emission impacts of carsharing in North America. IEEE Trans. Intell. Transp. Syst. **12**, 1074–1086 (2012)
6. D. Jorge, G. Correia, Carsharing systems demand estimation and defined operations: a literature review. Eur. J. Transp. Infrastruct. Res. **13**(3), 201–220 (2013)
7. L. He, H.-Y. Mak, Y. Rong, Z.-J. Max Shen, Service region design for urban electric vehicle sharing systems. Manuf. Serv. Oper. Manag. **19**(2), 309–327 (2017)
8. G.H. de Almeida Correia, A.P. Antunes, Optimization approach to depot location and trip selection in one-way carsharing systems. Transp. Res. Part E: Logist. Transp. Rev. **48**(1), 233–247 (2012)
9. Y. Zhang, M. Lu, S. Shen, On the values of vehicle-to-grid electricity selling in electric vehicle sharing. SSRN Electron. J. (2018)
10. A. Ben-Tal, B. Do Chung, S.R. Mandala, T. Yao, Robust optimization for emergency logistics planning: risk mitigation in humanitarian relief supply chains. Transp. Res. Part B: Methodol. **45**(8), 1177–1189 (2011)
11. L. El Ghaoui, M. Oks, F. Oustry, Worst-case value-at-risk and robust portfolio optimization: a conic programming approach. Oper. Res. **51**(4), 543–556 (2003)
12. W. Chen, M. Sim, J. Sun, C.-P. Teo, From CVaR to uncertainty set: implications in joint chance-constrained optimization. Oper. Res. **58**(2), 470–485 (2010)
13. Xu Min, Qiang Meng, Zhiyuan Liu, Electric vehicle fleet size and trip pricing for one-way carsharing services considering vehicle relocation and personnel assignment. Transp. Res. Part B: Methodol. **111**, 60–82 (2018)
14. https://www.zap-map.com/charge-points/smart-ed-charging-guide/

Research on the Method of Identifying Opinion Leaders Based on Online Word-of-Mouth

Chenglin He, Shan Li, Yehui Yao and Yu Ding

Abstract Opinion leaders are attracting increasing attention on practitioners and academics. Opinion leaders' online Word-of-Mouth (WOM) plays a guiding and decisive role in reducing risks and uncertainty faced by users in online shopping. It is of great significance of businesses and enterprises to effectively identify opinion leaders. This study proposes an integrated method by looking at not only essential indicators of reviewers but also the review characteristics. The RFM model is used to evaluate the activity of reviewers. Four variables L (text length), T (period time), P (with or without a picture) and S (sentiment intensity) are derived to measure review helpfulness from review text. And two effective networks are built using the Artificial Neural Network (ANN). This study utilizes a real-life data set from Dianping.com for analysis and designs three different experiments to verify the identification effect. The results show that this method can scientifically and effectively identify the opinion leaders and analyze the influence of opinion leaders.

Keywords Online WOM · Opinion leader · ANN · RFM · WOM content

1 Introduction

With the rapid development of the Internet, traditional reviews have become a greater influence and spread a broader range of online Word-of-Mouth (WOM). Compared with traditional reviews, two distinctive points of online WOMs have been well-documented [1]. Firstly, online WOM will not disappear as time and space change. Secondly, Online WOM can break the constraints on time and space. It will be fully presented in front of the public. Online WOM becomes a critical source of information because they can reshape the behavior of online consumers by reducing products information asymmetry between firms and consumers [2]. Service providers and manufacturers attach great importance on online WOM, they may show genuine

C. He · S. Li (✉) · Y. Yao · Y. Ding
College of Economics and Management, Nanjing University of Aeronautics and Astronautics,
Nanjing 211106, Jiangsu, China
e-mail: lishan@nuaa.edu.cn

© Springer Nature Switzerland AG 2020
H. Yang et al. (eds.), *Smart Service Systems, Operations Management, and Analytics*, Springer Proceedings in Business and Economics,
https://doi.org/10.1007/978-3-030-30967-1_19

concerns of online customers and provide useful market intelligence [3]. Online WOM is a major source of the evaluations of products and services, so retailers and marketers are utilizing online WOMs to understand the consumers' purchasing trends and attitudes. In a neutral sense, online WOM is a huge opportunity or thread for merchants and enterprises [4].

The consumers whose opinions are representative and authoritative are called opinion leaders. Opinion leaders play an essential role in online WOM and could greatly affect online consumption decisions [5]. These opinion leaders are the most concerned customers of the enterprise [6]. They have a n influential position in the fields of system recommendation [7], information security [8], network public opinion dissemination [9] and information sharing [10]. Therefore, it's very important to identify online opinion leaders. However, how to identify them correctly is a big challenge. The present paper aims to identify online opinion leaders more effective.

2   Related Work

For a long time, many scholars have been studying how to identify online opinion leaders. In this field, most scholars build an evaluation system around the nature or characteristics of opinion leaders to measure their influence, even if they use different methods. Yu and Wei [11] mining opinion leaders based on sentiment analysis. Cho and Hwang [12] using the Social Network Analysis (SNA) and thresholds method to examine which opinion leaders can quickly spread information. Their method helps companies make marketing choices. Ellero and Sorato [13] considered a new model, where opinion leaders are introduced as special agents, and study its specific properties which significantly recast some conclusions worth. Ma and Cai [14] measured the influence based on the RFM model to distinguish opinion leaders from general users. There are many other ways to identify opinion leaders, but the focus is not the same. For example, Weng [15] identified topic-sensitive opinion leaders classified by topics. Zhang et al. [16] put forward a network opinion leader recognition method, which integrates a network opinion leader recognition system and an opinion leader discovery module based on Markov Logic Networks. Miao [17] identified opinion leaders and explore the mechanism of how they influence others in public opinion events.

Consumers can get much information about products and services to reduce consumption risk and find out popular trends through online WOM [18]. Online WOM has a significant impact on the helpfulness of online WOM perception of online users. There is no unified view on the helpfulness of online WOM currently [19]. Some researchers have studied the helpfulness of online WOM in terms of reviewer level [20], text length [21], text recency [22], and sentiment intensity and so on [23]. It is feasible to study the credibility of the reviews of these aspects [24].

The text of online WOM inevitably involves some subjective description. Consumers are eager to get objective and realistic information. Now many users upload pictures in a review. They believe more about 'without a picture, there is no truth'.

Some studies have proved this point. Yang and Zhu [25] showed that text affects consumer attitude more for the previous review. And picture influences the recent one more. Chen [26] concluded that picture form has a positive effect on adopting the online WOM by consumers. More and more web users post pictures on the internet. Comments with pictures have become the key to the adoption of others information.

There are three disadvantages of the above literature. First, indicators are not comprehensive in these papers. Researchers pay more attention to users but did not focus on the content of online WOM. Second, the constructed model is not universal, it can only be applied to specific fields. Third, the methods hang behind because people believe 'without a picture, there's no truth', but the study did not consider the impact on the picture comments. This paper sheds new light on the content of online WOM to identify opinion leaders. In this paper, we propose a three-stage approach to rank opinion leaders. Specifically, we first use the RFM model to measure the activity of users because these three variables are effective against representing users' influence. Then, four variables of L (text length), T (period time), P (with or without a picture) and S (sentiment intensity) are proposed to measure the helpfulness of online WOM. At last, we use a ranking method which ranks opinion leaders according to not only the activity ranking but also the helpfulness ranking.

This paper makes the following contributions. First, we have developed an integrated approach to rank opinion leaders, which takes into account both user characteristics and online WOM content characteristics. Second, we use the ratio of interaction with the total number of reviews to represent M, which is a novel way. Third, we derive the P values in the evaluation of review helpfulness. Fourth, we have conducted evaluation studies using real-word data set to assess the effectiveness of the proposed method.

3   Method

This study derives a novel mechanism including the activity of users and the helpfulness of online WOM to identify opinion leaders. The weights of seven variables are assigned using the ANN method. This study first applies the RFM model to measure the activity of users. In the activity measurement network, three variables of R (recency), F (frequency) and M (monetary) are input, and the degree of centrality (DC) is output. Then, this study measures the helpfulness of online WOM. S is measured by SnowNLP package after segmentation and POS tagging. In the helpfulness measurement network, four variables of L, T, P and S are input, and the degree of agreement (DH) is output. DC is the total number of friends and fans. DH is the statistical number of likes and response. Next, this study calculates the activity and helpfulness score using the obtained ANN network. We divide the calculation results into active and inactive, helpful and unhelpful, identifies the opinion leaders. The influence ranking of each user by sorting the sum of activity and helpfulness ranking. Finally, this paper uses data from Dianping.com to verify the proposed method. The

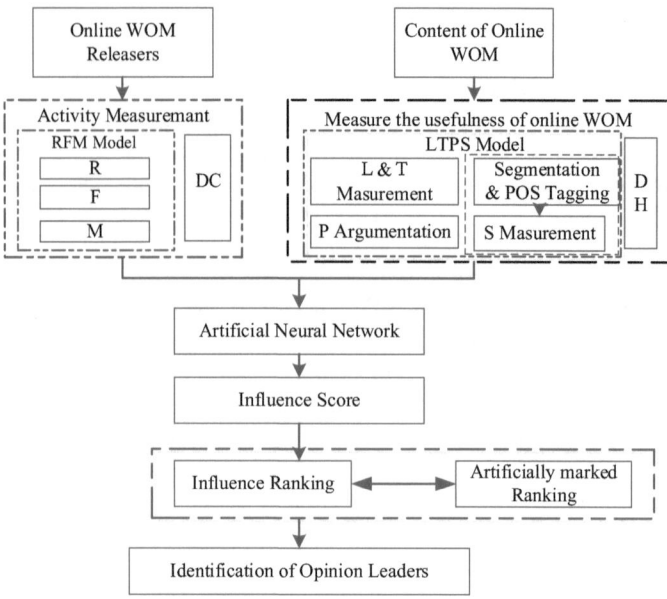

Fig. 1 The method of identifying opinion leaders

results will be contrasted with the artificially marked ranking to verify its validity (as shown in Fig. 1).

3.1 RFM Measurement

The RFM model is a practical method of weighing customer value and customer profitability in the marketing category. It has three indicators of recency, frequency and monetary, respectively indicating a customer recent consumption behavior, the frequency of consumption over a period and the total amount of consumption during the period. RFM model to measure customer value is based on the following three assumptions: (1) R is smaller the better. That is, a customer last spending time of the current time is shorter, he is the more feasible to consume again. (2) F is larger the better. That means the more a customer consumes in the unit time, the more feasible he consumes again. (3) M is larger the better. In a certain period, the greater the total amount of customer spending, the more feasible he is to consume again. Depending on the needs of different business scenarios, R, F, M will be given different weights W_r, W_f, W_m, so each customer's value score is expressed as:

$$W = W_r \times R + W_f \times F + W_m \times M \tag{1}$$

This paper adopts RFM model to measure the R, F, and M values of network users and evaluate the activity of network users in the network activities, to obtain which online WOM users have the qualifications to become opinion leaders. Combined with the characteristics of online WOM and the difference in application scenarios, the three indicators were appropriately modified.

Recency. Recency refers to the interval between the last time the user released online WOM to the current in days. For the online WOM user i, recency $R_i = N - l_i$. N represents the current date and l_i represents the last time the user released online WOM. Since the units of R, F, and M are different, all the data needs to be normalized before calculation, and they are respectively represented by R', F' and M':

$$R'_i = 1 - \frac{R_i - R_{min}}{R_{max} - R_{min}} \tag{2}$$

R_{max} represents the maximum recency of all online users, R_{min} represents the minimum one.

Frequency. Frequency means the number of times that online WOM is published by users within a specific period. This article set three months as a time interval. For the online WOM user i, frequency F_i is the total times of online WOM user published WOM in three months. The normalization formula is as follows:

$$F'_i = \frac{F_i - F_{min}}{F_{max} - F_{min}} \tag{3}$$

F_{max} represents the maximum frequency of all online users, F_{min} represents the minimum one.

Monetary. Online WOM has an influential impact on others' purchasing decision. It's difficult to evaluate how the review influences others accurately. As many websites allow consumers to interact with each other, this paper utilizes the interaction value. Users generate reviews and allow other users to estimate and reply. This study represents the interaction by the indicator of 'M'. For higher precision, we use the ratio of interaction with the total number of reviews to indicate M value. For example, the number of interaction is 1056 and the total number of review is 1129 for the user i, so the value M_i is 1056/1129. Similarly, M_i needs to be normalized as follows:

$$M'_i = \frac{M_i - M_{min}}{M_{max} - M_{min}} \tag{4}$$

M_{max} represents the maximum monetary of all online users, M_{min} represents the minimum one.

3.2 LTPS Measurement

This study uses four variables of L (text length), T (period time), P (with or without a picture) and S (sentiment intensity) to measure the helpfulness of online WOM. L is the number of words of an online WOM. L is proportional to the amount of information it contains. The more content it contains, the more feasible it is that the impact on consumer choice is greater, and the more feasible it is to be helpful. T indicates the timeliness of an online WOM. Instant online WOM is closer to the actual state that consumers are easier to accept it. P allows users to intuitively understand the real situation of products and services because pictures make the online WOM more convincing. In this study, if there are pictures recorded as 1, otherwise recorded as 0. S represents sentiment intensity. The explosive WOMs are more truthful to reflect the true feelings of the user experience compared to the plain one. It can provide more in-depth information on others. For each online WOM j, L, T and S were normalized consistent with RFM measurement as follows:

$$L'_j = \frac{L_j - L_{min}}{L_{max} - L_{min}} \tag{5}$$

$$T'_j = 1 - \frac{T_j - T_{min}}{T_{max} - T_{min}} \tag{6}$$

$$S' = \frac{S_j - S_{min}}{S_{max} - S_{min}} \tag{7}$$

L_{max} represents the maximum length of review, L_{min} represents the minimum length, T_{max} represents the maximum recency of review, T_{min} represents the minimum recency, S_{max} represents the maximum value of sentiment intensity and S_{min} represents the minimum value of sentiment intensity.

Before measuring the value of S, the text of online WOM needs to be segmented and tagged. This article selects Jieba to fulfill Chinese word segmentation and tagging because some Chinese scholars achieved the task and obtained a good result. Word segmentation and POS tagging play a key role in the following calculation. The precision of word segmentation directly affects the correction of S. To improve the accuracy of word segmentation, this paper extracts keyword and word frequency before word segmentation applying a tool called NLPIR. We build a personal dictionary suitable for food reviews. This study selects SnowNLP to get S. SnowNLP has high recognition of the shopping text and it can adapt to jagged, diverse forms of text. At the same time, Jieba and SnowNLP are programmed in Python 3.6 which achieved a unified.

4 Experiment

4.1 Experiment Data

In this article, the data used is from Dianping.com. Dianping.com is a typical third-party platform. It guides customers to provide an online WOM. Data acquisition is using a third-party collector 'Octopus'. For each online WOM, this study collected users name, release time, online WOM text, and number of pictures. For each user, this study collected user name, number of friends, number of fans, number of interaction, total number of online WOM, release time, number of likes, number of responses, and the frequency. Octopus collected a total of 5658 pieces of comment data and user data until March 30th, 2017 in Tuolejia business district. The original user data is cleaned by four steps: (1) delete 1239 pieces of duplicate data; (2) delete 899 pieces of data with M being zero; (3) delete 379 pieces of data with total number of online WOM being no more than 3; (4) clean the comment data according to the user data. After completing these steps, 3141 pieces of data are left.

In this paper, the data preprocessing uses Excel to calculate the time interval T, the sum of the number of likes and responses DH and the sum of the number of friends and fans DC. Among them, DH is the artificial neural network output calculated by the helpfulness of online WOM, and DC is the output calculated by user activity.

4.2 Sentiment(S) Measurement

This paper uses Python language and some tools of Chinese text mining to measure S value. First, NLPIR is used to extract keyword and word frequency statistics firstly to build a dedicated dictionary suitable for this study. According to keyword extraction and word frequency statistics, some special words have been added to the dictionary, such as '李公主卷(Princess Lee volume)', '棒棒哒(awesome)', '麻麻(mom)', '寿喜锅(Shouxi pot)'. Second, 'posseg' component is loaded with Jieba to segment word and POS tagging based on the dedicated dictionary. Third, some stopwords such as '好(nice)', '一般(general)', '哈哈(haha)' are removed from the HIT's Chinese stopwords data set. These words can express sentiment and have high frequency. In the stopwords corpus, words such as '是(is)', '的(of)', '有(have)', '也(also)', '了(have done)' which are auxiliary word even though have a relatively high frequency and some noun words such as '大众点评(Dianping)', '评论(review)', '商店(store)', '托乐嘉(Tuolejia)', '乐尚天地(Leshangtiandi)', '后街(Hou street)', '苏果(Suguo)' and so on are added. After these three steps, SnowNLP package in Python translator to automatically calculate S for each piece of text. S is the probability of positive sentiment, and the value range is [0, 1]. The calculation results are shown in Fig. 2.

User Name	Text	S
无肉不欢的小珠	点了鸡爪炒饭 想说挺一般的……炒饭可能又	0.993530248
魏茜_9607	还不错，口味还行，盒子挺有创意	0.999668226
天天山水	还行吧，有点干。辣点更好吃！！！！！	0.422235478
kitty靓儿	特意去打包外卖回家，发现点评的名字换了	1
熹范2店	店面太小，环境不是很好。饭太硬了	0.299472712
小可爱8282	环境:清爽干净，富有特色，老板还是花了	1
人类的好盆油	地铁通道里的一家小店……味道不错	0.970615237
短毛的羊	位置不是很好找但是挺好吃的 包装盒也很	0.990776656
dpuser_181964	油烟味超级重！在乐尚天地负一楼外卖生意	0.999996793
爱文静D	地点在乐尚天地，靠近南航，位置挺好找白	1
小九葵	我要给这家炒饭满分！真的很惊喜哇！口B	1
julylast	位置在地铁站出来的负一层。炒饭味道不错	0.998777854
Melilotus	大中午找吃的，想吃米饭便用点评寻到了i	0.999744618
咕咕快吃	好久之前团购的了，突然翻到，想建议一	0.871033331
wxm梅梅梅	差不多有小半年没有去过了，今天中午特别	1
小傻呆呆猪	外卖送的有点慢啊，味道还是不错的	0.921054345
想不起来昵称所	位置在乐尚那边，挺好找的，店面不大。	0.999999999
lisawang是我	上周末跑开托乐嘉吃饭，无意间发现的一	1
铁道部乐队_鼓	服务很好，还帮我水杯加了水，地方也不	0.999675308
sureyangyang	一般的小店，专门做外卖生意，口味还行	0.991600864

Fig. 2 The calculation results of S

4.3 Identifying Opinion Leaders

The relationship between R, F and M is very complex and the relation among L, T, P and S is uncertain. This complex relationship cannot be accurately described by linear or static weights, so we choose the Artificial Neural Network (ANN). Fortunately, ANN is suitable to fix the complex problem with multiple parameters, and the uncertainty of individual behaviors, emotions and attitudes will gradually clear through a lot of training and simulations. In this paper, the three-layer feed-forward neural network is used to measure the activity of the online WOM publishers and the helpfulness of its content, using the neural network tool in MATLAB2014b. Each network concludes the input layer, the hidden layer, and the output layer as shown in Fig. 3. The data is divided into training data and test data. This study inputs R, F and M in the input layer, and the output is DC. The four indexes of L, T, P and S are taken as input, DH is taken as output. ANN will adjust the weights adaptively and keep on learning in training in the hidden layer. Its adaptive weighting mechanism can determine the weight of each index, and then can calculate the influence score, get rankings.

This paper takes the calculation of the user's activity as an example. There are three variables of R, F, and M in the input layer, so the number of input nodes is set to be three. At the output layer, there is only a value DC, so the number of nodes is set to one. In the hidden layer, choosing the number of nodes is a complex problem. In this paper, the empirical formula $2^x > n$ is referenced, where x denotes the number of hidden layer nodes and n denotes the number of samples. If the number of training times is too huge or the result cannot converge on a given number of training times, then the training will stop out and a lot of new nodes will be selected and trained. The number of nodes in the hidden layer sets from 12, constantly changing the value

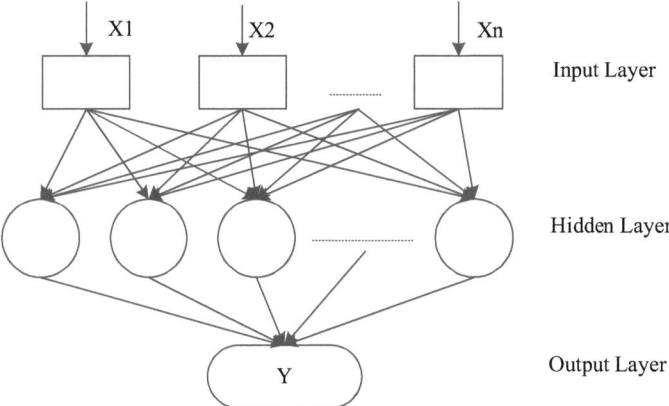

Fig. 3 Three-layer feed-forward neural network

Table 1 Training parameters

Parameter	Value	Parameter	Value
Type of ANN	feed-forward neural network	Performance function	MSE
Input nodes	3	Goal	0
Hidden nodes	190	Epochs	1000
Output nodes	1	lr	0.01
Training function	TRAINGDX	Mu_max	100,000,000
Activation function	LOGSIG	hitNum	100

of x, observing the performance of the training, estimating the range of nodes until the algorithm converges. The training parameters are shown in Table 1.

For 3141 pieces of user data, select 2800 pieces as training data, 341 pieces as test data. In this paper, the number of nodes in the hidden layer changed from 12 to 200 by observing the training performance with using least square method, and when the number of training nodes was 190, the training error and training steps both achieve the minimum value, so the neural network model structure is $3 \times 190 \times 1$. When the number of the hidden layer nodes is 190, the best training is obtained when iteration is 976, and the algorithm converges, as shown in Fig. 4.

4.4 Results and Analysis

Based on the trained ANN model, this paper calculates the activity and the helpfulness score, and then ranks the score. According to the ranking, score will be artificially divided into active and inactive, helpful and unhelpful. As a result, 26 active and

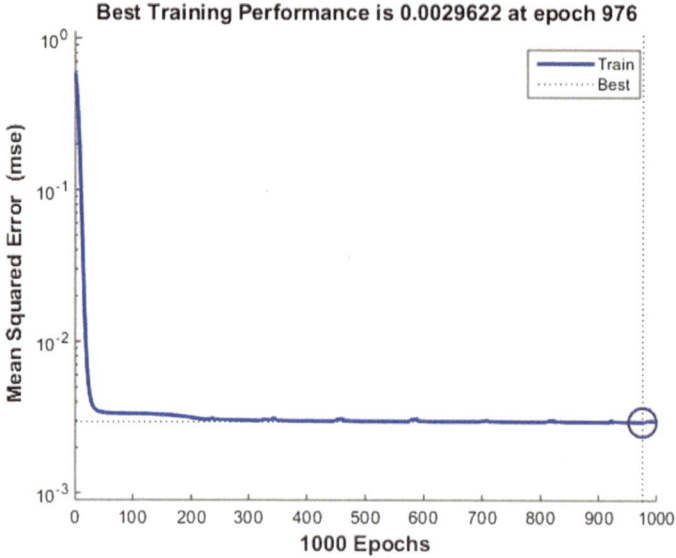

Fig. 4 Training performances

helpful users were obtained. There are 26 opinion leaders, 139 potential opinion leaders, 183 suspicious opinion leaders and 2,793 ordinary users as shown in Fig. 5.

The contribution of users can indicate the impact of the user on other consumers to a certain degree. In this research, the user's contribution was obtained from Dianping.com. This study ranks 26 users who are both active and helpful according to the artificial ranking, and then compares the contributions of user with the artificial ranking.

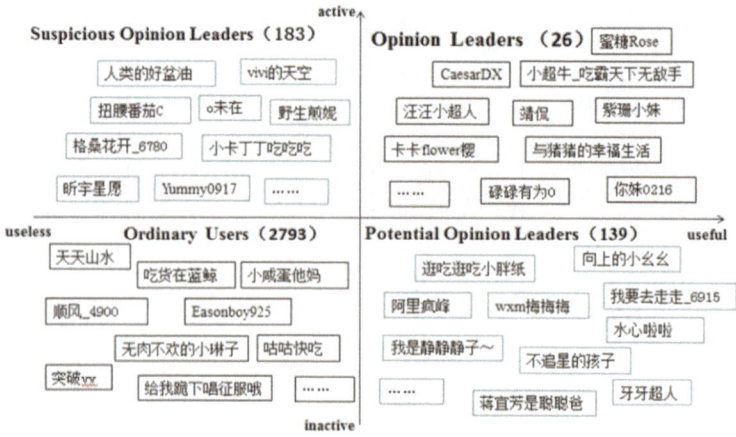

Fig. 5 Classification results

The validity of the model is verified as shown in Table 2. The consistency of the activity ranking and the helpfulness ranking with the artificial ranking is 5 and 4 respectively. It indicates that single dimension either activity or helpfulness measurement has little effect on identifying opinion leaders. The artificial ranking is compared with the contribution ranking and the influence ranking, the consistency is 10 and 14 respectively, which means the results of this research obtained the largest

Table 2 Results comparison

User name	Artificial ranking	Activity ranking	Helpfulness ranking	Influence ranking	Contribution ranking
蜜糖Rose	1	2	1	1	1
小超牛_吃霸天下无敌手	2	4	4	2	2
紫珊小妹	3	12	2	4	11
阿布好棒	4	10	25	12	4
靖侃	5	5	16	8	18
你妹0216	6	9	10	6	26
碌碌有为0	7	14	7	9	25
CaesarDX	8	1	9	3	12
汪汪小超人	9	3	12	5	9
与猪猪的幸福生活	10	6	15	10	10
蓝鲸树下的小阿菲	11	16	27	15	5
木易冰客	12	21	23	17	13
易足差华	13	22	17	13	14
一只自由行走的花	14	20	22	14	6
卡卡flower樱	15	15	5	7	17
莜LOVE美	16	24	20	16	7
开卫星的土豆	17	8	21	11	8
kitty靓儿	18	18	26	18	16
爱旅行超会吃的小疯子	19	26	19	19	19
萨克有声	20	28	18	20	20
木小西爱吃肉	21	27	33	26	21
全球最Ging的女人	22	19	31	22	22
小茜茜要瘦成一道闪电	23	23	28	23	15
陈小鱼逛吃逛吃	24	31	24	24	24
shark_口口	25	25	30	25	23
黑椒糖	26	17	29	21	3
Consistency	–	5	4	14	10

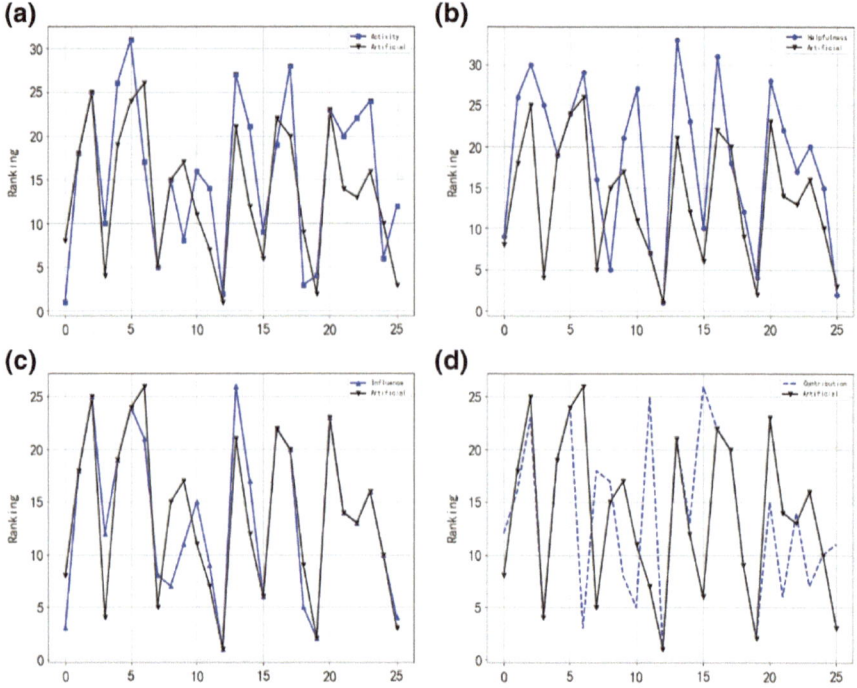

Fig. 6 Comparison of different ranking with the artificial ranking

consistency. It indicates that the proposed method has some advantages over the contribution value of Dianping.com, which is more in line with the actual situation.

In another way, it can be seen from Fig. 6. The influence ranking has the best fit with the artificial ranking. The contribution ranking fits better than the activity ranking, and the activity ranking fits better than the helpfulness ranking. But, there were no significant differences between the fitting effect of the activity ranking and the fitting effect of the helpfulness ranking. The single most striking observation to emerge from the data comparison was the combination of activity and helpfulness ranking can effectively identify opinion leaders. The results indicate that the proposed method can effectively identify the online opinion leaders. It is interesting to note that the activity of user is more important than the helpfulness of WOM content in identifying opinion leaders.

5   Conclusion

As stated in the Introduction, our main aim was to identify opinion leaders more effective. In this paper, the RFM model and the helpfulness of online WOM are combined to identify the opinion leaders. Two aspects of users' activity and the helpfulness of WOM are organically combined, which includes the most important indicators performing the reputation for users. To obtain accurate and reasonable results, this article uses a large amount of data to train the ANN model to make it closer to the real situation. The data experiments show that the proposed method is effective. It has the best fitting with the artificial ranking. Our results share number of similarities with previous findings. Three variables of R, F, M are helpful to identify opinion leaders. The results have enriched the method of identifying opinion leaders and provided a practical idea for the businesses to manage different clients. Besides, this research recorded as 1 with a picture and recorded as 0 without a picture. It did not consider the content of the picture on calculating the helpfulness of online WOM. The follow-up study could consider the opinion leaders by studying the changes of the number of friends and fans in continuous state, also could explore the impact on picture content of consumers. These issues deserve further study in the future research.

References

1. X. Lu, S. Ba, L. Huang, Y. Feng, Promotional marketing or word-of-mouth? Evidence from online restaurant reviews. Inf. Syst. Res. **24**(3), 596–612 (2013)
2. B. Gu, J. Park, P. Konana, The impact of external word-of-mouth sources on retailer sales of high-involvement products. Inf. Syst. Res. **23**(1), 182–196 (2012)
3. A.H. Huang, K. Chen, D.C. Yen, T.P. Tran, A study of factors that contribute to online review helpfulness. Comput. Hum. Behav. **48**(C), 17–27 (2015)
4. X. Yan, J. Wang, M. Chau, Customer revisit intention to restaurants: evidence from online reviews. Inf. Syst. Front. **17**(3), 645–657 (2015)
5. C. Wu, H. Che, T. Chan, X. Lu, The economic value of online reviews. Mark. Sci. **34**(5), 739–754 (2015)
6. H. Choi, S. Kim, J. Lee, Role of network structure and network effects in diffusion of innovations. Ind. Mark. Manage. **39**, 170–177 (2010)
7. S. Mohammadi, A. Andalib, Using the opinion leaders in social networks to improve the cold start challenge in recommender systems, in *2017 3th International Conference on Web Research (ICWR)*, (IEEE, Tehran, Iran, 2017), pp. 62–66
8. S.B. Yudhoatmojo, I. Budi, F.K. Dewi, Identification of opinion leader on rumor spreading in online social network twitter using edge weighting and centrality measure weighting. in *2017 Twelfth International Conference on Digital Information Management (ICDIM)* (IEEE, Fukuoka, Japan 2018), pp. 313–318
9. Y. Ma, X. Shu, S. Shen, J. Song, G. Li, Q. Liu, Study on network public opinion dissemination and coping strategies in large fire disasters. Procedia Eng. **71**, 616–621 (2014)
10. M. Shinde, S. Girase, Identification of topic-specific opinion leader using SPEAR algorithm in online knowledge communities, *2016 International Conference on Computing*, Analytics and Security Trends (CAST) (IEEE, Pune, India, 2017), pp. 144–149
11. X. Yu, X. Wei, X. Lin, Algorithms of BBS opinion leader mining based on sentiment analysis. in *WISM 2010: Web Information Systems and Mining* (Springer, Sanya, China 2010), pp. 360–369
12. Y. Cho, J. Hwang, D. Lee, Identification of effective opinion leaders in the diffusion of technological innovation: A social network approach. Technol. Forecast. Soc. Chang. **79**, 97–106 (2012)
13. A. Ellero, A. Sorato, G. Fasano, A new model for estimating the probability of information spreading with opinion leaders. Department of Management at Università Ca' Foscari Venezia working Paper No. 13, Italy (2011)

14. Y. Ma, S. Cai, R. Wang, Study on the method of identifying opinion leaders based on online customer reviews. in *2011 International Conference on Management Science & Engineering* (IEEE, Rome, Italy 2011), pp. 10–17

15. J. Weng, E.P. Lim, J. Jiang, Q. He, TwitterRank: finding topic-sensitive influential twitterers. in *Proceedings of the third ACM International Conference on Web Search and Data Mining* (ACM, New York 2010), pp. 261–270

16. W. Zhang, X. Li, H. He, X. Wang, Identifying network public opinion leaders based on markov logic networks. Sci. World J., 1–8 (2014)

17. Q. Miao, Y. Meng, J. Sun, Identifying the most influential topic-sensitive opinion leaders in online review communities. in *2016 IEEE International Conference on Cloud Computing and Big Data Analysis* (IEEE, Chengdu, China 2016), pp. 330–335

18. Y. Liu, J. Jin, P. Ji, A.H. Jenny, Y.K.F. Richard, Identifying helpful online review: a product designer's perspective. Comput. Aided Des. **45**(2), 180–194 (2013)

19. M.S.I. Malik, A. Hussain, An analysis of review content and reviewer variables that contribute to review helpfulness. Inf. Process. Manage. **54**, 88–104 (2018)

20. H. Hong, D. Xu, G. Wang, W. Fan, Understanding the determinants of online review helpfulness: a meta-analytic investigation. Decis. Support Syst. **102**, 1–11 (2017)

21. S.M. Mudambi, D. Schuff, What makes a helpful review? A study of customer reviews on Amazon. Com. MIS Q. **34**(1), 185–200 (2010)

22. Y. Hu, K. Chen, P. Lee, The effect of user-controllable filters on the prediction of online hotel reviews. Inf. Manag. **54**, 728–744 (2017)

23. A.Y. Chua, S. Banerjee, Helpfulness of user-generated reviews as a function of review sentiment, product type and information quality. Comput. Hum. Behav. **54**, 547–554 (2016)

24. Q. Yan, Y. Meng, Factors affecting the perceived helpfulness of online reviews—an empirical study based on online film reviews. Chin. J. Manag. Sci. S1, 26–131 (2013). 闫强, 孟跃: 在线评论的感知有用性影响因素——基于在线影评的实证研究.中国管理科学 (S1), 26–131 (2013)

25. Y. Yang, Y. Zhu, No picture, no truth? The effect of pictorial and verbal service online reviews on consumer attitudes. Psychol. Explor. **34**(1), 83–89 (2014). 杨颖, 朱毅: 无图无真相?图片和文字网络评论对服务产品消费者态度的影响.心理学探新 34(1), 83–89 (2014)

26. S. Chen, Research on the relationship between picture form word-of-mouth features and adoption effect. Huazhong University of Science and Technology, Hubei, China (2013). 陈珊珊: 图片形式网络口碑信息特性与其采纳效果的关系研究.华中科技大学, 湖北 (2013)

People Analytics in Practice: Connecting Employee, Customer, and Operational Data to Create Evidence-Based Decision Making

Fiona Jamison

Abstract People analytics is a rapidly growing field and one that can be daunting for many HR professionals. The sophistication and capability of organizations vary considerably. Some organizations are just starting to get a handle on their data and improve its quality while others are on the cutting edge of predictive analytics. This paper outlines an analytics maturity model and various case studies sharing how organizations have moved from data to action. Examples will demonstrate how companies have embarked on a journey to connect data on employee attitudes, customer loyalty and satisfaction, operations, and financial performance to make more informed evidence-based business decisions.

Keywords People analytics · Operational data · Decision making · Human capital

1 Introduction

The largest fixed cost for most organizations is their human capital the workforce that keeps their operations humming. Their most pressing and vexing challenge is determining the correct type and number of people to hire and retain to achieve optimum performance and profitability. Despite the vital significance of this task few organizations know how to accurately measure both the cost of their human capital and how their employees impact their bottom line.

The recent explosion of data analytics and data management has begun to transform how organizations understand, motivate, recruit, promote, and nurture their most critical resource, human capital. As human resources technology advances to better capture, measure, and monetize this critical data businesses can capitalize on this information for a competitive advantage. Indeed, companies that excel at people analytics are 3.1 times more likely to outperform their peers financially, according to a recent global leadership forecast [1].

F. Jamison (✉)
Spring International, Conshohocken, PA 19428, USA
e-mail: FJamison@springitl.com

© Springer Nature Switzerland AG 2020   223
H. Yang et al. (eds.), *Smart Service Systems, Operations Management, and Analytics*, Springer Proceedings in Business and Economics, https://doi.org/10.1007/978-3-030-30967-1_20

Business leaders are getting this message albeit slowly: 53% of CEOs agree they must invest significantly more in talent analytics to quantify the value of human capital and 43% of CEOs want to significantly improve human capital metrics reporting to help investors accurately measure their companies' value [2]. But those leaders continue to struggle with how to best use their human capital. About 60% of businesses merely react to basic data they collect rather than refining it and deploying it for strategic or predictive purposes.

Fewer than one in 10 businesses say they understand what factors drive employee performance—and just 8% believe they possess data they can use. Interestingly, a 2017 survey of global attitudes about people analytics found that 81% of respondents in China rated this area as "important" or "very important"—behind only Brazil and India, and above the UK and United States [3]. Business leaders' familiarity with human capital analytics varies greatly depending on their field, their background, their industry, their location, and their company's size.

2   Maturity Model

Through its work with hundreds of companies Spring International has identified four levels of maturity in assessing and leveraging human capital analytics: reactive, proactive, strategic, and predictive as shown in Fig. 1. Based on a recent study of 105 companies in the Northeast United States, we identified that only 11% of those surveyed were using advanced or predictive analytical models to make more informed business decisions [4]. Most companies react using generic surveys and basic analytics resulting in few or even no behavior changes based on practical

Fig. 1 Human capital analytics maturity model

evidence. Proactive organizations deploy custom targeted surveys and assessments and more refined analytics to identify vital issues and then plan to act based on those findings. The strategic approach to human capital analytics seeks to shape a business's decision-making by using sophisticated tools to solve challenges and anticipate—and prevent—future problems. Finally, the predictive level in the model incorporates analytics data, tools, and models that shape business strategy and enable evidence-based decision-making.

Spring International's research and extensive consulting work have shown again and again that when organizations leverage data on a regular proactive basis to make informed and strategic decisions about their people, their business, their customers, and their operations they enjoy a significant competitive advantage. This can manifest as lower turnover, higher job satisfaction, less shrinkage (employee theft of merchandise or materials), reduced accident rates, increased quality leading to higher revenues, and profitability.

2.1 Using a Human Capital Analyst

Leaders typically seek help from human capital analysts only after failing to correct missteps themselves which prolongs their problems. Human capital analysts and consultants are at their best unbiased parties who provide a fresh perspective on business operations and challenges. Using deliberately crafted data collection, data mining, surveys, interviews, observations, and focus groups to gather new data and information they put this information into context and suggest clear actionable remedies.

Many companies start their effort with a small pilot project that directly addresses a "pain point" that is troubling an organization's leadership. That was the case with a Pennsylvania healthcare system that suffered from a significant turnover of its nursing staff. The churn was beginning to affect the patients' quality of care and forced the human resource staff into an exhausting and frustrating pattern of continually recruiting, hiring, and training new nurses only to see them leave. Top executives didn't understand the root causes of why this was happening, but they knew they had to act because nurse turnover was costing them about $3.1 million per year.

Using data drawn from several years of nurse engagement surveys and the hospital HR information system (HRIS), Spring International built a model that predicted potential turnover with 90% accuracy. The model identified the demographics of the nurses most at risk of leaving in the next three months. Deploying 2016 Engagement and HRIS data, Spring International first built a model using logistic regression to determine the strongest predictors of voluntary leave. The model was then applied to the 2018 Engagement and HRIS data to identify nurses with the greatest likelihood of quitting. Figure 2 outlines the final model.

The Nurse Turnover model was able to identify that nurses younger than 35 with less than four years of tenure and employed at a particular location were at the greatest risk of leaving. The model further identified that nurses were on average

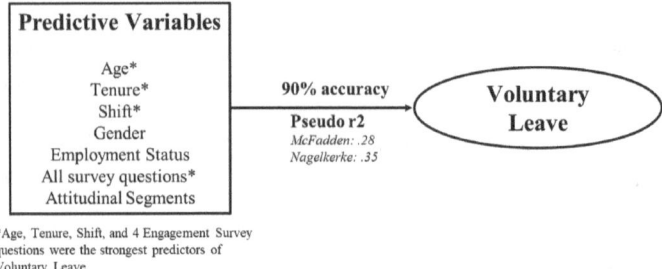

Fig. 2 Nurse turnover model (Summary Visual)

twice as likely to stay if they received regular feedback on their performance, felt recognized for their work, and described themselves as not planning any major life changes (stated intent to leave, planning to retire, move or go back to school full time in the next three years). While this model helped identify key areas for the health organization to focus on it was limited in the type of data it could include. First, attitudinal engagement data was collected only every two years. Ideally, the model would be built on more current data which makes the case for collecting attitudinal data more frequently through surveys. Further, the HRIS data was limited to basic demographic data such as age, tenure, gender, and shift. Additional metrics such as distance to work, time since last promotion, and performance ratings would provide stronger models.

Additional analysis, including in-depth one-on-one interviews homed in on exactly why these nurses were unhappy. Data showed that nurses were stressed and burned out. They felt overworked and unappreciated. Once the leadership knew precisely what work conditions were triggering the nurses' dissatisfaction, it could directly address those issues and stem the tide of resignations. As a result, the health system began a targeted recognition and feedback system to support nurses and adjusted its scheduling system.

Those solutions were specific to that healthcare company and those nurses. Each office, factory, and business is unique with its own quirks and particular strategies for providing service, handling operations, and customer interactions and managing growth. Each organization also faces its own limitations in the data available to them which illuminates the need to ensure that a solid data collection plan is in place so an analyst can pull current and accurate data that can be linked between systems. There are no one-size-fits-all solutions to solving service, personnel or management problems but data can help identify the right solution.

2.2   *Targeting Efforts to Drive Tactical Solutions*

Gathering data for its own sake is pointless. A valuable study is one that drills deep into a specific problem area and provides practical and tactical solutions. In the case of involving nurse attrition Spring's analytics team learned that nurses wanted meaningful feedback about their performance and recognition. The team then took this information a step further to learn exactly what "meaningful" feedback looked like to these nurses. Was it a supervisor's "thank you" at the end of a shift? A day off? A pin for their uniform? An award? The team learned that nurses wanted both feedback on their performance and recognition when they went above and beyond. The only way to impact the bottom line is to translate detailed analysis and survey findings into actions an organization can take.

In another case, the leadership team for a chain of convenience stores asked Spring to analyze why the employee recognition program it had created wasn't motivating its workers. Company leaders hoped the program would boost morale but instead the data showed that leaders were relying too much on the recognition program and not managing employee performance. Over a two-year period, Spring collected data from more than 15,000 employees across 500 stores. The survey measured employee engagement, job satisfaction, leadership, and operational effectiveness. This data was compiled into a centralized data set with the recognition award system and customer and operational data. Using a combination of structural equation modeling, regression, and correlation analysis Spring uncovered a surprising truth: The organization's best managers rarely made use of the employee recognition program—but their workers didn't care. These employees were happy and engaged and felt appreciated and respected. The stores' lowest-performing managers, however, handed out many awards to their workers, yet morale was low and the management was ineffective. What was going on?

In-depth interviews revealed a strong link between disengaged employees and ineffective managers. When an employee called in sick or failed to show up for a shift the manager would ask another worker to cover the hours and then "recognize" the hardworking employee with a management sanctioned trinket. The unreliable worker, however, was never confronted, disciplined or fired. Bad managers exhibited poor leadership skills and an over-reliance on meaningless employee recognition "awards."

The company's executive team was initially surprised by these findings. But they couldn't argue with the findings of an analysis crafted specifically to understand their particular problem and suggest solutions. Executives took the corrective action: They changed how they trained managers and reduced their investment in the employee recognition program.

2.3 Analytics is a Journey, Not a Destination

Human capital analytics enables leaders and managers to transition from decisions made on the fly or on a hunch to analysis based recommendations grounded in data and evidence. This approach, however, requires an ongoing commitment, and that takes time and money.

All organizations are dynamic. The workforce's demographics change. The economy changes. The industry changes. Management changes. Technology changes. Correct one problem and something else will crop up in a month or a year. But when leaders have a clear understanding of how these variables affect the intersection of employee engagement, employer costs, and profitability they can better use a predictive approach to manage operations to keep customers happy and staff engaged.

That was the case for big-box retail chain that experiences seasonal sales fluctuations. The retailer asked Spring's team of analysts to probe the relationship between leadership engagement, employee engagement and turnover, customer loyalty and satisfaction, sales volume, average ticket, and profitability. Spring collected and studied quarterly data from more than 200,000 employees across 1,800 stores—data from attitudinal surveys, HRIS, operational metrics, and financial data [5]. Spring used structural equation modeling to build several models answering a variety of business, HR, and service-related questions. Specifically, when the analysts calculated the level of staffing the stores needed to ensure the highest quality service the retailer realized it needed to change its staffing ratio and supervisor support structure and leadership training. Leadership reorganized its staffing model to provide better customer service which ultimately improved the chain's profitability.

3 Summary

The pressure on HR executives to adopt and optimize human capital analytics has intensified over the last several years. Human resource leaders are beginning to use analytics a s aroute to greater strategic influence. Our research shows that the longer an organization collects, analyzes, and deploys data the more precise and meaningful that information becomes. Over time, the process becomes a roadmap to making better decisions, more effective management, more engaged workers, better overall service, and for businesses greater profitability. Making a long-term commitment to a predictive model for human capital analytics pays rich dividends for everyone.

References

1. E. Sinar, R. Wellins, A. Canwell, R. Ray, S. Neal, A. Abel, "Global Leadership Forecast" Development Dimensions International Inc., The Conference Board, EYGM Limited (2018)
2. CEB Talent Analytics Quarterly, Q1 (2018)
3. https://www2.deloitte.com/insights/us/en/focus/human-capital-trends/2017/people-analytics-in-hr.html
4. Spring International & Philadelphia Society of People & Strategy, 'Human Capital Analytics Landscape Study' (2018)
5. C. Coco, F. Jamison, H. Black, Connecting people investments and business outcomes at Lowe's: using value linkage analytics to link employee engagement to business performance, people & strategy. **34**(2) (2011)

Multiple-Disease Risk Predictive Modeling Based on Directed Disease Networks

Tingyan Wang, Robin G. Qiu and Ming Yu

Abstract This paper studies multiple-disease risk predictive models to assess a dis-charged patient's future disease risks. We propose a novel framework that combines directed disease networks and recommendation system techniques to substantially enhance the performance of multiple-disease risk predictive modeling. Firstly, a directed disease network considering patients' temporal information is developed. Then based on this directed disease network, we investigate different disease risk score computing approaches. We validate the proposed approaches using a hospital's dataset. Promisingly, the predictive results can be well referenced by healthcare professionals who provide healthcare guidance for patients ready for discharge.

Keywords Directed disease network · Predictive modeling · Multiple-disease risk assessment

1 Introduction

Identification of disease risks and intervention at the earliest stage can lead to better medical results and lower medical cost [1]. Identifying disease risks for hospitalized patients in time helps not only improve their received healthcare services [13] but also reduce their readmission rates [2]. However, currently millions of individuals suffer from late-diagnosed chronic diseases [7], causing heavy burdens to the society

T. Wang
Nuffield Department of Medicine, University of Oxford, South Parks Road, Oxford OX1 3SY, UK
e-mail: tingyan.wang@ndm.ox.ac.uk

R. G. Qiu (✉)
Big Data Lab, Division of Engineering and Information Science, The Pennsylvania State University, Malvern, PA 19355, USA
e-mail: robinqiu@psu.edu

T. Wang · M. Yu
Department of Industrial Engineering, Health Care Services Research Centre, Tsinghua University, Beijing 100084, China
e-mail: mingyu@tsinghua.edu.cn

© Springer Nature Switzerland AG 2020   229
H. Yang et al. (eds.), *Smart Service Systems, Operations Management, and Analytics*, Springer Proceedings in Business and Economics,
https://doi.org/10.1007/978-3-030-30967-1_21

[12]. Recently, the assessment of health risks has drawn much attention in academia and practice, yet the accuracy of health risk assessment remains one of the main challenges [1].

As well-known, there is a tremendous amount of health-related information in electronic health records (EHRs). For example, EHRs contain medical information with respect to disease correlations and progression [10]. By relying on EHRs researchers can exploit comorbidity relationships among diseases using multiple-disease risk predictive modeling [5]. However, they had focused on disease prediction modeling through investigating concurrence of diseases with the same individuals while excluding their temporal relations among occurred diseases. Recently, researchers started to consider disease temporal information to explore directed disease networks [10] or directly incorporate disease temporal relations into disease risk predictive modeling [6, 12]. However, using directed disease networks while fully leveraging disease temporal relations in EHRs for multiple-disease prediction modeling has not been well studied yet.

In this study, we formulate the multiple-disease prediction as a medical recommendation system problem and propose a novel framework that combines directed disease networks and recommendation system techniques for multiple-disease risk predictive modeling. Firstly, diseases are grouped and ranked according to the metrics of disease burdens designed by the World Health Organization (WHO). Secondly, we construct a directed network to express temporal flows among these disease groups. Based on the constructed network, we then explore different risk score computing methods and generate recommendations of disease risks for individuals. Correspondingly, performance evaluation metrics for recommendation systems are adopted to evaluate the proposed methods. Finally, we apply the proposed framework to an archived dataset provided by a large hospital in Beijing to validate the applicability of the framework.

The remaining paper is organized as follows: Section 2 briefly reviews the literature. Section 3 details the proposed multiple-disease risk modeling. Analytical results based on a hospital dataset is presented. Lastly, Sect. 4 discusses the conclusion of this study and highlights the future studies.

2 A Brief Study Background

In general, there are two different ways to study disease prediction modeling. One is a single disease prediction modeling, which only involves one type of disease risk assessment. The other is multiple-disease prediction modeling, which aims to identify all new diseases an individual might develop. The former is disease-centric. That is, a model predicts the likelihood of an individual getting a disease, regardless of whether other diseases this individual might suffer. The latter is patient-centric. In other words, a model can systematically assess multiple-disease risks for an individual simultaneously. In this study, we concentrate on the latter as it is more realistic than the former.

Table 1 A sample of medical records of a patient

Patient ID	nth Visit	Diagnosis codes
P23	1	I20.001, I25.101, I09.901, I05.201, I48, Z95.001, I10, N18.901
P23	2	I20.001, M94.082, K26.901, I05.201, I48, Z95.001, I10, N18.901, D35.101, I69.301
P23	3	I20.001, I25.101, I05.101, I48, Z95.001, I10, N18.901, D35.101, K26.901, I69.301

Each hospital visit by a patient produces a set of medical diagnoses. Patients' information in our studied dataset includes a patient's identification, the sequence number of a hospital visit, and a set of medical diagnoses. For example, as shown in Table 1, each row represents a hospital visit of a patient. The first column indicates the patient's identification and the second column indicates the sequence number of a hospital visit, and the third column shows the patient's diagnoses on that visit. Each medical diagnosis in this study is standardized and encoded as a disease code according to the 10th Revision of International Statistical Classification of Diseases and Related Health Problems (ICD-10) [15]. The question becomes how to assess disease risks that a patient might have on the next hospital visit.

Single disease prediction modeling has been well studied, such as cardiovascular risk prediction [14] and diabetes [16]. Future disease risk assessments for an individual using recommendation algorithms have also been explored [2]. With the support of recommendation algorithms researchers have been trying to directly incorporate temporal information among diseases for multiple-disease prediction modeling [4, 12]. In fact, many scholars started to try to bring big data techniques and deep learning algorithms to the field of multi-disease risk prediction modeling [11]. To explore an alternative approach, we explore disease risk predictive modeling using directed disease networks while fully leveraging disease chronological orders in EHRs.

3 Multiple-Disease Risk Predictive Modeling

Figure 1 shows our proposed modeling framework. According to ICD-10, there are over thousands of diseases with four digits in total. In this study, diseases are systematically classified into categories at the levels for easy interpretations and management in hospitals. Therefore, clustering diseases into groups that are easy to understand and interpret is essential, i.e., mapping patients' diagnoses to the ranked 125 disease groups of Disability Adjusted Life Year (DALY) provided by WHO [3].

An example of patients' diagnostic data transformation is briefly illustrated in Fig. 2, where the column named *id* denotes the identification of a patient; the column named nth indicates the nth hospital visit of the patient. For the column D_i, $i = 1, 2, \ldots, 125$, the value is 1 if a patient suffers from a disease D_i, and 0 otherwise.

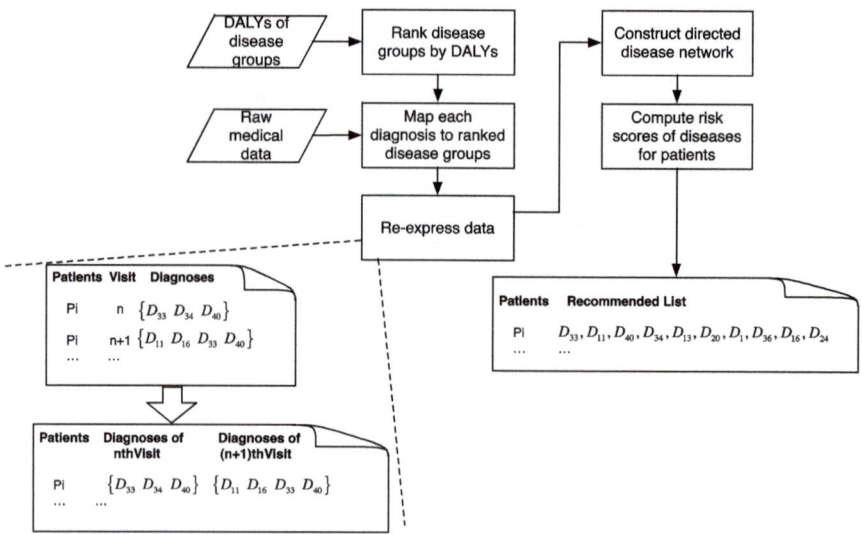

Fig. 1 The process overview of multiple-disease predictive modeling

$$
U = \begin{array}{c|cccccc}
id & n^{th} & D_1 & D_2 & D_3 & \cdots & D_{125} \\
\hline
1 & 1 & 0 & 1 & 1 & \cdots & 0 \\
1 & 2 & 1 & 1 & 0 & \cdots & 1 \\
1 & 3 & 1 & 1 & 0 & \cdots & 0 \\
2 & 1 & 0 & 1 & 0 & \cdots & 1 \\
2 & 2 & 1 & 0 & 1 & \cdots & 0 \\
\vdots & \vdots & \vdots & \vdots & \vdots & \ddots & \vdots
\end{array}
$$

$$
V = \begin{array}{c|cccccc|cccccc}
id & n^{th} & D_1 & D_2 & D_3 & \cdots & D_{125} & id & n^{th} & D_1 & D_2 & D_3 & \cdots & D_{125} \\
\hline
1 & 1 & 0 & 1 & 1 & \cdots & 0 & 1 & 2 & 1 & 1 & 0 & \cdots & 1 \\
1 & 2 & 1 & 1 & 0 & \cdots & 1 & 1 & 3 & 1 & 1 & 0 & \cdots & 0 \\
2 & 1 & 0 & 1 & 0 & \cdots & 1 & 2 & 2 & 1 & 0 & 1 & \cdots & 0 \\
\vdots & \vdots & \vdots & \vdots & \vdots & \ddots & \vdots & \vdots & \vdots & \vdots & \vdots & \vdots & \ddots & \vdots
\end{array}
$$

Fig. 2 Transforming data for developing a directed disease network

To construct a directed disease network, we must transform the data in a way of showing the temporal relationship between a patient's two consecutive hospital visits. As shown in Fig. 2, matrix V can be derived from matrix U after the transformation is applied.

3.1 Proposed Approach to Predictive Modeling

A directed disease network can be then easily developed. A vertex x in the network denotes a disease. A directed arc (x, y) indicates disease x pointing to disease y, which is constructed from patients' two consecutive visits. The flow of the directed arc (x, y), called the outflow of the vertex x to the vertex y, shows that the number of patients with disease x for a hospital visit and with disease y diagnosed in the corresponding subsequent visit. The total outflows of vertex x are equal to the flows coming out from this vertex to other vertices.

To make it easy to understand, we use a simplified network to explain the developed disease network under study. As displayed in Fig. 3, there are three vertices, which represent stroke, liver cancer, and diabetes mellitus. The numbers beside directed arcs indicate flows among various vertices. For example, the outflows from stroke to liver cancer are 11 which means that there are 11 patients who had a stroke in their last visits and liver cancer in their current visits in the studied dataset. Note that the total outflow of stroke i s 2201, which includes flows to other vertices that are not shown in this simplified network. It is worth mentioning that in this study the directed arcs in the network are not causal links, i.e., the flows among disease nodes simply stand for the correlations from the temporal perspective rather than their causal relationships. Obviously, it surely gets extremely complicated and difficult to read if we draw all the disease groups to build a complete network.

Herein, we explore two approaches to compute disease risk scores. An ordered list of diseases that are sorted by the risk scores is then assessed for a patient.

The first approach directly leverages the temporal relations between single nodes in the network which is denoted as Approach 1. Based on the diseases in a patient's medical history we can then compute the likelihoods of the next developing diseases of the patient. Considering patient i with M diseases in the medical history denoted as $H = \{h_1, h_2, \ldots, h_M\}$, the basic idea of Approach 1 is that the risk score of the target patient i developing a disease k is computed according to the relations between disease k and each disease in the patient's medical history, i.e.,

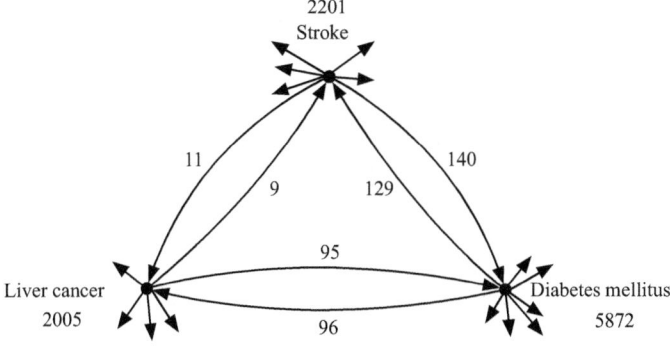

Fig. 3 A simplified view of three diseases with outflows

$$Score_k^{(i)} = g\left(r_{h_m \to k}\right) \tag{1}$$

where $h_m \in H$, $r_{h_m \to k}$ indicates a defined temporal relation between disease h_m and disease k, and $g(\cdot)$ represents a function that combines disease temporal relations.

In Approach 1, we use the flows between each disease and other diseases in a patient's medical history to define the temporal relations between network nodes. The total outflow of the vertex $h_1, h_2, \ldots, h_M$ is $O_{h_1}, O_{h_2}, \ldots, O_{h_M}$, respectively, and their outflows to the vertex k is $O_{h_1 \to k}, O_{h_2 \to k}, \ldots, O_{h_M \to k}$. We compute the risk score of vertex k through using the outflow of each single node in the set H directing to vertex k. Formally, Eq. (1) can be rewritten as the average risk score of patient i with disease k in the patient's next hospital visit, i.e.:

$$Score_k^{(i)} = \frac{1}{M} \sum_{m=1}^{M} \frac{O_{h_m \to k}}{O_{h_m}} \tag{2}$$

where $h_m \in H$, and $O_{h_m \to k}$ indicates the outflow of the disease h_m to the disease k.

Different from incorporating the temporal relations between network nodes, we consider the temporal relations between each subset (i.e., a group of similar patients) of the target patient's medical history and disease k, and then derive the risk score of developing disease k by aggregating risk scores from all the subsets. We denoted this approach as Approach 2.

Once again, we consider target patient i with M diseases in the medical history, denoted as $H = \{h_1, h_2, \ldots, h_M\}$. In Approach 2, we first figure out all the subsets of the target patient's medical history and then calculate the score for each subset. Finally, we aggregate the scores of subsets with different weights and obtain the risk score of the target patient developing the disease k, i.e.,

$$Score_k^{(i)} = f\left(w_{H_{sub}} \cdot SubScore_{H_{sub} \to k}^{(i)}\right) \tag{3}$$

where $H_{sub} \subseteq H$, $H_{sub} \to k$ indicates a defined temporal relation between the subset H_{sub} and disease k, $w_{H_{sub}}$ are weights derived according to the sizes of subsets of the target patient's medical history, $SubScore_{H_{sub} \to k}^{(i)}$ indicates the score derived from each subset and $f(\cdot)$ represents a function that combines scores from different subsets.

Note that weights for subsets are introduced in Approach 2. The bigger size of a subset indicates that more diseases that other patients "share" with the target patient in their medical histories. Accordingly, we define a similarity metric based on patients' medical histories, called "medical history similarity". The "medical history similarity" between two patients is thus quantified as the size of the set of the same diseases in these two patients' medical histories. Without question, the more diseases "shared" by two patients in the medical histories, the more similar their medical histories are. In other words, the larger the medical history similarity between the target patient and other patient is, the greater the likelihood that the target patient will develop an identical disease with other patient. Therefore, we assign different weights to subsets with various sizes in Approach 2 based on the calculated medical history similarity among patients.

To assign different weights to subsets with various sizes, it is necessary to group medical history subsets with the same size which are clustered into the same category based on their sizes. Given patient i with the medical history $H = \{h_1, h_2, \ldots, h_M\}$, we get the subsets of the medical history, such as $\{h_1\}, \{h_2\}, \ldots, \{h_1, h_2\}, \{h_1, h_3\}, \ldots, \{h_1, h_2, h_3\}, \{h_1, h_2, h_4\}, \ldots, \{h_1, h_2, \ldots, h_M\}$. Then these medical history subsets are classified into multiple categories $J_1, J_2, \ldots, J_M$ according to their sizes. Mathematically, we have

$$J_m = \{A \subseteq H \,|\, Card(A) = m\} \tag{4}$$

where $m = 1, 2, \ldots, M$, $Card(A)$ is the number of elements in the set A.

For each subset $sub_j \in J_m$, $j = 1, 2, \ldots, sub_j \in Card(J_m)$, we calculate the total outflow and their outflow directing to vertex k. Thus, the average score of each category is

$$SubScore_{J_m \to k}^{(i)} = \frac{1}{Card(J_m)} \sum_{j=1}^{Card(J_m)} \psi(sub_j \to k) \tag{5}$$

$$\psi(sub_j \to k) = \begin{cases} \dfrac{O_{sub_j \to k}}{O_{sub_j}}, & O_{sub_j} \neq 0 \\ 0, & otherwise \end{cases} \tag{6}$$

where $sub_j \in J_m$, O_{sub_j} indicates the total outflow of the subset sub_j, $O_{sub_j \to k}$ denotes the outflow from subset sub_j to disease k, $Card(J_m)$ is the numbers of subsets in category J_m and $m = 1, 2, \ldots, M$.

Note that we have classified medical history subsets with an identical number of elements into the same category. In this study, the weights are calculated based on an exponential function of the sizes of subsets. Formally, given m denotes the size of a medical history subset, then the weight of the category is e^m. Correspondingly, the normalized weights of each category are

$$w_{J_m} = \frac{e^m}{\sum_{m=1}^{M} e^m} \tag{7}$$

where $m = 1, 2, \ldots, M$. The risk score of patient i with disease k in the next hospital visit becomes:

$$Score_k^{(i)} = \sum_{m=1}^{M} \left(w_{J_m} \cdot SubScore_{J_m \to k}^{(i)} \right) \tag{8}$$

Based on Eqs. (5), (6), and (7), Eq. (8) can be specifically expressed as:

$$Score_k^{(i)} = \frac{1}{\sum_{m=1}^{M} e^m} \cdot \sum_{m=1}^{M} \left[\frac{e^m}{Card(J_m)} \cdot \sum_{j=1}^{Card(J_m)} \psi\left(sub_j \to k\right) \right] \qquad (9\text{-}1)$$

$$\psi\left(sub_j \to k\right) = \begin{cases} \dfrac{O_{sub_j \to k}}{O_{sub_j}}, & O_{sub_j} \neq 0 \\ 0, & otherwise \end{cases} \qquad (9\text{-}2)$$

Through the above discussed risk score computing schemes, we have a list of descending disease scores for patient i. Let $Score_{(1)}^{(i)}, Score_{(2)}^{(i)}, \ldots, Score_{(K)}^{(i)}$ denote the sequence of the scores $Score_k^{(i)}, k = 1, 2, \ldots, K$, then we have the corresponding sequence of diseases, denoted as $\varphi^{(i)} = \varphi_1^{(i)}, \varphi_2^{(i)}, \ldots, \varphi_K^{(i)}$. As mentioned earlier, only top Q diseases in the sequence $\varphi^{(i)} = \varphi_1^{(i)}, \varphi_2^{(i)}, \ldots, \varphi_K^{(i)}$ of patient i will be finally recommended to healthcare professionals.

To evaluate the list-based prediction results, we propose an accuracy evaluation approach based on half-life accuracy evaluation [9]. Given that there is a set of G diseases denoted as $\theta^{(i)} = \left\{\theta_1^{(i)}, \theta_2^{(i)}, \ldots, \theta_G^{(i)}\right\}$ that patient i will have when visiting the hospital next time, we have the whole predicted list $\varphi^{(i)} = \varphi_1^{(i)}, \varphi_2^{(i)}, \ldots, \varphi_K^{(i)}$. As only top Q diseases on the list will be recommended to medical experts, the assessment value of the recommended list for patient i can be calculated as follows:

$$List_k^{(i)} = \begin{cases} 1 \text{ if } \varphi_k^{(i)} \in \theta^{(i)} \\ 0 \text{ otherwise} \end{cases}, \qquad (9)$$

where $k = 1, 2, \ldots, Q$. In other words, $List_k^{(i)}$ is 1 if a disease will develop on the next hospital visit and 0 otherwise.

Note that the half-life accuracy score for a recommended list is as following:

$$ListAccuracy^{(i)} = \sum_{k=1}^{Q} 2^{-k/c} \cdot List_k^{(i)} \qquad (10)$$

where k represents the position on the recommended list, $2^{-k/c}$ represents the weight of the kth positions of the recommended list, and c is a constant that can be adjusted by the user. We further introduce normalization for the half-life accuracy score. Given that the significance of top Q positions on the list changes based on exponential function the adapted normalized weight function becomes:

$$ListWeight_k = \frac{2^{-k/c}}{\sum_{g=1}^{G} 2^{-g/c}} \qquad (11)$$

where $k = 1, 2, \ldots, Q$, G is the number of diseases that patient i has on the subsequent hospital visit. Thus, the prediction accuracy of patient i will be:

$$ListAccuracy^{(i)} = \sum_{k=1}^{Q} List_k^{(i)} \cdot ListWeight_k \tag{12}$$

where $List_k^{(i)}$ represents the value of the kth position on the recommended list and $ListWeight_k$ represents the weight of the kth position on the recommended list. Hence, the mean accuracy score of all the test samples is:

$$MeanAccuracy = \frac{1}{N} \sum_{i=1}^{N} ListAccuracy^{(i)} \tag{13}$$

where N is the size of the test samples.

3.2 Validation of the Proposed Approach

To validate the applicability of the proposed framework, we use a dataset in which data were collected from a hospital in Beijing, China. The dataset includes medical records of 50,499 patients, covering their 66,901 visits in total. Each record includes patient identification, the sequence number for a hospital visit, gender, and diagnosis codes. The diagnosis codes contain one principal diagnosis and several secondary diagnoses. Because our goal is to predict what diseases a patient might have on the next hospital visit, a patient that only had onetime hospital visit would be removed from the dataset. As a result, 7989 patients with 24,391 visits in total remain in the final dataset that are used in this study.

To compare the overall accuracy of the different proposed methods, 10-fold cross-validation is implemented using the transformed dataset. First, the two proposed approaches are, respectively, performed on the training dataset and various recommended lists for the test dataset are achieved. In this study, we set $Q = 10$, that is, only top 10 diseases of a scored list will be recommended to medical experts. Therefore, all the metrics were derived using recommendation lists with a length of 10.

As shown in Table 2, the mean accuracy score of Approach 2 is higher than the one of Approach 1. One score is defined as a recommendation list that is 100% correct. More impressively, the percentage of one score for Approach 2 is 71.39%, which gets improved by almost 7% compared with Approach 1. Overall, the aggregated

Table 2 Performances of the proposed approaches with 10-fold cross-validation

Methods	Mean accuracy score (%)	Ratio of 1 score (%)
Approach 1: disease temporal link	88.31	64.42
Approach 2: aggregated disease temporal links	89.87	71.39

disease temporal links approach is superior to the disease temporal link approach. The insight derived from this conclusion is that if a prediction model can leverage more information, it will improve the performance. Therefore, Approach 2 should be adopted in a real-life implementation.

We compare the proposed Approach 2 with some baseline methods that are frequently used in a recommendation problem. Three baseline methods include ranking algorithms, recommendation algorithms, and classifiers. As for ranking algorithms and recommendation algorithms, page rank [8] and collaborative filtering [4] are used in the comparisons below. As discussed earlier, the problem in this study is a multi-label classification problem [17]. Therefore, two classical multi-label classifiers are used below, i.e., K-nearest neighbor and decision tree. Since the output of a classifier is the probabilities of developing diseases, we first sort these probabilities in descending order and then generate an ordered recommendation list. Note that all the results for various methods are obtained from 10-fold cross-validation experiments. The recommendation performances of these different methods are shown in Fig. 4. Promisingly, the proposed Approach 2 is superior to all the baseline methods with respect to all the measured performances.

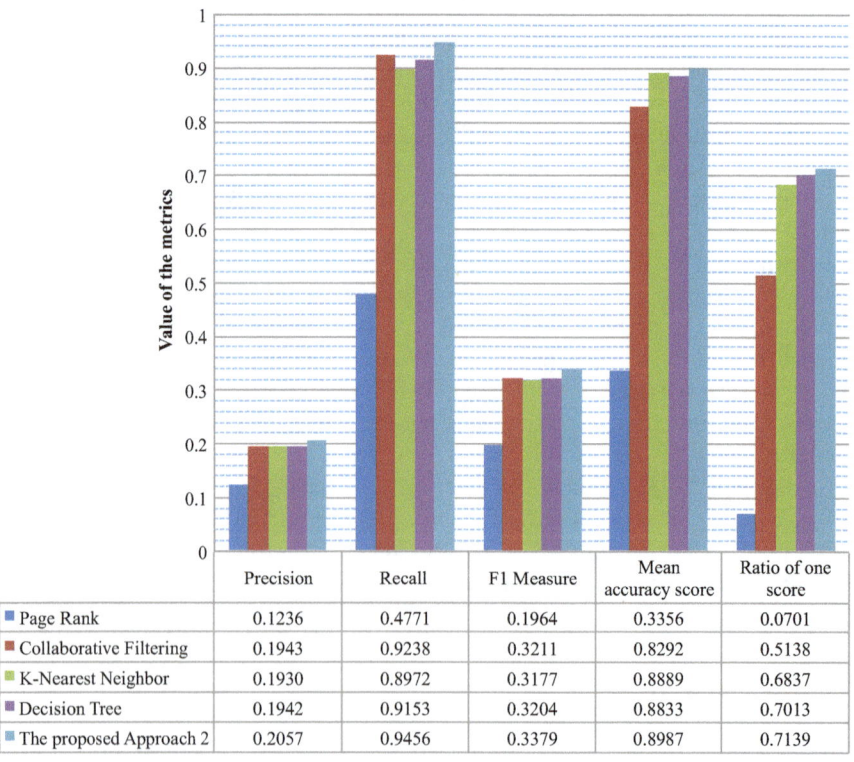

	Precision	Recall	F1 Measure	Mean accuracy score	Ratio of one score
■ Page Rank	0.1236	0.4771	0.1964	0.3356	0.0701
■ Collaborative Filtering	0.1943	0.9238	0.3211	0.8292	0.5138
■ K-Nearest Neighbor	0.1930	0.8972	0.3177	0.8889	0.6837
■ Decision Tree	0.1942	0.9153	0.3204	0.8833	0.7013
■ The proposed Approach 2	0.2057	0.9456	0.3379	0.8987	0.7139

Fig. 4 Performances of various methods with 10-fold cross-validation

4 Brief Final Remarks

In this study, we proposed a novel framework that integrates directed disease networks with recommendation system techniques for multiple-disease risk predictive modeling. By leveraging patients' medical diagnoses between successive hospital visits, the proposed directed disease networks successfully incorporated diseases' temporal relations into the proposed disease risk prediction models. The proposed framework has been well validated using a hospital dataset. In the future, we can study how other information of individual patients could be fully incorporated into the proposed models for better performances.

Acknowledgements A significant part of this work from Tingyan Wang and Robin Qiu was done with the support from the Big Data Lab at Penn State. This project was partially supported by IBM Faculty Awards (RDP-Qiu2016 and RDP-Qiu2017).

References

1. M. Bayati, S. Bhaskar, A. Montanari, Statistical analysis of a low cost method for multiple disease prediction. Stat. Methods Med. Res. **27**(8), 2312–2328 (2018)
2. N.V. Chawla, D.A. Davis, Bringing big data to personalized healthcare: a patient-centered framework. J. Gen. Intern. Med. **28**(3), 660–665 (2013)
3. A. Chen, K.H. Jacobsen, A.A. Deshmukh, S.B. Cantor, The evolution of the disability-adjusted life year (DALY). Socio-Econ. Plann. Sci. **49**, 10–15 (2015)
4. D.A. Davis, N.V. Chawla, N. Blumm, N. Christakis, A.L. Barabasi, Predicting individual disease risk based on medical history. in *Proceedings of the 17th ACM conference on Information and knowledge management*, pp. 769–778 (2008)
5. F. Folino, C. Pizzuti, Link prediction approaches for disease networks. in *International Conference on Information Technology in Bio-and Medical Informatics*, (Springer, Berlin, Heidelberg, 2012), pp. 99–108
6. F. Folino, C. Pizzuti, A recommendation engine for disease prediction. IseB **13**(4), 609–628 (2015)
7. A.J. Frandsen, *Machine Learning for Disease Prediction, Master thesis* (Brigham Young University, 2016)
8. T.H. Haveliwala, Topic-sensitive pagerank: a context-sensitive ranking algorithm for web search. IEEE Trans. Knowl. Data Eng. **15**(4), 784–796 (2003)
9. J.L. Herlocker, J.A. Konstan, L.G. Terveen et al., Evaluating collaborative filtering recommender systems. ACM Trans. Info. Syst. **22**(1), 5–53 (2004)
10. V. Kannan, F. Swartz, N.A. Kiani, G. Silberberg, G. Tsipras, D. Gomez-Cabrero, K. Alexanderson, J. Tegnèr, Conditional disease development extracted from longitudinal health care cohort data using layered network construction. Sci. Rep. **6**, 26170 (2016)
11. R. Miotto, L. Li, B.A. Kidd, J.T. Dudley, Deep patient: an unsupervised representation to predict the future of patients from the electronic health records. Sci. Rep. **6**, 26094 (2016)
12. M. Nasiri, B. Minaei, A. Kiani, Dynamic recommendation: Disease prediction and prevention using recommender system. Int. J. Basic Sci. Med. **1**(1), 13–17 (2016)

13. J.A. Paul, L. MacDonald, G. Hariharan, Modeling risk factors and disease conditions to study associated lifetime medical costs. Serv. Sci. **6**(1), 47–62 (2014)
14. S. Selvarajah, G. Kaur, J. Haniff, K.C. Cheong, T.G. Hiong, Y. van der Graaf, M.L. Bots, Comparison of the Framingham risk score, SCORE and WHO/ISH cardiovascular risk prediction models in an Asian population. Int. J. Cardiol. **176**(1), 211–218 (2014)
15. World Health Organization (2019) International statistical classification of diseases and related health problems, 10th Revision. Retrieved 8 Jan 2019. http://apps.who.int/classifications/icd10/browse/2016/en
16. C. Willi, P. Bodenmann, W.A. Ghali, P.D. Faris, J. Cornuz, Active smoking and the risk of type 2 diabetes: a systematic review and meta-analysis. JAMA **298**(22), 2654–2664 (2007)
17. M.L. Zhang, Z.H. Zhou, A review on multi-label learning algorithms. IEEE Trans. Knowl. Data Eng. **26**(8), 1819–1837 (2014)

Service Performance Tests on the Mobile Edge Computing Platform: Challenges and Opportunities

Ishtiaque Hussain, Qiang Duan and Tiffany Zhong

Abstract Mobile Edge Computing (MEC) is a fast growing research area that may soon offer alternate service platforms to the clients over today's Cloud Computing vendors (e.g., Amazon AWS, Google Cloud, or Microsoft Azure, etc.). And when the MEC services are ready and available, just like when Cloud Computing services were presented to the "On-Premises" server clients, despite available research data on compatibility, business owners and engineers will still have to practically implement and host their applications to the MEC servers, measure performance and be convinced first that MEC services can withstand the real users load without any performance degradation or disruption before finally making the switch. In this paper, we first discuss how today's "On-Premises" server–clients conduct performance analysis before switching to the Cloud computing vendors, discuss the differences between the Cloud computing and MEC paradigms and then we try to envision the challenges and opportunities that may unfold for the performance tests on the future MEC service platforms.

Keywords Mobile Edge Computing (MEC) · Service performance · Testing

I. Hussain (✉)
Computer Science Department, Pennsylvania State University,
Abington, PA 19001, USA
e-mail: ihussain@psu.edu

Q. Duan
Information Sciences & Technology Department, Pennsylvania State University,
Abington, PA 19001, USA
e-mail: qxd2@psu.edu

T. Zhong
Germantown Academy, Fort Washington, PA 19034, USA
e-mail: tzhon20@germantownacademy.org

© Springer Nature Switzerland AG 2020
H. Yang et al. (eds.), *Smart Service Systems, Operations Management,
and Analytics*, Springer Proceedings in Business and Economics,
https://doi.org/10.1007/978-3-030-30967-1_22

1 Introduction

Cloud computing is a large-scale distributed computing paradigm driven by economies of scale, in which a pool of abstracted, virtualized, dynamically scalable computing resources are delivered on-demand as services to external customers over networks. The rapid advances in Cloud computing technologies in the past decade have enabled a wide spectrum of Cloud services that have been adopted as indispensable ingredients in the information systems of numerous organizations. Various methods have been developed for evaluating Cloud service performance, among which measurement-based performance testing is one of the main categories [7].

With the wide adoption of the Cloud service paradigms, some Cloud applications demand high-throughput data transmissions between user devices to Cloud data centers and require low latency for service response, which not only increases traffic loads in networks but also introduces performance bottleneck in Cloud service delivery. On the other hand, recent developments in networking technologies have enabled deployment of virtualized compute and storage resources at network edges, often in the wireless mobile networks. This triggered a trend of extending Cloud-like capabilities to the edge of networks in close proximity to end users, which is often referred to as Mobile Edge Computing (MEC) [15].

MEC is expected to play a crucial role in provisioning a wide variety of Cloud-like services in some emerging areas such as Internet of Things (IoT) and 5G networks. One of the main objectives of MEC is to enhance the service performance by leveraging the distributed compute and storage capacities in users' proximity at network edges. Therefore, testing performance of the services hosted upon the MEC platform becomes an important topic. However, MEC has some unique features that differentiate it from the traditional Cloud computing model, for example, distributed computational resources embedded inside network infrastructures, a large number of servers with constrained capacities, collaboration between edge servers and cloud data centers, etc. These special features of MEC make performance testing a new challenge that deserves more thorough investigation.

In this paper, we attempt to study the problem of service performance test on the MEC platform. We first review related work and discuss the motivation of this paper. We then review the typical approach to testing Cloud service performance in current practice, then we present a comparison between MEC and the traditional Cloud computing paradigm, based on which we identify a set of new challenges brought in by MEC to service performance testing, and discuss some opportunities for future research.

2 Related Work and Motivation

There have been numerous studies on Cloud computing paradigm evaluating the performance of networks and computing resources [8, 9, 13]. However, MEC has some special features that differentiate it from traditional Cloud computing and bring in new challenges to service performance evaluation. Researchers have developed different types of simulators and conducted simulation-based performance evaluations studies on MEC. For example, CloudSim is commonly used for conventional cloud computing scenarios [5]. Running on top of CloudSim, iFogSim [10] is another simulator that simulates IoT and Fog environments by modeling components like the sensor, actuator, fog devices, and the cloud. As reported by Sonmez et al. [16], one major limitation of iFogSim is that it assigns static values to these components and does not consider any mobility model.

Sonmez et al. proposed a dynamic simulator called EdgeCloudSim [16] that meets the specific demands of Edge Computing research and supports necessary functionality in terms of computation and networking abilities. GlobeTraff [11] is another traffic workload generator tool that promises highly parameterizable generation of realistic, synthetic mixture of network traffic to support performance evaluation of existing Cloud Computing and future Fog/Edge network architectures.

Zhu et al. [20] discussed performance optimization for MEC paradigm by combining existing techniques [17, 18] for web optimization and unique knowledge that is only available at the Fog nodes. Server nodes in a MEC paradigm are in a unique position that could utilize a clients' resource status information (e.g., CPU, memory, and graphics processing units usage, network congestion, display size etc.) and mobility information (e.g., proximity to an access point (AP), roaming statistics) and apply relevant optimization techniques to render web responses faster [20]. However, they do not discuss how to evaluate their optimization technique's performance compared to the core network or central server optimization techniques.

The main motivation of this paper is to envision the challenges and opportunities researchers and practitioners will face once they attempt to evaluate MEC performance in the real world and discuss some opportunities for future research. To the best of our knowledge, no other related work has studied these before. For example, for conducting performance testing in MEC, as we discuss in Sect. 5.1 in details, we need to come up with algorithms, protocols, and approaches to handle— among many other things, scenarios like participating server node's availability, clients mobility, server–clients' resource utilization, network congestion and quality, establishing MEC performance benchmarking criteria, etc. We believe, MEC will be able to offer solutions to some unique problems and create opportunities that will be worth researchers' and practitioners' effort in developing tools and techniques in this emerging field of MEC performance evaluation.

3 Current Practice for Cloud Service Testing

Studies show that more and more companies are adopting Cloud Computing (CC) vendors (e.g., Amzon AWS, Google Cloud, Mircrosoft Azure, etc.) to host their services [2, 12]. They are preferring these CC services over their own 'on-premises', company-owned infrustructure because of mainly the following reasons: CC services allow "pay-as-you-use" model, they are scalable, secure, fault tolerant, and reliable. Moreover, there is significant amount of cost and overhead involved in maintaining a company-owned, "on-premises" infrustructre since companies need to invest on updating, upgrading to the newer and better hardware and network equipment as they become available, maintain workforce to keep them running in a secure and effective way. Since CC vendors provide all these as part of their services, the overall cost, overhead, and man-hours saved in the long run by switching to CC vendors are huge incentives for companies as they can now focus on their business more than the infrustructre on which they used to run their applications. To reap these benefits, not only large enterprises, many small and medium enterprises are also making the move toward the CC vendors.

However, before making the final switch, companies are often curious to know how well these CC service providers would handle their customer load and prefer to conduct a study by first deploying their system (or subsystems) to the CC platform and then performing a load test to validate against their production performance metrics. In the rest of this section, we discuss the algorithm or the current practices on how companies perform these performance or load tests studies.

Any software system, even the small-sized businesses, may have many use case scenarios for its users. Each scenario can consists of simple to complex business logic, invole different modules of the system including database and network. However, not all the scenarios are heavily used at all the time. Often time there are few Use Cases (UCs) that are primarily used and are the main contributing factors to the overall load to the system. When a business is trying to move from its own "on-premesis" hosting servers to a CC vendor, at first they want to ensure that their primary use cases would be available and functioning properly under the regular user load in the Cloud. If the CC platform can handle these primary use cases under substantial user load, then they are convinced, confident, and feel comfortable moving forward with the switch.

Figure 1 shows the overview of the steps involved in performing a load test for some primary use cases in the Cloud. A test engineer, with the help of the business analyst and software developers has to first identify the subsystem of the application that the users or clients use the most (step 1). Once that is established, she has to find out the use cases in that subsystem and know the detailed, end-to-end (E2E) steps involved in the use cases from a user's perspective (step 2). She also has to know what database tables, web pages, etc., these E2E steps are accessing. At the next steps (steps 3–4), with the help of the DevOps, she has to enable different performance monitors (e.g, page hits counter, page response times, memory usage, network bandwidth, database I/Os, etc.) in the production environment and keep

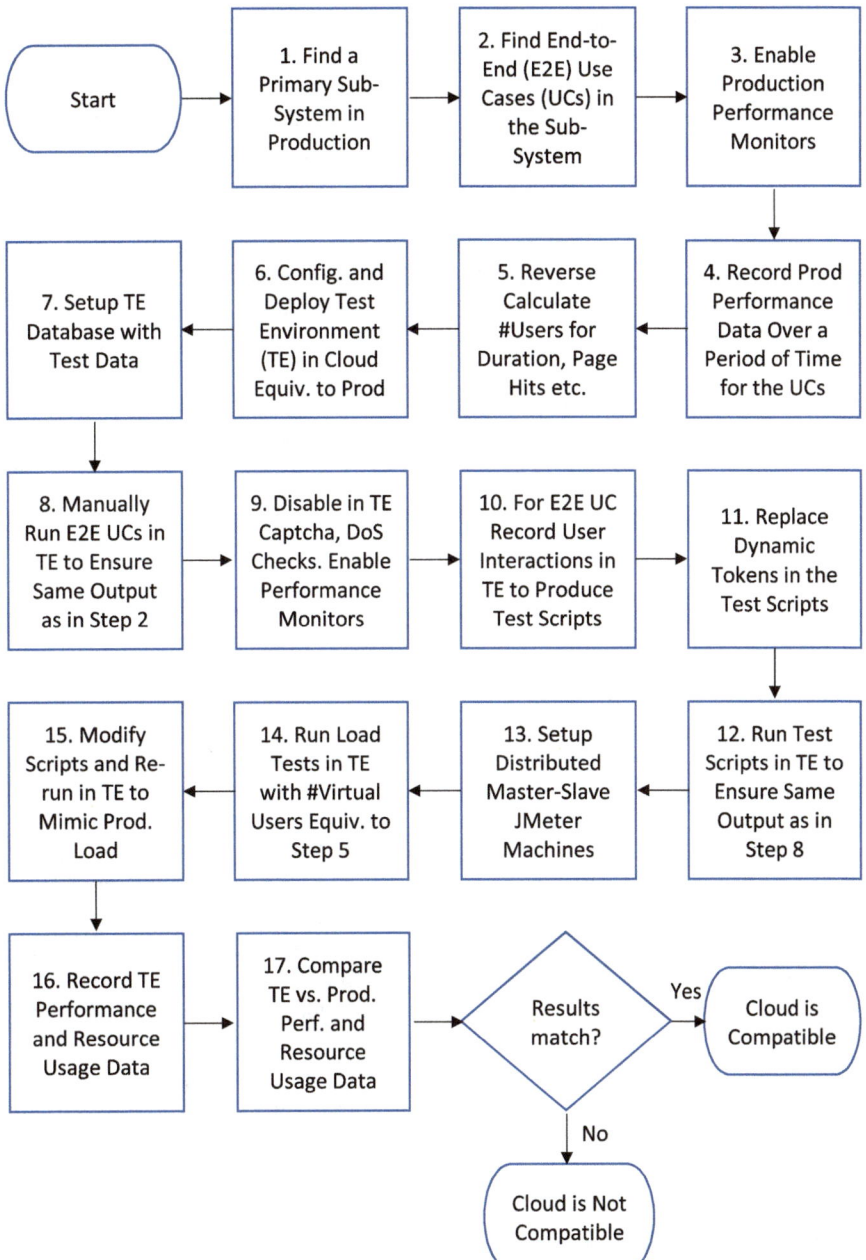

Fig. 1 Overview of the steps involved, from setting up to performing a load test and comparing the results in cloud

recording the data over a period of time. Once the user and system performance data are collected, she has to figure out (step 5), in a reverse engineering fashion, for the given UCs, what are the number of users at any given point of time is hitting the system, what are the page response time for those web pages, how long a user spends in these E2E UCs, etc. Basically, this step helps set the benchmark for the performance of the subsystem.

At this point, the test engineer has gathered all she needs from the production environment and needs to setup the test environment. With the help of the DevOps, she now has to configure the test environment that matches the production hardware and software configurations and when it is ready, deploy the application into the Cloud (step 6). After the successful deployment, she now has to setup the test environment with test data (step 7). Next step is to manually verify the E2E UCs in the test environment to ensure that they produce the same output as in Production in step 2.

Any web application has to consider security vulnerabilities and apply defense mechanisms to protect against known forms of attacks. For example, Denial of Service (DoS) attack is the most common attacks and websites usually deploy CAPTCHA, 2-step verification and other counter measures to protect against these attacks. However, when load testing a web application with virtual users using a tool like JMeter [1], since these tools cannot handle CAPTCHA or other image-based verification mechanisms, they can severely restrict a meaningful performance or load test. Therefore, for testing purpose only, a test engineer with the help of the developers and DevOps has to disable these checks in the test environment (step 9).

As the next step (step 10), a test engineer has to record, using a tool like JMeter, all the interactions between a user and the application during executing an end-to-end use case to produce a test script. This is the script that will be used in a load test, running it multiple times, simultaneously mimicking actual users' interaction with the application that will generate users load. To protect against attacks, misuse and to maintain users authenticity, web applications usually utilize dynamic token mechanism in their web page request and responses. A test engineer has to modify the test script (step 11) so that it extracts these dynamic tokens in real time from the web response and use it in its following requests to the application. Otherwise, the website would simply deny any service for a request, not work as intended and block the user.

After all these steps, the test engineer is now ready with her script to run the load tests. At this point, she can run the script with few virtual users to ensure that it produces the same desired output (step 12). However, to create users load matching the production data gathered at step 5, she may need to use multiple machines. Usually, a single conventional desktop machine cannot produce the required user load that would match the production. Therefore, a test engineer has to configure multiple machines (as master–slave distributed system) and use the setup as the test load generator (step 13). Once she runs the load test with this setup (step 14), she may have to update the scripts one more time to match the ramp-up time for the virtual users, duration of the test execution, etc. Finally, (in steps 16–17) she can record the application performance along with resource usage and compare with the data collected from the production in step 4. If they match, she can declare that the

Cloud is compatible to switch from the "on-premises" hosting servers. Otherwise, it is not compatible and may need further investigation, development, and optimization before making the switch.

4 Comparison Between MEC and Cloud Computing

Traditional Cloud infrastructures are based on centralized data centers in which a large cluster of servers are interconnected through a dedicated high-speed network. The data centers are typically located inside backbone network so often have to be accessed through long-distance communications via wide area networks. MEC infrastructures comprise highly distributed Edge servers located at different sites scattered across the entire mobile networks. The Edge servers are at network edges in proximity of end users thus may be accessed without long-distance networking.

Cloud data centers are composed of a large number of servers with high compute, storage, and network capacities; therefore, the pool of resources in a Cloud infrastructure is often assumed to have unlimited service capacity. In addition, the servers in a Cloud data center are typically homogeneous—the same typical of hardware hosting the same operating system; and the data center networks are also often have a certain typical of topology such as fat tree. In contrast, Edge servers are typically realized using the extra resources available on the servers deployed at network edge by network service providers, e.g., gateway servers in wireless mobile networks, therefore have limited compute and storage capacities. Also, these Edge servers are implemented with various types of hardware platforms and software systems. The network bandwidth among Edge servers are determined by the capacities of the wireless mobile networks where the Edge servers are located and are often constrained. Therefore, comparing to high-capacity homogeneous Cloud data centers, MEC infrastructures comprise much more heterogeneous servers and networks with constrained capacities.

The heterogeneous hardware and software in MEC infrastructures make an MEC platform much more dynamic and less reliable compared with traditional Cloud computing. Since an MEC infrastructure essentially shares computing and networking resources with the network infrastructure in which it is embedded, both its server capacities and network bandwidth may be influenced by multiple factors such as the traffic loads in networks thus varying with time. Also, compared with Cloud servers hosted on dedicated hardware and network, Edge servers are much less reliable.

The data center infrastructures of a service provider can be assumed in a single administrative domain, in which the system management function has a global view of the entire domain and can directly control the operation of the entire system. On the other hand, MEC infrastructures often comprise multiple domains cooperating with each other for service provisioning. For example, the access network, backbone network, and cloud data center that are owned and operated by different service providers may need to cooperate for end-to-end service provisioning in an MEC environment. Therefore, inter-domain service orchestration becomes an important issue particularly for MEC.

Networking in Cloud computing provides the required data transmissions among servers within a data center and supports communications between user devices and cloud data centers. Due to the highly distributed computing resources deployed inside networks for MEC, networking becomes a much more crucial component of the MEC infrastructure. Typical MEC use cases involve multiple MEC servers interconnected through network connections, sometimes also connected with cloud data centers, for service provisioning. In addition, mobility of both user devices and servers in MEC calls for more agile network control for supporting dynamic and elastic services.

A high-level comparison between Edge computing and Cloud computing is given in Table 1.

5 Challenges and Opportunities

In this section, we discuss the challenges and opportunities that researchers and practitioners may face when conducting actual performance tests on Mobile Edge Computing platforms.

5.1 Challenges

Recently there have been some studies [5, 10, 11, 16] that utilized simulators in analyzing performance of the Fog or Mobile Edge Computing (MEC) paradigms. But to the best of our knowledge, no studies have shown or discussed what are the challenges practitioners and researches might face when conducting an actual performance test on the MEC platforms. Below are some of the challenges and questions that we envision researchers and practitioners have to deal with or answer before moving forward with the actual performance tests on the MEC platforms:

1. *Participating Nodes as Servers*: Like Cloud Computing, in MEC, the nodes that will serve as web servers, should be able to handle client web requests. However, since any node in an MEC platform will be resource constrained, at the times of peak user requests load, there should be a protocol that will decide when and to which participating nodes to delineate server responsibilities. This could be done either by spinning up new server nodes in real time or by forecasting requests load and performing a load balancing as a preemptive step.

2. *Server Availability*: Web servers in MEC platforms may not be dedicated servers. They may be ad-hoc servers that are allowing its clients to utilize their spare computing, network, and cache memory resources. It is possible that these servers may be switched off and become unavailable for service. To maintain performance, there should be enough available servers to fill-in in such cases.

Table 1 High-level comparison between edge computing and cloud computing

Aspects	Fog/Edge computing	Cloud computing
Decentralized versus centralized	Servers scattered across the network at different geographical locations	Centralized servers, data centers located at a few sites
Proximity	Servers typically at network edges closer to user devices	Servers inside the Internet, accessed through long-distance communication via WAN (backbone network)
Infrastructure capacity	Large numbers of servers with constrained computing and storage capacity, interconnected with limited network bandwidth (e.g., wireless mobile networks)	Cluster of servers with high computing, storage and network bandwidth capacity. Compared to Fog/Edge, they are considered as unlimited capacity
Hetero- versus Homogeneity	Heterogeneous infrastructure resources (e.g., different hardware and software devices) interconnected with heterogeneous network components (e.g., radio access network, WLAN, WAN, etc.)	Homogeneous infrastructure resources (e.g., single or few types of server) interconnected via DCN which is also considered as homogeneous network
Inter- versus Intra-domain	Inter-domain service delivery may span across multiple service providers' domains, including cloud and edge computing	Intra-domain in nature since a cloud data center typically owned by a single cloud service provider
Dynamic versus relatively static	Servers are hosted on network devices even user devices, therefore, are more dynamic in terms of availability (nodes switched on-off), device mobility, capacity sharing etc.	Server states, resources, network connections are typically more static in nature since they are owned by single provider
Roles of networking	Roles of networking is much more crucial since configuration, availability and mobility of servers may change frequently, virtualization, and communication between heterogeneous systems pose challenges	Main roles of networking includes two aspects: data communication between servers and users for cloud services, communication between cloud data centers

3. *Mobility of Clients*: Web servers might be required to be static (i.e., hosted in an MEC access point) to ensure better performance, but clients in a MEC paradigm are bound to be mobile. If clients are roaming and are moving away from an initial access point or web server, there need to be protocols in place that would handle switching service over to a participating nearby web server node and communicating client's current status (e.g., authentication, state of service, etc.) so that the services are seamless and uninterrupted.

4. *Difficulty in Defining the Performance Benchmark*: In MEC platforms, since there are many variable factors involved (e.g., server availability, client mobility, network congestion, proximity, etc.), defining a stable performance benchmark will be challenging. Unlike Cloud Computing, where the web servers are placed in dedicated data centers behind load balancers (to maintain performance), since servers and clients in the MEC platforms are heterogeneous in nature, performance may vary at any point of time. It might be necessary to define multiple benchmarks depending upon these variable factors. For example, a benchmark of less than two seconds for a page response time in a usual best case scenario may have to be set to less than 5 s when the network is congested or when not many participating server nodes are available in the MEC platform.

5. *Deployment of Application Performance Management Tools (APM)*: Each client in an MEC platform must include some sort of APM tools to monitor resource usage and report to the server nodes. Clients may include these reports as part of their subsequent web requests. Server nodes, on the other hand, have to analyze these reports and decide on their response material (e.g., size of the image files, CSS stylesheets or required JavaScripts etc.). As discussed in [20], consider a case where a client initially has enough network bandwidth, CPU, memory, and graphics processing resources to process a multimedia-rich website. After collecting the data from the embedded APM tools, the client sends this information along with the web request to the participating server node. Servers, while processing a previous client's request, may have already cached these requested materials from the core web server and after analyzing the new client request and realizing that the client has available resources, may send all the required materials to the client in their original, uncompressed form. However, after some time, the situation may change and the client (again from its embedded APM tool data) may report that it has now limited resources along its subsequent web requests. Server now needs to adjust and send a lower resolution graphics and restrict itself from sending unimportant scripts to the client. Deployment of APM tools will be particularly required for latency-sensitive applications such as gaming, video streaming, or augmented reality.

6. *Privacy Concerns*: Use of APM tools to monitor clients' resource usage status will be a key factor in the successful MEC paradigm implementation. Moreover, clients' location, mobility, and proximity to an MEC access point, etc., are properties that will need to be shared to determine the quality of service. This will be true for a successful performance testing in MEC as well. However, sharing these data need to be anonymous, i.e., these cannot be used to uniquely identify, contact, or locate a single person. Otherwise, it will be considered as Personally Identifiable

Information (PII) sharing and could be a privacy concern, even a breach to the privacy regulations [3, 19] with a severe penalty.

7. *Heterogeneous Client Workload*: In MEC platforms, the clients may not be simply PC or laptops. They can be of heterogeneous types (e.g., different devices of different hardware, OS, display size, computation, memory, and network capacities, etc.). An actual performance testing has to consider these spectra of clients' perspectives. In Cloud Computing, some performance test vendors, e.g., BlazeMeter [4], now offers different actual and virtualized devices to generate the user load, even from different geographic locations. In conducting performance tests in MEC platforms, to the best of our knowledge, there are no such service providers yet. Researchers and industry practitioners need to come forward and collaborate in building similar frameworks and tools that can utilize these MEC clients and use them as actual workload generators instead of just simulation-based analysis.

5.2 Opportunities

Despite all the listed challenges in Sect. 5.1, we believe MEC platforms offer some unique opportunities. These potential, unique opportunities are worth researchers' and industries' effort in solving or working on the abovementioned challenges. Some researchers have already pointed out the potential applications of MEC platforms over Cloud Computing [6, 14, 20]. We are listing some more opportunities below:

1. *Increased Client Satisfaction and Engagement*: Previously it was believed that page load-time was mainly determined by the servers, however, researches as reported in [17, 20] show that 80–90% of the load-time issues actually occur at the front end, i.e., on the client's web browser. These research describes guidelines on how to optimize web page responses to ensure pages can load into a browser faster. One of the key contributors to the page load-time is the lengthy round-trip of web requests and responses, from the client to the server and vice-versa, through Wide Area Networks (WAN). In MEC, when some participating node works as a proxy server, it can cache these time expensive responses from the code web servers and render them to the subsequent clients' requests. Since both the client and server remain in close proximity, users can experience a better service that can result in greater user satisfaction and more engagement with the website and generate revenue.

2. *The EU General Data Protection Regulation (GDPR)*: The European Union (EU) parliament approved the General Data Protection Regulation (GDPR) on April 14, 2016 [3]. It went effective from May 25, 2018 and has significant impact on the businesses that store, process, and report on the Personal Identifiable Information (PII) of their EU citizen clients. Among other things, this regulation allows EU citizens to request businesses to purge (also known as: right to be forgotten) their PII data, halt processing their data, update their PII data and even request a copy of their entire information in a digital format. Businesses that do not conform

to this regulation may face up to 20 million Euro or 4% of their annual profit, whichever is higher—as a penalty [3, 19]. Some EU business clients also require that companies that offer services have to host the service and store user data in EU locations. In general, EU has stricter law on privacy than most other developed countries. MEC platforms can offer efficient solution to these locations and PII-based regulations. Since participating MEC nodes are in close proximity of the user, storing data and hosting the service in MEC nodes will meet the requirement and will be easier to maintain.

3. *Cost-Effective Solution*: Once MEC starts offering computation, hosting, and storage services to their clients in an ad-hoc basis, existing Cloud Computing platform vendors will have to reevaluate their pricing and may reduce it to match the competition. Moreover, since network bandwidth usage will reduce due to the local access points communication (i.e., instead of a long round-trip with the core data centers through WAN), it will improve the overall energy consumption and lower the cost.

6   Conclusion

The Mobile Edge Computing (MEC) paradigm, which extends the Cloud computing into wireless mobile networks in proximity of the end users, is expected to be widely deployed for supporting various performance-sensitive applications such as IoT and 5G services. Therefore, performance evaluation for the MEC services are critical and testing service performance on the MEC platform becomes an important research topic. In this paper, we first reviewed the current practices for testing Cloud service performance and then compared MEC against traditional Cloud computing. Our analysis shows that the currently available methods for Cloud performance testing may not be directly applicable to MEC due to its unique features. Then, we identified some new challenges brought in by the MEC to service performance testing and listed some opportunities for future research with the hope to evoke the research community's interest in this emerging field.

References

1. The Apache JMeter: designed to load test functional behavior and meseaure performance. https://jmeter.apache.org/
2. Cloud computing trends: 2018 state of the cloud survey. https://www.rightscale.com/blog/cloud-industry-insights/cloud-computing-trends-2018-state-cloud-survey. Accessed 13 Feb 2018
3. The EU general data protection regulation (gdpr). https://eugdpr.org/
4. BlazeMeter. API, website, app testing leveraging open source tools jmeter, gatling, selenium and taurus. https://www.blazemeter.com/ (2011)

5. R.N. Calheiros, R. Ranjan, A. Beloglazov, C.A. De Rose, R. Buyya, CloudSim: a toolkit for modeling and simulation of cloud computing environments and evaluation of resource provisioning algorithms. Softw. Pract. Exp. **41**(1), 23–50 (2011)
6. A.V. Dastjerdi, R. Buyya, Fog computing: helping the internet of things realize its potential. Computer **49**(8), 112–116 (2016)
7. Q. Duan, Cloud service performance evaluation: status, challenges, and opportunities-a survey from the system modeling perspective. Dig. Commun. Netw. **3**(2), 101–111 (2017)
8. R.R. Expósito, G.L. Taboada, S. Ramos, J. Touriño, R. Doallo, Performance analysis of HPC applications in the cloud. Future Gener. Comput. Syst. **29**(1), 218–229 (2013)
9. M. Femminella, M. Pergolesi, G. Reali, Performance evaluation of edge cloud computing system for big data applications, in *2016 5th IEEE International Conference on Cloud Networking (Cloudnet)*, pp. 170–175. IEEE (2016)
10. H. Gupta, A. Vahid Dastjerdi, S.K. Ghosh, R. Buyya, iFogSim: a toolkit for modeling and simulation of resource management techniques in the internet of things, edge and fog computing environments. Softw. Prac. Exp. **47**(9), 1275–1296 (2017)
11. K.V. Katsaros, G. Xylomenos, G.C. Polyzos, GlobeTraff: a traffic workload generator for the performance evaluation of future internet architectures, in *2012 5th International Conference on New Technologies, Mobility and Security (NTMS)*, pp. 1–5. IEEE (2012)
12. A. Khajeh-Hosseini, D. Greenwood, I. Sommerville, Cloud migration: a case study of migrating an enterprise IT system to IaaS, in *2010 IEEE 3rd International Conference on Cloud Computing*, pp. 450–457. IEEE (2010)
13. R. Ledyayev, H. Richter, High performance computing in a cloud using openstack. Cloud Comput. 108–113 (2014)
14. T.H. Luan, L. Gao, Z. Li, Y. Xiang, G. Wei, L. Sun, Fog computing: focusing on mobile users at the edge. arXiv:1502.01815 (2015)
15. P. Mach, Z. Becvar, Mobile edge computing: a survey on architecture and computation offloading. IEEE Commun. Surv. Tutor. **19**(3), 1628–1656 (2017)
16. C. Sonmez, A. Ozgovde, C. Ersoy, EdgeCloudSim: an environment for performance evaluation of edge computing systems. Trans. Emerg. Telecommun. Technol. **29**(11), e3493 (2018)
17. S. Souders, High performance web sites. Queue **6**(6), 30–37 (2008)
18. S. Souders, Even Faster Web Sites: Performance Best Practices for Web Developers. O'Reilly Media, Inc. (2009)
19. P. Voigt, A. Von dem Bussche, The EU general data protection regulation (GDPR), in *A Practical Guide*, 1st edn. (Springer International Publishing, Cham, 2017)
20. J. Zhu, D.S. Chan, M.S. Prabhu, P. Natarajan, H. Hu, F. Bonomi, Improving web sites performance using edge servers in fog computing architecture, in *2013 IEEE Seventh International Symposium on Service-Oriented System Engineering*, pp. 320–323. IEEE (2013)

Study on an Argumentation-Based Negotiation in Human-Computer Negotiation Service

Mukun Cao and Gong Jing

Abstract With the rapid development of e-commerce, the automatic transaction has been a potential demand of enterprises. Concentrating on B2C e-commerce scenario, this paper designs a human-computer negotiation model and algorithm, through which an agent can negotiate with a human using natural language. To validate the model and algorithm, we conducted a between-group experiment based on a prototype system comparing the negotiation effect between the groups with and without arguments. The experimental results show that adding arguments into price negotiation can significantly increase the success rate of the human-computer negotiation. Moreover, there is a significant positive impact on the buyer's subjective feelings on system using, as well as the seller agent's economic utility, so that it can finally help both sides to reach a win-win situation in the negotiation. The contribution of our study can apply to B2C e-commerce platforms for improving the performance of their intelligent customer service agent. Our study is also meaningful for helping the two parties to increase their trading efficiency and decrease their trading cost.

Keywords E-commerce · Automated negotiation · Argumentation-Based negotiation · Human-Computer interaction · Agent · Service system

1 Introduction

E-commerce has had a substantial impact on the traditional retail industry [1, 2], which is profoundly changing people's purchasing behavior, and at the same time, it has brought substantial business opportunities to merchants [3]. In 2017, on the day of "Double Eleven," Ali Tmall's trading volume reached 168.2 billion yuan. The turnover exceeded 10 billion yuan after this online shopping event began only three minutes [4]. Some merchants' trading volume was dozens of times of the usual, even hundreds of times. Such huge concurrent trading volume requires more intelligent system support. However, the current mainstream B2C e-commerce platforms (such

M. Cao (✉) · G. Jing
School of Management, Xiamen University, Fujian, Xiamen 361005, China
e-mail: mkcao@xmu.edu.cn

© Springer Nature Switzerland AG 2020   255
H. Yang et al. (eds.), *Smart Service Systems, Operations Management,
and Analytics*, Springer Proceedings in Business and Economics,
https://doi.org/10.1007/978-3-030-30967-1_23

as Tmall and Jingdong) mainly rely on fixed price transactions. If customers want to negotiate the price with sellers they need to bargain with the back-office customer service personnel through instant messaging tools (such as Ali Want). This method is costly and slow to respond and is easily affected by personal emotional factors. Nevertheless, in the future, e-commerce does need a human-computer interaction system that can automatically negotiate the price with customers, namely, "human-computer negotiation system", referred to as "human-Agent negotiation system" [5]. The human-computer negotiation system relies on agent technology, free from time and geographical restrictions, and can conduct a negotiation anytime and anywhere. With broad application prospect in B2C e-commerce, it can alleviate the pressure of insufficient customer service personnel, significantly improve negotiation efficiency, and reduce selling costs [6]. An agent can not only bargain but also give reasons for offer to increase persuasiveness [7]. It can automatically identify the bidding characteristics of consumers and provide accurate and reasonable argument support to quick and smooth deals between buyers and sellers, consequently improving e-commerce efficiency [8, 9].

2   Literature Review

The research of this paper depends on the theory of argumentation-based negotiation. Argumentation-based negotiation allows the negotiator to fully express his own belief in the negotiation process allowing opponent to better understand his position [10]. Arguments include the reasons for accepting or rejecting the opponent's bid which facilitates the opponent to adjust strategy timely; the defense of one's own bid to persuade his opponent to accept the offer; warning the opponent the price of refusing the bid, or the reward of accepting it. In the study of automated negotiation system in which computers replace humans the common practice of employing arguments is as follows: agent first evaluates arguments, generates arguments, and selects arguments in a way similar to predicate logic reasoning; and then, according to the predefined interaction rules the communication is completed in a standard communication language format [11]. However, these formalized arguments based on logical reasoning are only applicable to machines and cannot meet the demand of human-machine negotiation in e-commerce [12], because people cannot understand them.

The latest representative outcomes of argumentation-based negotiation are summarized below. El-sisi and Mousa believe that argument negotiation can be an alternative method based on game theory and heuristic negotiation model. The experiment shows that the effect of argumentation-based negotiation is superior [13]. However, their experiment is only a simulation experiment between agents without any human participation. Most studies tend to address issues such as the generation, evaluation, and exchange of arguments [14, 15] and how to design appropriate negotiation protocol to support the communication between agents. However, a crucial issue is how to determine the uncertainty risk of arguments directly affecting the quality of Agent decision-making. But there are a few related studies. Recent representative achievement is to construct a model identifying the uncertainty of Agent to guide the choice of arguments by using Shannon's information entropy theory and hypothesis testing method [16]. In order to improve the effectiveness of argument selection in negotiation between agents the reinforcement learning mechanism is used to update the strategy of argument selection, introduce dynamic learning, and update the method of argument choice preference [17]. Despite this, most studies of argumentation-based negotiation stay in the stage of theoretical construction and simulation experiment but are far from practical application. Real e-business negotiation requires natural language communication close to human habits.

Scholars, such as Sun et al., (2014), have studied argumentation-based negotiation from the perspective of agent persuasion classified as complaint type, explanation type, and analogy type, and combined concession strategy with persuasion argument making agent negotiation step forward to personification. However, the argument form is too single to handle complex human-computer negotiation [18]. Closer to human-computer communication is an intelligent sales agent designed by Huang and Lin (2007). By using intensive learning technology to train an abstract argument framework and selecting appropriate arguments according to the keywords spoken by the opponent, it can satisfy the demand for simple human-computer dialogue [12]. However, the model has artificially divided the negotiation into two stages of persuasion and negotiation. It is unable to persuade the opponent with arguments while negotiating the price which is inconsistent with the real negotiation. Meanwhile, the bidding strategy is too simple to cope with the challenges from human.

To sum up, most of the existing argumentation-based negotiation models have not been tested by human-computer negotiation and remain in the stage of theoretical research. Some models have passed the human-computer negotiation tests but are unable to deal with people's complex negotiation behaviors because of the simple design and lacking research on users' psychological feelings, so they are hard to be applied in real life. To solve this problem, we introduce argumentation-based negotiation into the real human-computer negotiation, combines the arguments with the concession bidding strategy, and uses economic and psychological metrics to examine the validity of the model to provide a feasible solution to practical application in this field.

3   Negotiation Model Based on Similarity Matching of Argument Vector

The design goal of the human-computer argument negotiation system is the seller Agent offers and gives its reasons for offering and makes the whole negotiation process closer to the real "bargaining" through textual expression, in a similar way of QQ chat. The specific approach is that the seller agent calculates human buyer's behavior vector $\vec{X}$ according to his or her bids, searches for the most similar semantic vector $\vec{Y}$ in vector statement library and obtains three statements according to the value of the vector in each dimension, then merges these statements with bid attached, finally generates arguments and returns to the buyer.

Arguments come from a predesigned statement library. Multidimensional statement library is one innovation of this paper, which can cope with complex negotiation situations. According to different negotiation situations, it can dynamically combine the statements of each dimension in the statement library to form a complete sentence and give a response. This function is not available in other automated question-answer systems. The dimension of the multidimensional statement library can be expanded to n dimensions according to the designer's requirements. In this paper, all statements are divided into three dimensions (x_1, x_2, x_3) according to their semantics.

(1) Dimension x_1: perception of concession, making a differentiated response to the opponent's concession in this round;
(2) Dimension x_2: perception of intention, judging to what extent both parties intended to make a deal according to the bidding gap between the seller and the buyer, and then make response, such as encouragement or persuasion;
(3) Dimension x_3: perception of negotiation time-consuming, making a reminder or urge according to the length of negotiation time.

Without loss of generality, in the example of multidimensional statement library given in this paper, there are 11 optional statements for each dimension (see appendix). According to the tone from weak to strong and priority of the situation, the score of each statement is set from small to large as 0, 0.1, 0.2, 0.3, 0.4, 0.5, 0.6, 0.7, 0.8, 0.9, and 1, 11 grades in all. In this way, theoretically, the multidimensional statement library can produce a combination of $11^3 = 1331$ statements ensuring the diversity of negotiation terms. For example, $\vec{X} = (0, 0.8, 0.4)$, when the seller offers a price of P, the corresponding argument is: "Dear, your concession is too small. I hope you are more sincere. This product has a large capacity of 20,000 mAh, which provides enough power to your phone when traveling. You can charge your phone five times if you charge it fully every time. I sincerely offer you a price of P yuan."

The behavior vector is a quantitative tool to describe an opponent's negotiation behavior. The core issue of argumentation-based negotiation is to judge the negotiation situation according to the opponent's bid and humanly respond with appropriate language, namely, the arguments must conform to the current negotiation context. The negotiation situation is judged by observing the bidding behavior of the opponent. Corresponding to the semantic vector the behavior vector $\vec{Y} = (y_1, y_2, y_3)$ is used to describe the negotiator's quantifiable negotiation response in three dimensions. Among them, y_1 calculates the difference of the opponent's bid in this round compared with the previous round, y_2 calculates the bidding gap between the two parties in the previous round and y_3 calculates the negotiation time consumed up to the current round. In order to unify the dimension with the value of the semantic vector, the values of y_1, y_2, y_3 are limited to the interval of $[0, 1]$ in the above algorithm. In this way, they are in the same interval with the values of x_1, x_2, and x_3.

It is essentially a matching problem of behavior vector and semantic vector to find the corresponding negotiation statement according to specific negotiation behavior. However, the calculation results of the behavior vector in each dimension are continuous values on the interval of [0, 1], while the values of the semantic vector in each dimension are discrete on the same interval. They are unable to be directly matched. It is only advisable to match behavior vector with the semantic vector having the most significant similarity in multidimensional statement library.

For any semantic vector $\vec{X} = (x_1, x_2, \ldots, x_n)$ and behavior vector $\vec{Y} = (y, y_2, \ldots, y_n)$, the overall similarity γ can be expressed as:

$$\gamma\left(\vec{X}, \vec{Y}\right) = \alpha \cdot \beta \tag{1}$$

where α is the length similarity, and β is the direction similarity. According to the above definition, after the seller agent obtains the buyer's behavior vector $\vec{Y}$, the system will traverse all the semantic vector $\vec{X}$ in the multidimensional statement library and calculate γ, which represents the overall similarity of vector $\vec{X}$ and vector $\vec{Y}$, then find the largest value of γ. The semantic vector $\vec{X}$, corresponding to γ, has the highest overall similarity. The argument, corresponding to $\vec{X}$, is what will be returned to the buyer from the seller.

The negotiation protocol and bidding strategy used by the computer in this paper, respectively, adopt the alternate bidding and multi-strategy selection mechanism [5]. The core is by judging the concession rate change of the opponent's offer price, the Agent dynamically adjusts its bidding strategy in real time. For example, when the opponent accelerates or decelerates the concession, that is, the range of concession increases or decreases compared with the previous one, the Agent will imitate the change rate of the opponent's concession and select an appropriate bid from the alternative strategy curve to deal with it. If the opponent keeps the same concession rate, Agent will respond by keeping the current bidding strategy curve unchanged. Since negotiation agreement and negotiation strategy are not the focus of this paper, only a brief introduction is made here. For details, please refer to Cao et al., (2015) [6]. An essential task of this paper is to combine the argument selection algorithm based on vector similarity with negotiation agreement and bidding strategy model.

4 Experimental Evaluation

In order to verify the validity of the established argumentation-based negotiation model and to explore the impact of arguments on human-computer negotiation, we have designed two groups of an experiment for between-group comparison. For the same kind of commodity, people and computer negotiate on the price. With people as a buyer and agent as a seller, both parties aim to maximize their economic utility. In these two groups, one group of the seller agent has argument supporting, while the other group does not.

In this human-computer negotiation experiment, among the questionnaires collected from the group with the argument, the settlement rate was 87.43%. However, among the questionnaires collected from the group without argument, the settlement rate was 82.58%. The settlement rate of the argument supporting group is higher than the group without arguments.

One-way Multivariate Analysis of Variance (MANOVA) was conducted to compare the differences in various evaluation metrics between the two groups. The analysis of social psychology metrics shows that, in terms of the buyer's satisfaction with the negotiation results, the group with arguments is significantly higher than the group without arguments, namely, the human buyer is more satisfied with the negotiation results when the seller agent provided arguments. In the matter of human buyer's perception of the seller agent's cooperativeness, the experimental group is significantly higher than that of the control group, in other words, the buyer is more likely to perceive the cooperativeness of the seller agent when the arguments exist. However, there is no significant difference in the buyers' perception of the negotiating process between the two groups.

The results of the economic metrics analysis show that the average deal price is significantly higher in the group with arguments than that in the non-argument group. The average utility of the human buyers in the argument group is significantly lower than that of the non-argument group. Correspondingly, the average utility of the seller agent in the argument supporting group is significantly higher than that of the other group. Therefore, the economic benefit of the seller agent significantly increases when there are arguments in the negotiation. On the other hand, the average utility product of the two sides was significantly higher in the argument group than in the non-argument group. The mean of utility difference of the two parties was significantly lower in the group with arguments than in the group without arguments, indicating that the mutual economic benefit of buyers and sellers significantly increased when arguments are involved.

5   Conclusions

The experimental results show that it is feasible for the system to automatically generate negotiation arguments according to the current negotiation context through similarity matching of verbal behavior and semantic vector. In the price negotiation between human and computer, argumentation-based negotiation can increase the success rate of negotiation, improve negotiating efficiency, and dramatically improve user's psychological feelings. Meanwhile, it can positively make a significant influence on the economic benefits of the seller agent and increase the confidence of the merchants using the system.

In the next step of the study, the negotiation model will be extended to the multi-attribute context, so that Agent can negotiate for other items in addition to the price. Consequently, the practical value of the human-computer negotiation system will be further enhanced.

Acknowledgments This work was supported by the China Scholarship Council (Grant# 201706315032), and the Natural Science Foundation of China (Grant# 71671154), and the Fundamental Research Funds for the Central Universities of China (Grant# 20720161052).

References

1. eMarketer, Worldwide Retail Ecommerce Sales Will Reach $1.915 Trillion This Year. 2016. https://www.emarketer.com/Article/Worldwide-Retail-Ecommerce-Sales-Will-Reach-1915-Trillion-This-Year/1014369
2. V. Zeng et al., China Online Retail Forecast, 2014 to 2019. Forrester Research (2015)
3. Y. Li, D. Wei, R. Cao, Negotiation support system for e-commerce. J. Manag. Sci. **4**(6), 13–17 (2001)
4. C. Team, China Double 11 Shopping Festival Sales Statistics 2017. China Internet Watch 2017 [cited 2018; https://www.chinainternetwatch.com/22791/double-11-2017/
5. Y.P. Yang, S. Singhal, Y.J. Xu, Alternate strategies for a win-win seeking agent in agent-human negotiations. J. Manag. Inf. Syst. **29**(3), 223–255 (2012)
6. M. Cao et al., Automated negotiation for e-commerce decision making: a goal deliberated agent architecture for multi-strategy selection. Decis. Support Syst. **73**, 1–14 (2015)
7. R. Lin, S. Kraus, Can automated agents proficiently negotiate with humans? Commun. ACM **53**(1), 78–88 (2010)
8. P. Chen, T. Gao, Y. Li, Research on agent-oriented Markov multi-issue automated negotiation model. J. Manag. Eng. **28**(3), 203–208 (2014)
9. H. Li, N. Cui, X. Xu, Negotiation mode of dual-channel supply chain wholesale price based on bargaining power. J. Manag. Eng. **29**(4), 124–132 (2015)
10. I. Rahwan et al., Argumentation-based negotiation. Knowl. Eng. Rev. **18**(04), 343–375 (2003)
11. A. Carrera, C.A. Iglesias, A systematic review of argumentation techniques for multi-agent systems research. Artif. Intell. Rev. **44**(4), 509–535 (2015)
12. S.L. Huang, F.R. Lin, The design and evaluation of an intelligent sales agent for online persuasion and negotiation. Electron. Commer. Res. Appl. **6**(3), 285–296 (2007)
13. A. El-Sisi, H. Mousa, Argumentation-based negotiation in multi-agent system. Int. Arab J. e-Technol. **3**(3), 155–162 (2014)
14. A. Monteserin, A. Amandi, Agents that learn how to generate arguments from other agents. New Gener. Comput. **32**(1), 31–58 (2014)
15. A. Beigi, N. Mozayani, A new dialogue strategy in multi-agent systems. J. Intell. Fuzzy Syst. **27**(2), 641–653 (2014)
16. O. Marey et al., Decision making under subjective uncertainty in argumentation-based agent negotiation. J. Ambient Intell. Humaniz. Comput. **6**(3), 307–323 (2015)
17. A. Monteserin, A. Amandi, A reinforcement learning approach to improve the argument selection effectiveness in argumentation-based negotiation. Expert Syst. Appl. **40**(6), 2182–2188 (2013)
18. H. Sun, W. Jinghua, T. Huang et al., A concession model based on classification of agent persuasion and its implementation. Oper. Res. Manag. **23**(6), 252–256 (2014)

On the Uncertain Accuracy of Seller-Provided Information in the Presence of Online Reviews

Tian Li and Zeting Chen

Abstract With the development of the Internet, online reviews play a more important role than ever. We study the impact of online reviews on a seller's information provision strategy regarding product quality. The accuracy of the seller-provided information can be uncertain. Yet the presence of online reviews can resolve such uncertainty. We find that the seller can benefit from a greater capability in supplying more accurate information. We also find that online reviews do not always improve the seller's expected profit and the uncertainty of the signal accuracy is not always unfavorable to the seller.

Keywords Online reviews · Seller reputation · Information provision

1 Introduction

With the ubiquity of the Internet and the prevalence of user-generated information, online reviews play a more important role than ever in impacting consumer market. In fact, an increasing number of consumers read online reviews before making purchase decisions [1, 2]. Online reviews are easy to access as almost all online platforms (e.g., Amazon.com, eBay.com, Epinions.com, TripAdvisor.com) have introduced online systems eliciting verbal or grading opinions from consumers [3]. Therefore, sellers adjust their various strategies at both the marketing and the operational levels to actively respond to online reviews.

When consumers face products with uncertain attributes (e.g., quality), online reviews are not the only source of information for them. Usually sellers also provide product information on the online platforms; for example, all sellers on Tmall.com describe their product via text, pictures, videos, etc. Hence, consumers aggregate both consumer-generated and seller-generated product information when they decide whether to buy. The two sources of information complement each other in the sense that they both are informative to the uncertain product attribute. Thus, in the presence

T. Li (✉) · Z. Chen
Business School, East China University of Science and Technology, Shanghai 200237, China
e-mail: litian@ecust.edu.cn

© Springer Nature Switzerland AG 2020　　　　　　　　　　　　　　　　263
H. Yang et al. (eds.), *Smart Service Systems, Operations Management, and Analytics*, Springer Proceedings in Business and Economics,
https://doi.org/10.1007/978-3-030-30967-1_24

of consumer product reviews, a seller may not always describe its product as accurate as possible, and he may strategically increase or decrease the accuracy of his provided product information.

In this paper, we study the impact of online reviews on a seller's information provision strategy regarding product quality. The accuracy of the seller-provided information can be uncertain. Yet the existence of online reviews can resolve such uncertainty. We find that given favorable product quality, the seller prefers high signal accuracy and benefits from a greater capability in supplying the signal accuracy; given unfavorable product quality, the seller prefers low signal accuracy. We also find that when the signal accuracy is uncertain, the seller's expected profit is improved by online reviews given favorable product quality and be reduced given unfavorable product quality.

The literature reviews include three research streams: online word of mouth, online reviews, and seller reputation.

Compared with the traditional word of mouth, the emergence of Internet-based communication has multiplied the dissemination of information [4]. Online Word of Mouth receives great attention and represents a relatively new research area [1]. Chen and Xie use two-stage signaling model to examine an online seller's strategic decision to supply consumer-created versus seller-created product information [1]. Kwark et al. consider online reviews as additional information for consumers to mitigate the uncertainty about the quality of a product and about its fit to consumers' needs [2]. Feng and Liu develop a model to investigate optimal dynamic pricing for a profit-maximizing firm selling a product in a large social network, where agents can only get informed about the product via word of mouth of previously informed friends [4].

This paper also extends research into online reviews, which thus far has been primarily empirical. Previous literature has studied the relationship between online reviews and product sales [5], investigated the influencing factors of this relationship [6], and identified factors that might influence review posting behavior [7]. Sun has highlighted how rating variance can influence subsequent demand, beyond the effect of the mean rating value, such that when the mean is low, greater variance in the ratings has a positive impact on sales [8]. Kwark et al. study how firms can use the upstream pricing scheme as a strategic tool to benefit from online reviews [9]. Li et al. find that information provided by online product reviews influences firms' pricing strategies for products purchased repeatedly [10].

Because of the uncertainty and risk of online transactions, seller reputation is considered to be an important factor in delivering unobservable product quality to consumers and influencing the buyer's decision-making [11]. Yoo and Kim investigate the effects of online product presentation on consumer responses from a mental imagery perspective and the moderating effect of the style of processing (SOP) [12]. Wang et al. take seller's reputation as a high-weight product information clue and consumer product evaluation as a low-weight product information clue to study the interaction on consumer product reviews in different involvement situations [13].

Aperjis and Johari use word-of-mouth (WAM) model to study how the seller ratings matter and establish an effective reputation mechanism to encourage sellers to disclose product information honestly [14].

The main purpose of this paper is to study the impact of online reviews on a seller's information provision strategy regarding product quality. Most of the studies [1, 2, 8, 9] focus on that online reviews can provide consumers with product information and reduce product uncertainty. We consider that online reviews can not only provide consumers with product information but also reduce the uncertainty of the seller-provided information.

2 Model

An online seller sells a product to a unit of consumers. The quality of the product v_t is uncertain, and takes two possible values, v_H and v_L, with a half probability for each value. The seller observes the realization of v_t, but the consumers do not observe it. The seller can provide imperfect information about the product quality, say by offering an online description of the product. In particular, the seller can provide a signal v_d of product quality v_t to the consumers, with v_d also taking the two possible values v_H and v_L. The conditional probability is $P(v_d = v_H | v_t = v_H) = P(v_d = v_L | v_t = v_L) = \alpha$, where $\alpha \in (1/2, 1)$ is the accuracy of signal v_d (i.e., the seller-provided information), and a larger α means that v_d is a more accurate signal of v_t. We call α the signal accuracy hereafter.

The seller may not always be able to deliver a signal of high accuracy, even if he would like to. For example, the seller's product description team makes an effort to fully reveal the product quality, but the description may backfire such that it turns out to be less informative. That is, the accuracy of the seller-provided information is *uncertain*, i.e., α is a random variable. The seller-provided information (the signal v_d) is of high accuracy ($\alpha = \alpha_H$) with probability θ (i.e., $P(\alpha = \alpha_H) = \theta$), and it is of low accuracy ($\alpha = \alpha_L$) with the complement probability. Here θ can be interpreted as the seller's ability to deliver an accurate signal, and it is common knowledge. Note that although the seller cannot control the realization of the signal accuracy α, he does have a bigger chance to offer a more accurate signal ($\alpha = \alpha_H$) if he has a greater capability in supplying it (i.e., greater θ). The signal v_d is informative, and hence $1 \geq \alpha_H > \alpha_L \geq 1/2$. We assume α and v_t are independent, that is, the signal accuracy does not depend on the quality of the product.

The consumers' utility is given by

$$U = v_t - p. \tag{1}$$

The consumers buy the product if and only if the utility is nonnegative. Because they are uninformed of the true product quality v_t, the consumers make their purchasing decisions based on the signal v_d, which is ironical of random accuracy. Online reviews can help resolve the uncertainty of the signal accuracy. Especially, in the presence of online reviews, both the seller and the consumers observe the realized value of α. Clearly, the seller's optimal selling price is always equal to the consumers' expected product quality conditional on the quality signal.

The sequence is as follows. First, the seller observes the true value of v_t, and a signal v_d is generated with unknown accuracy. Second, the realized signal accuracy α is made public if online reviews exist; otherwise, both the seller's and the consumers' beliefs about α remain prior. Third, the seller sets a price and the consumers make their purchasing decisions based on v_d and its accuracy α (if available). The seller collects his profit.

To study the impact of online reviews and the uncertain signal accuracy, we first discuss the following three cases separately and then compare them to generate managerial insights. Case A, certain signal accuracy; Case B, uncertain signal accuracy without online reviews; Case C, uncertain accuracy with online reviews.

3 Analysis and Results

3.1 Case A. Certain Signal Accuracy

In Case A, the signal accuracy α is certain. It is worth noting that the exact value of α is common knowledge at the beginning of the game; this is quite different from the case of uncertain signal accuracy with online reviews (i.e., Case C discussed later) in which α is revealed through online reviews after the signal v_d is provided. In fact, Case A can be deemed as the special cases of $\theta = 1$ ($\alpha = \alpha_H$ for sure) or $\theta = 0$ ($\alpha = \alpha_L$ for sure).

Given a signal $v_d = v_i$ with accuracy $\alpha = \alpha_j$ ($i, j \in \{H, L\}$), the consumers update their belief on product quality

$$P_A\left(v_t = v_k | v_d = v_i, \alpha = \alpha_j\right) = \begin{cases} \alpha_j & if\ k = i \\ 1 - \alpha_j & if\ k \neq i \end{cases}. \tag{2}$$

Hence their expected willingness to pay is

$$E_A\left[v_t | v_d = v_H, \alpha = \alpha_j\right] = \alpha_j v_H + \left(1 - \alpha_j\right)v_L \tag{3}$$

when the signal is $v_d = v_H$, and it is

$$E_A\left[v_t | v_d = v_L, \alpha = \alpha_j\right] = (1 - \alpha_j)v_H + \alpha_j v_L \tag{4}$$

when the signal is $v_d = v_L$. Anticipating the consumers' willingness to pay, the seller's optimal price (also the optimal profit conditional on signal v_d) is given by

$$p_i^j = E_A\left[v_t | v_d = v_i, \alpha = \alpha_j\right], \tag{5}$$

where the superscript $j \in \{H, L\}$ represents the signal accuracy and the subscript $i \in \{H, L\}$ represents the signal of product quality. For example, $p_L^H = (1 - \alpha_H)v_H + \alpha_H v_L$ is the seller's optimal price if the signal is low ($v_d = v_L$) and its accuracy is high ($\alpha = \alpha_H$).

Lemma 1 *Suppose α is certain and known to both the seller and the consumers at the very beginning. The seller's optimal prices satisfy $p_H^H > p_H^L > p_L^L > p_L^H$.*

Lemma 1 states that a favorable signal ($v_d = v_H$) about the uncertain product quality is always desirable to the seller ($p_H^j > p_L^j$ for any signal accuracy $j \in \{H, L\}$). However, a higher signal accuracy may lead to a lower price. For example, if the signal is unfavorable ($v_d = v_L$), it lowers the consumers' willingness to pay if it is of high accuracy compared to if it is of low accuracy ($p_L^L > p_L^H$).

Given the true quality of the product $v_t = v_H$ and the signal accuracy α, the seller's expected profit before the generation of signal v_d is

$$A_H^j = \alpha_j p_H^j + (1 - \alpha_j) p_L^j, \qquad (6)$$

and given the true quality of the product $v_t = v_L$, it is

$$A_L^j = \alpha_j p_L^j + (1 - \alpha_j) p_H^j, \qquad (7)$$

where the superscript $j \in \{H, L\}$ represents the signal accuracy and the subscript represents the true product quality.

Proposition 1 *Suppose α is certain and known to both the seller and the consumers at the very beginning. The seller's optimal expected profits satisfy $A_H^H > A_H^L > A_L^L > A_L^H$.*

Proposition 1 states that favorable product quality ($v_t = v_H$) is always desirable to the seller ($A_H^j > A_L^j$ for any signal accuracy $j \in \{H, L\}$). However, a higher signal accuracy may lead to a lower expected profit. Because a higher signal accuracy can make the consumers know more information about the product quality which can benefit the seller given the true product quality $v_t = v_H$ but hurt the seller given the true product quality $v_t = v_L$. Therefore, given favorable product quality ($v_t = v_H$), the seller prefers high accuracy; given unfavorable product quality ($v_t = v_L$), the seller prefers low accuracy.

3.2 Case B. Uncertain Signal Accuracy Without Online Reviews

In Case B, the signal accuracy α is uncertain, and it cannot be revealed either, due to the absence of the online review system. Thus, both the seller and the consumers remain uninformed of α throughout the game.

Given a signal $v_d = v_i$ ($i \in \{H, L\}$), the consumers update their belief on product quality

$$P_B(v_t = v_k | v_d = v_i) = \begin{cases} \bar{\alpha} & if\ k = i \\ 1 - \bar{\alpha} & if\ k \neq i \end{cases}, \tag{8}$$

where $\bar{\alpha} = \alpha_H \theta + \alpha_L(1 - \theta)$. Hence, the consumers' expected willingness to pay is

$$E_B[v_t | v_d = v_H] = \bar{\alpha} v_H + (1 - \bar{\alpha}) v_L \tag{9}$$

when the signal is $v_d = v_H$, and it is

$$E_B[v_t | v_d = v_L] = (1 - \bar{\alpha}) v_H + \bar{\alpha} v_L \tag{10}$$

when the signal is $v_d = v_L$. Anticipating the consumers' willingness to pay, the seller's optimal price is

$$p_i = E_B[v_t | v_d = v_i], \tag{11}$$

where the subscript $i \in \{H, L\}$ represents the signal of product quality. Given favorable product quality ($v_t = v_H$), the seller's expected profit before the generation of signal v_d is

$$B_H = \theta[\alpha_H p_H + (1 - \alpha_H) p_L] + (1 - \theta)[\alpha_L p_H + (1 - \alpha_L) p_L], \tag{12}$$

and given unfavorable product quality ($v_t = v_L$), it is

$$B_L = \theta[\alpha_H p_L + (1 - \alpha_H) p_H] + (1 - \theta)[\alpha_L p_L + (1 - \alpha_L) p_H]. \tag{13}$$

Proposition 2 *Suppose α is uncertain and cannot be revealed. A greater capability in supplying α (i.e., greater θ) can benefit the seller given $v_t = v_H$, i.e., $dB_H/d\theta > 0$, but hurt the seller given $v_t = v_L$, i.e., $dB_L/d\theta < 0$.*

Proposition 2 states that suppose the signal accuracy α is uncertain and cannot be revealed due to the absence of online reviews, a greater capability in supplying the accuracy can benefit the seller given favorable product quality ($v_t = v_H$). Proposition 2 also states that the seller prefers high accuracy given $v_t = v_H$ and low accuracy given $v_t = v_L$ no matter the signal accuracy is certain or uncertain. Because a greater θ means that the seller has a bigger chance to offer a more accurate signal and a more accurate signal makes the consumers know more information about the product quality which raises the consumers' willingness to pay given $v_t = v_H$ but lowers the consumers' willingness to pay given $v_t = v_L$. Thus, a greater θ can improve the seller's expected profit given favorable product quality ($v_t = v_H$) and reduce the profit given unfavorable product quality ($v_t = v_L$).

Proposition 3 *Compared with the expected profits in Case A, the seller's optimal expected profits in Case B satisfy* $A_H^H > B_H > A_H^L > A_L^L > B_L > A_L^H$.

Proposition 3 states that favorable product quality ($v_t = v_H$) is always desirable to the seller no matter the signal accuracy is certain or uncertain and the uncertainty of the signal accuracy makes the seller collect medium expected profit (i.e,. $A_H^H > B_H > A_H^L$, $A_L^L > B_L > A_L^H$). Because the uncertainty of the signal accuracy makes the uninformed α between α_H and α_L. As stated in Proposition 2, the seller prefers high accuracy given $v_t = v_H$ and low accuracy given $v_t = v_L$ no matter the signal accuracy is certain or uncertain. Hence, the seller's expected profit B_H is greater than A_H^L and smaller than A_H^H, B_L is greater than A_L^H and smaller than A_L^L.

3.3 Case C. Uncertain Signal Accuracy with Online Reviews

In Case C, the signal accuracy α is uncertain, but it can be observed by both the seller and the consumers through online reviews, after the signal v_d is generated.

Given a signal $v_d = v_i$ with accuracy $\alpha = \alpha_j$ ($i, j \in \{H, L\}$), the consumers' updated belief on product quality is identical to Case A. The consumers update their belief on product quality

$$P_C(v_t = v_k | v_d = v_i, \alpha = \alpha_j) = \begin{cases} \alpha_j & if\ k = i \\ 1 - \alpha_j & if\ k \neq i \end{cases}. \tag{14}$$

Hence their expected willingness to pay is

$$E_C[v_t | v_d = v_H, \alpha = \alpha_j] = \alpha_j v_H + (1 - \alpha_j)v_L \tag{15}$$

when the signal is $v_d = v_H$ and it is

$$E_C[v_t | v_d = v_L, \alpha = \alpha_j] = (1 - \alpha_j)v_H + \alpha_j v_L \tag{16}$$

when the signal is $v_d = v_L$. Anticipating the consumers' willingness to pay, the seller's optimal price is

$$p_i^j = E_C[v_t | v_d = v_i, \alpha = \alpha_j]. \tag{17}$$

Although the consumers' updated belief on product quality is identical to Case A, the seller's expected profit is different from Case A. Before the generation of the signal and online reviews, the seller doesn't know the realization value of the signal accuracy. The seller calculates the profit based on the uninformed signal accuracy. Given the true quality of the product $v_t = v_H$, the seller's expected profit before the generation of signal v_d is

$$C_H = \theta \alpha_H p_H^H + \theta(1 - \alpha_H) p_L^H + (1 - \theta)\alpha_L p_H^L + (1 - \theta)(1 - \alpha_L)p_L^L, \quad (18)$$

and given the true quality of the product $v_t = v_L$, it is

$$C_L = \theta \alpha_H p_L^H + \theta(1 - \alpha_H) p_H^H + (1 - \theta)\alpha_L p_L^L + (1 - \theta)(1 - \alpha_L)p_H^L. \quad (19)$$

Proposition 4 *Suppose α is uncertain but can be observed through online reviews. A greater capability in supplying α (i.e., greater θ) can benefit the seller given $v_t = v_H$, i.e., $dC_H/d\theta > 0$, but hurt the seller given $v_t = v_L$, i.e., $dC_L/d\theta < 0$.*

As the same as Proposition 2, Proposition 4 states that the seller's expected profit in Case C is increasing with θ given the true quality of the product $v_t = v_H$ and decreasing with θ given the true quality of the product $v_t = v_L$. Although online reviews resolve the uncertainty of the signal accuracy after the signal v_d is generated, the seller still does not know the signal accuracy before the generation of the signal. Therefore, a greater θ can reduce the uncertainty of the seller to the signal so that improve the seller's expected profit before the generation of the signal given the true quality of the product $v_t = v_H$ and reduce the profit given $v_t = v_L$.

Finally, we compare the seller's expected profit in three cases, as shown in Figs. 1 and 2.

Figure 1 *shows that given favorable product quality ($v_t = v_H$), the higher the signal accuracy, the higher the seller's expected profit. Figure 1 also shows that the seller's expected profit is increasing with θ in both Case B and C, yet the profit in Case C is always larger than that in Case B. Because the signal uncertainty is resolved by the presence of online reviews in Case C. $C_H - B_H$*

Fig. 1 Seller's profit given favorable product quality

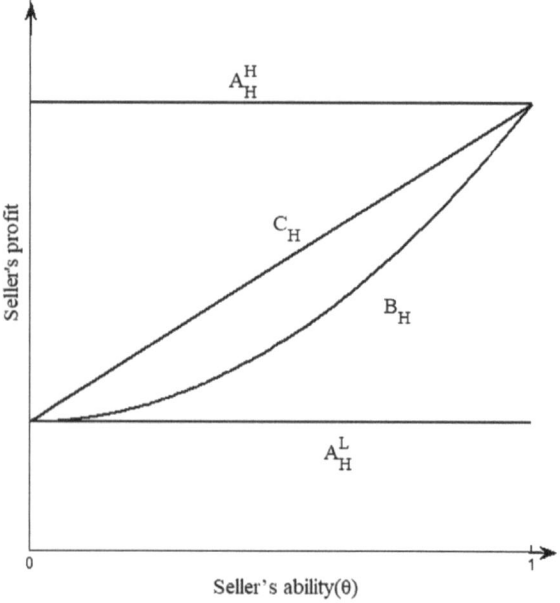

Fig. 2 Seller's profit given
unfavorable product quality

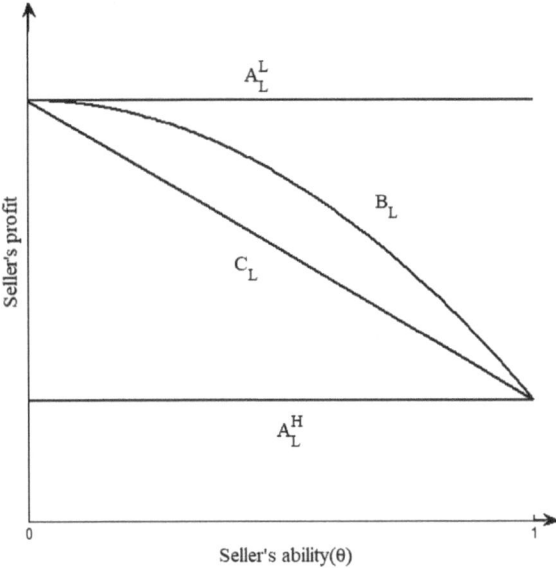

represents the increased profits by the resolved signal uncertainty through online reviews. Since the signal uncertainty is low when θ is quite large or small, the seller's profit increased by the presence of online reviews is low. When the signal uncertainty is high, online reviews play a more significant role in reducing the signal uncertainty, so that the seller's profit increased by online reviews is relatively high. Therefore, $C_H - B_H$ increases first and then decreases with the increase of θ.

Figure 2 *explains something similar to* Fig. 1. *Different from* Figs. 1, and 2 *shows the seller's expected profit in three cases given unfavorable product quality* ($v_t = v_L$). *As can be seen from* Fig. 2, *the seller's expected profit is decreasing with* θ *in both Case B and C, yet the profit in Case C is always lower than that in Case B. The explanation of* Fig. 2 *is the same as that of* Fig. 1. *As Proposition 2 states, the seller prefers low accuracy given* $v_t = v_L$. *A greater* θ *means that the seller has a bigger chance to offer a more accurate signal, hence* B_L *and* C_L *are both decreasing with* θ. *Meanwhile, due to the presence of online reviews, the uncertainty of the signal accuracy is resolved, the consumers know more information about product quality, which hurts the seller given* $v_t = v_L$. *Consequently, the profit in Case C is always lower than that in Case B.*

Proposition 5 *Compared with the expected profits in Case A and Case B, the seller's expected profits in Case C satisfy* $A_H^H > C_H > B_H > A_H^L > A_L^L > B_L > C_L > A_L^H$.

Proposition 5 states that online reviews do not always improve the seller's expected profit and the uncertainty of the signal accuracy is not always unfavorable to the seller. Given favorable product quality ($v_t = v_H$), *online reviews can improve the seller's expected profit. However, given unfavorable product quality* ($v_t = v_L$), *online reviews can resolve the uncertainty of the signal accuracy, which helps the consumers know more about the product quality and reduce willingness to pay, so that hurt the seller's expected profit.*

4 Discussion and Future Research

In this paper, we study the impact of online reviews on a seller's information provision strategy regarding product quality. The accuracy of the information provided by the seller may be uncertain. However, the presence of online reviews can resolve such uncertainty. We establish a game model and discuss the following three cases separately, Case A, certain signal accuracy; Case B, uncertain signal accuracy without online reviews; Case C, uncertain accuracy with online reviews. Then we compare the seller's profit in three cases to generate managerial insights. We find that favorable product quality is always desirable to the seller while a higher signal accuracy may lead to a lower expected profit. No matter the signal accuracy is certain or uncertain, given favorable product quality, the seller prefers high signal accuracy; given unfavorable product quality, the seller prefers low signal accuracy. We also find that online reviews do not always improve the seller's expected profit and the uncertainty of the signal accuracy is not always unfavorable to the seller. When the signal accuracy is uncertain, the seller's expected profit is improved by online reviews given the true quality of the product is high and be reduced given the true quality of the product is low.

In this paper, the presence of online reviews means that consumers know the realization value of the signal accuracy and we don't model the online reviews generation process. In future research, we may establish the online reviews generation process.

Acknowledgements The study is supported by the National Natural Science Foundation of China (71671067, 71301050) and the Fundamental Research Funds for Central Universities.

References

1. Y. Chen, J. Xie, Online review: word-of-mouth as a new element of marketing communication mix. Manag. Sci. **54**(3), 477–491 (2008)
2. J. Chen, Y. Kwark, S. Raghunathan, Online reviews: implications for retailers and competing manufacturers. Inf. Syst. Res. **25**(1), 93–110 (2014)
3. O. Ivanova, M. Scholz, How can online marketplaces reduce rating manipulation? A new approach on dynamic aggregation of online ratings. Decis. Support Syst. **104**, 64–78 (2017)
4. F. Jian, L. Bin, Dynamic impact of online word-of-mouth and advertising on supply chain performance. Int. J. Environ. Res. Public Health **15**(1), 69–85 (2018)
5. C. Dellarocas, G. Gao, R. Narayan, Are consumers more likely to contribute online reviews for Hit or Niche products. J. Manag. Inf. Syst. **27**(2), 127–158 (2010)
6. X. Lu, S. Ba, L. Huang, Y. Feng, Promotional marketing or word-of-mouth? Evidence from online restaurant reviews. Inf. Syst. Res. **24**(3), 596–612 (2013)

7. P.B. Goes, M. Lin, C. Yeung, Popularity effect in user-generated content: evidence from online product reviews. Inf. Syst. Res. **25**(2), 222–238 (2014)
8. M. Sun, How does the variance of product ratings matter. Manag. Sci. **58**(4), 696–707 (2012)
9. Y. Kwark, J. Chen, S. Raghunathan, Platform or wholesale? A strategic tool for online retailers to benefit from third-party information. MIS Q. **41**(3), 763–785 (2017)
10. X. Li, L.M. Hitt, Z.J. Zhang, Product reviews and competition in markets for repeat purchase products. J. Manag. Inf. Syst. **27**(4), 9–42 (2011)
11. M. Luca, G. Zervas, Fake it till you make it: reputation, competition, and yelp review fraud. Manag. Sci. **62**(12), 3393–3672 (2016)
12. J. Yoo, M. Kim, The effects of online product presentation on consumer responses: a mental imagery perspective. J. Bus. Res. **67**(11), 2464–2472 (2014)
13. Q. Wang, X. Cui, L. Huang, Y. Dai, Seller reputation or product presentation? An empirical investigation from cue utilization perspective. Int. J. Inf. Manag. **36**(3), 271–283 (2016)
14. C. Aperjis, R. Johari, Optimal windows for aggregating ratings in electronic marketplaces. Manag. Sci. **56**(5), 864–880 (2010)

Route Planning for Vehicles with UAVs Based on Set Covering

Qiuchen Gu and Tijun Fan

Abstract Due to the complexity of the last mile delivery, the emerging business called "just-in-time delivery" often has a strict service range limit. As a newly received method for last mile delivery, the UAV can expand the service range of just-in-time delivery with the assistance of vehicles. This paper proposes a model based on the set covering model to locate the UAV take-off point. After that, an improved ant colony algorithm is used to obtain the location result. Finally, the customer's allocation method is proposed. Both methods are tested with randomly generated data and proved to be effective. It provides a reference for the future practice of the just-in-time delivery business.

Keywords UAV · Set covering · Location · Route planning

1 Introduction

Demand for just-in-time delivery grows rapidly as the lazy economy develops. Just-in-time delivery refers to a new form of logistics that delivery is carried out at once after the orders are placed online and it has high timeliness within certain space. According to reports, the number of users in China reached 355 million in 2018, an increase of 21.2% over 2017. It is foreseeable that the just-in-time delivery service will become a competitive strategy adopted by the logistics industry and will help enterprises to seize more market share.

As an example, Fresh Hema can provide delivery services of fresh products and consumer products for nearby customers in the fastest 30 min which covers residential areas around 3 km. However, once the customer's location exceeds the coverage, Fresh Hema cannot provide the just-in-time delivery service. Therefore, such service often has a strict service range limit.

Q. Gu (✉) · T. Fan
Business School, East China University of Science and Technology Shanghai, Shanghai 200237, China
e-mail: 871868947@qq.com

© Springer Nature Switzerland AG 2020   275
H. Yang et al. (eds.), *Smart Service Systems, Operations Management, and Analytics*, Springer Proceedings in Business and Economics, https://doi.org/10.1007/978-3-030-30967-1_25

The UAV is a new technology that is widely used in disaster management, military, and environmental monitoring [1]. So far, countries such as China and the United States have also begun to legally permit the use of civilian UAVs in logistics [2, 3], so it's believable that last mile delivery by UAV will one day be taken into practice due to its fast speed and portability. As "last mile delivery" is the chief difficulty of the just-in-time delivery business, the UAV can obviously be taken as a future practice of the business. Utilization of UAVs can expand the service range of just-in-time delivery with the assistance of vehicles.

However, how to effectively configure vehicles and UAVs, plan the take-off position of UAVs, and optimize the delivery route of vehicles and drones is a challenging problem. Mathew et al. [4] considered the scenario of one vehicle taking a single UAV to provide the service. They use map theory to select take-off points in the transportation network and plan the path of the UAV and the vehicle. Mourelo et al. [5] used k-means clustering and genetic algorithm to give the location and number of optimal take-off points, and then minimize total vehicle transit time. Chang and Lee et al. [6] use k-means to cluster customer points and adjust the location of cluster center as the take-off points, from which the UAVs were dispatched for delivery tasks. Although they give the solution algorithm of selecting the take-off points and the method of assigning the customers to vehicles, they do not give a way to assign the customers to the UAVs. In addition, the use of k-means clustering method to select the take-off position is not stable enough and is hard to take the capacity limit of vehicles into consideration. Therefore, this paper proposes the model for locating take-off points based on the set covering model and designs a heuristic algorithm to solve it effectively.

This paper is organized as follows: Sect. 2 proposes the model of the problem studied in this paper, Section 3 explains the algorithm for solving this problem, and Sect. 4 is the result and conclusion.

2   Problem Description and Mathematical Formulation

For a group of known customers, each vehicle carries a number of drones from the warehouse and selects a stopping point as a take-off point for the UAVs. The UAVs work parallel to send parcels to customers and then return to the vehicle to recharge. The above processes of UAVs are repeated until the parcels responsible by this vehicle are all delivered. The drone has a flight range limit and the vehicle has a capacity limit. The objective is to find the minimum number of take-off points (i.e., dispatch as few vehicles as possible to reduce cost), and on this basis, minimize the duration to complete the delivery tasks. The following operating conditions are assumed:

Battery charging of the UAVs is completed in an instant.

The number of UAVs carried by a single vehicle is the same.

The UAV can only carry one parcel at a time, but the weight of a single parcel does not exceed the UAV's capacity.

The UAV has a constant service time at each customer.

The problem is divided into three echelons to solve that are to decide locations and number of take-off points, to allocate customers to vehicles, and to allocate customers to UAVs.

2.1 Set Covering Based Model of Take-off Points Location Problem

The first echelon of the model is to determine the minimum number of take-off points to meet the demands of all customers. We take the model of set covering problem as the basis and additionally add a constraint to limit the total weight that a single vehicle can hold. When the number of these points is minimized, there may be various solutions of which alternative points are selected to be the group of take-off points. Different selections will bring out different results in the follow-up echelons, so all the feasible selections shall be saved as effective solutions. The programming of this echelon is as follows:

$$\min \sum_{i \in N} O_i \tag{1}$$

$$F_{ij} \leq \begin{cases} 1 & D_{ij} < \theta \\ 0 & \text{else} \end{cases} \quad \forall i \in N, j \in C \tag{2}$$

$$\sum_{j \in C} F_{ij} \cdot g_j \cdot O_i \leq q \quad \forall i \in N \tag{3}$$

$$\sum_{i \in N} O_i \cdot F_{ij} \geq 1 \quad \forall j \in C \tag{4}$$

$$O_i \in \{0 \ 1\} \tag{5}$$

$$F_{ij} \in \{0 \ 1\} \tag{6}$$

where set N contains the alternative points O that may be selected as a take-off point; set C is the customer set; D_{ij} is the distance between alternative point i and customer j by UAV; θ is the range limit of a UAV; g_j is the demand of each customer; q is the capacity limit of a vehicle; O_i and F_{ij} are 0–1 variables, O_i takes the value of 1 if alternative point i is selected as take-off point while F_{ij} equals 1 if customer j is covered by alternative point i.

2.2 Models of Allocating Customers to Vehicles and UAVs

Two linear programming models are presented to describe the allocation problems. In the second echelon, with one of the effective solutions obtained above, the customers are expected to be allocated to the nearest take-off points under the premise that vehicles' capacity limit should be met. The model is given as follows:

$$\min \sum_{\substack{j \in C \\ i \in SN}} D_{ij} \cdot belong_{ij} \tag{7}$$

$$belong_{ij} \leq F_{ij} \quad \forall i \in SN, j \in C \tag{8}$$

$$\sum_{j \in C} belong_{ij} \cdot g_i \leq q \quad \forall i \in SN \tag{9}$$

$$\sum_{i \in SN} belong_{ij} = 1 \quad \forall j \in C \tag{10}$$

$$belong_{ij} \in \{0 \quad 1\} \tag{11}$$

where set SN contains the points selected as take-off points in the first echelon with corresponding F_{ij} become known parameters; $belong_{ij}$ is a new 0–1 variable which takes the value of 1 if customer j is allocated to take-off point i.

The UAV routes are designed in the third echelon to minimize the duration of completing all delivery tasks. As vehicles and UAVs work parallel, each vehicle with its UAVs has a service duration, and the total duration can be equated to the maximum duration of them. The model is given as follows:

$$\min sevT \tag{12}$$

$$\sum_{k \in K} Z_{jk} = 1 \quad \forall j \in C \tag{13}$$

$$sevT \geq CD_i/vt + \sum_{j \in C} belong_{ij} \cdot Z_{jk} \cdot (2 \cdot D_{ij}/vd + con\_sev) \quad \forall i \in SN, k \in K \tag{14}$$

where set CD_i is the distance between take-off point i and the warehouse by vehicle; vt and vd stand for speed of vehicle and UAV, respectively; con_sev is the constant service time UAV need at each customer; $belong_{ij}$ becomes known after the second echelon, but 0–1 variable Z_{jk} which shows allocation of customers to UAVs shall be calculated in this echelon, and so that dependent variable $sevT$ which means the total duration can be obtained.

3 Solution Algorithm

We proposed an improved ant colony algorithm to solve the NP-hard location problem in the first echelon, and the allocation problems are solved simply by exact algorithms. Figure 1a shows the solution process and Fig. 1b shows the generation of a single ant.

The coverage capability of an alternative point refers to the number of "currently uncovered customers" whose distance from the alternative point is within its coverage radius. The coverage radius is limited by the range of UAV and the capacity of the vehicle. We set the flight range of the UAV as initial value for each alternative point, but if the total demand of the customers covered by an alternative point exceeds the capacity limit of vehicle, the coverage radius for this point has to be reduced until the capacity limit is satisfied. In addition, to increase the convergence speed, we suppose that alternative point covered by any selected take-off point in the same ant cannot be selected as another take-off point.

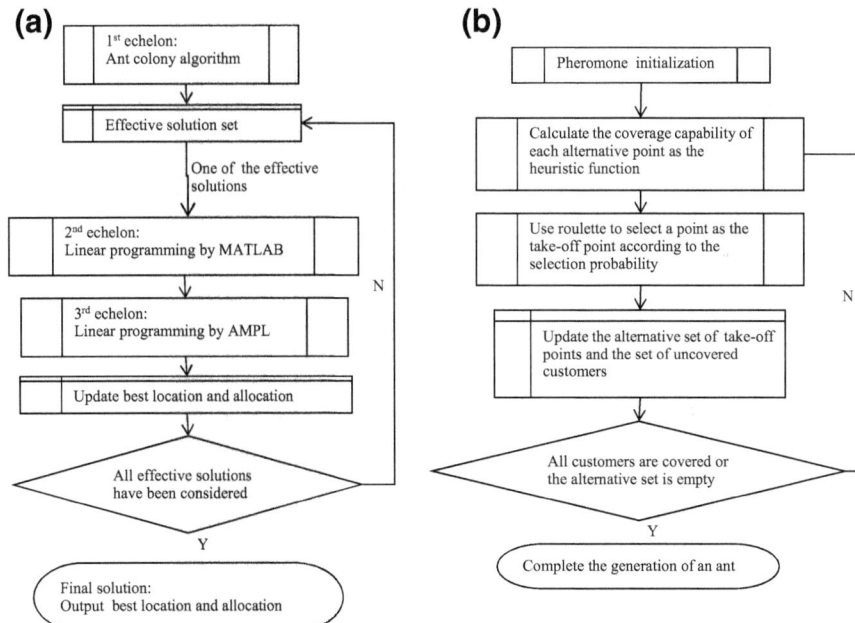

Fig. 1 The solution processes

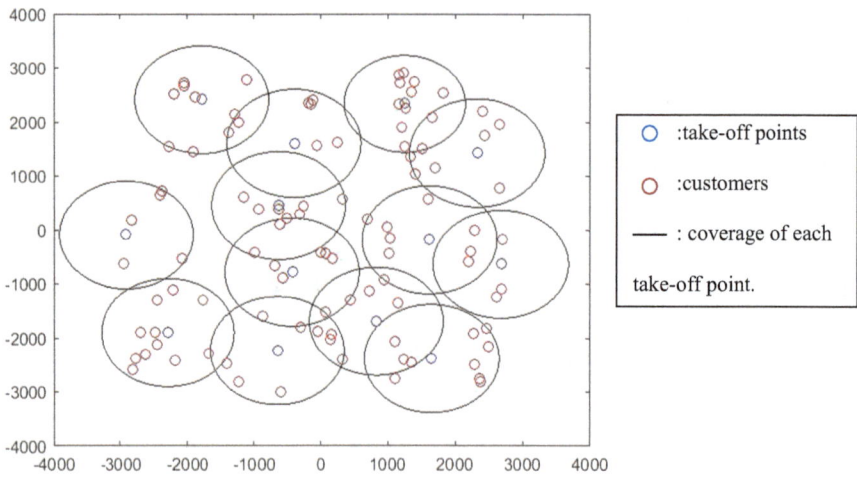

Fig. 2 The best selection of first echelon

4 Illustrative Example

An illustrative example is provided to test the algorithm. A set of 100 customers and a set of 400 alternative points are randomly generated in a square area with a side length of 3000 m. The other parameters are set as follows:

Demand of each customer: randomly generated within 0.5–5 kg.
Capacity of vehicle: 30 kg.
Flight range of UAV: 1000 m.
Speed of UAV and vehicle: 1000 and 500 m/min.

In the example, no feasible solution of the location problem in first echelon can be obtained by CPLEX in acceptable duration, but our ant colony algorithm can give its best result after 500 iterations that is only 13 take-off points are selected to serve the 100 customers, and it also gives 95 different selections of the 13 take-off points. It is a rather rational result since the same set covering problem without capacity limit is solved by CPLEX and the optimal solution is 11. After that, algorithms of second echelon and third echelon starts up. Figure 2 shows the best selection of the 95 options in which the minimum duration is 15.44 min.

5 Conclusion

In this paper, we have developed models and algorithms that can be used to design routes for vehicles and UAVs to deliver parcels to customers. It provides a reference for the future practice of the just-in-time delivery business. Future research will study the expansion to real data and consider adding the vehicle routing problem model to it.

References

1. A. Otto, N. Agatz, J. Campbell, B. Golden, E. Pesch, Optimization approaches for civil applications of unmanned aerial vehicles (UAVs) or aerial drones: a survey. Networks **72**, 411–458 (2018)
2. https://www.cnbc.com/2016/08/29/faas-new-drone-laws-go-into-effect-monday-allowing-us-companies-to-innovate.html
3. https://tech.sina.com.cn/2018-06-24/doc-iheirxye9538870.shtml
4. N. Mathew, S.L. Smith, S.L. Waslander, Planning paths for package delivery in heterogeneous multirobot teams. IEEE Trans. Autom. Sci. Eng. **12**(4), 1298–1308 (2015)
5. S. Mourelo, F.T. Harbison, T. Weber, R. Sturges and R. Rich, Optimization of a truck-drone in tandem delivery network using k-means and genetic algorithm, J. Ind. Eng. Manag. **9**(2), 374–388 (2016)
6. Y.S. Chang, H.J. Lee, Optimal delivery routing with wider drone-delivery areas along a shorter truck-route. Expert Syst. Appl. **104**, 307–317 (2018)

Index Frequency-Based Contour Selection of Gray Wave Forecasting Model and Its Application in Shanghai Stock Market

Xingyuan Li, Qifeng Tang and Shaojun Ning

Abstract Indexes reflect the mechanism of the stock market and the Gray Wave Forecasting Model (GWFM) which has been confirmed to be one of the most effective methods for forecasting. However, the previous method did not take into account the fact that the larger the index frequency is, the more likely this index is to appear in the future. According to the changing law of indexes, an index frequency-based contour selection of GWFM is put forward in this study where the classical uniformly spaced contour line is used twice to select the contour lines. Using this model, the fluctuation trend of Shanghai stock indexes is well predicted which demonstrated that this model has certain advantage over the original GWFM at forecasting stock indexes.

Keywords Gray wave predicting · Index frequency-based contours · Shanghai stock market indexes

1 Introduction

The security market is an integral part of the financial market in national economic development. And the stock index always is seen as "a barometer" of the security market. Therefore, forecasting stock index is extremely beneficial for national economy and individual investors. On the one hand, the stock index has been found to reveal the macroscopic economy and financial risks. On the other hand, investors always perceive market risks by forecasting the stock index. However, the change of stock market indexes is influenced by many factors, and there are complex relationships among them, so it is difficult to build an accurate predicting model. Through historical data of security indexes some trend changes of security index in the future

X. Li (✉)
School of Business, East China University of Science and Technology, Shanghai 200237, China
e-mail: lixy_sjp@126.com

Q. Tang · S. Ning
Shanghai Zamplus Technology Co., Ltd, Shanghai 200072, China

© Springer Nature Switzerland AG 2020
H. Yang et al. (eds.), *Smart Service Systems, Operations Management, and Analytics*, Springer Proceedings in Business and Economics,
https://doi.org/10.1007/978-3-030-30967-1_26

could be found, and forecasting time series of stock market index has always been the focus of investors and researchers [1].

There are two main aspects of the research related to this paper. One is about the prediction of the security indexes. Most of the existing security forecasting methods are based on the data of a single security market. Through cointegration test and Granger causality test of the stock market indices of China and the United States, Lee also found that there is a linkage between the two markets [2]. Lu uses the Nonlinear Independent Component Analysis (NLICA) to extract features from the dynamic stock index data and uses Support Vector Machine (SVM) to predict security indexes [3]. In recent years, with the continuous development of computer technology, management science, and operations research many new methods have been applied to the prediction of security index such as nonlinear dynamic model [4], support vector machine (SVM) [5], artificial neural network [6, 7], and fuzzy mathematics approach [8]. Different from those literature, this paper uses gray waveform prediction model which regards the sequence of security prices as waveform to predict security prices.

The second is about the application of gray prediction theory in the field of finance. Gray system prediction model has been favored by many scholars because it needs less data and does not need to consider the type of data distribution [9]. Zhou et al. used GM (1, 1) model to predict the casualties of high-speed traffic accidents, railway traffic accidents, and mine production accidents in China [10]. Wan et al. proposed an improved gray waveform predicting model for time series with irregular fluctuation amplitude [11]. Chen uses monthly prices of two metals—aluminum and nickel—to assess the performance of the novel gray wave forecasting model with a multistep-ahead prediction [12]. The classical model is mainly for behavior sequence data similar to the form of sine wave. However, for the stock market indexes, the time series is not fixed, the period is irregular, and the distribution is extremely uneven. If the classical model is used to select the contour line of stock indexes it is easy to cause the higher frequency indexes not to get selected as the contour line resulting in the lack of effective information, thus affecting the prediction error of the sequence fluctuation trend. Although Chen [12] used a non-equal interval contour method based on quantile in metal price forecasting. However, there are several problems in that method. Firstly, the prediction is greatly affected by the number of contour lines initially determined (if different number of contour lines are selected, the contour values of the determined contour lines are different). Secondly, that method does not consider the law of recurrence of the price index (that is, the more frequency a price index appears in the past, the more likely it is to appear in the future). In order to overcome the shortcomings of the above method, this paper proposes a contour selection method. Our method selects the contour line with twice according to the frequency rule of the indexes. At the first interpolation the number of contour lines is given externally and at the second time it is determined by the frequency which the price index appears, so the prediction is less affected by the number of contours initially determined. In addition, there is another difference between our model and Chen [12]. The contour selection method in Chen' model is based on the order of the price index sequence. However, we directly divide the index value, doing so, it will be more reasonable to select a higher frequency as a contour to predict.

The chapter of this paper is arranged as follows: The second part analyzes the shortcomings of existing improved models and proposes a new method. The third part is the empirical analysis and model comparison. The last part is the conclusion and discussion.

2 Basic Model and Existing Problems

The gray wave prediction model is a method in the gray system theory, which mainly predicts sequences with periodic fluctuations, and thus has the general advantage of the gray model. The stock indexes are a kind of time sequence data with certain periodic fluctuation characteristics, so it can be predicted by the gray waveform prediction model. The classical GWFM is uniformly spaced contour and can well predict sine-like time sequence. But unlike the sine waves, stock market indexes have irregular period and amplitude. The characteristics of stock market indexes can be described as follows: (1) peaks and valleys with less data, intermediate with mass data; (2) irregular period and amplitude. The classical GWFM may have a problem that contour lines in the peak and valley with less data are relatively more than contour lines near the middle. So many useful information gets lost in the contour lines near the middle. Therefore, the classical GWFM may not predict the trend of stock market indexes well. In this regard, it is reasonable that nonuniform spaced contours line replace uniform-spaced contours, because contour lines near the middle should add more while contour lines in the peak and valley reduce.

There are mainly four steps on the calculation of the GWFM. The first step is the contours selection of the stock indexes sequence. Most improvements (according to the author's knowledge) of the gray waveform prediction model are different on the contour selection, and this article is not an exception. The second step is to calculate contour time sequences. Then the third step is to forecast the contour sequences time using GM (1, 1) model. The last step is to rank the predicted indexes time sequences in chronological order, and the ranked predicted value is the predicted result.

This paper inspired by Chen [12] which uses the new GWFM to predict the metal price indexes. In that method, a percentile-based contour interpolation is adopted. To be specific, first, the metal price index is ranked by numerical value and the first point (that is, the smallest price index) is taken as the value of the first contour line, and the last point (that is, the largest price index) is taken as the s-th contour line where s is given by the externally. Secondly, using the percentile method, the new sequence after ranking is inserted into the contour lines at equal intervals according to the position of the sequence so that the inserted contour values are arranged at non-equal intervals. For instance, a set of price index sequences X and a sequence X' of ranked value indices are, respectively.

$$X = (x_0 = 2500, x_1 = 2300, x_2 = 2700, x_3 = 2200, x_4 = 1800, x_5 = 3200,$$
$$x_6 = 2300, x_7 = 2200, x_8 = 2300)$$

$$X' = \left(x'_0 = 1800, x'_1 = 2200, x'_2 = 2200, x'_3 = 2300, x'_4 = 2300,\right.$$
$$\left.x'_5 = 2300, x'_6 = 2500, x'_7 = 2700, x'_8 = 3200\right)$$

Suppose four contours are selected, the 0-th contour value 1800, and the 3rd contour value 3200 has been determined. The 1st contour' position is $9 \times (1/3) = 3$, that is the second contour value is 2300. And the 2nd contour' position is $9 \times (2/3) = 6$, that is the value is 2500. Finally, the contours are selected as $\{\xi_1 = 1800, \xi_2 = 2300, \xi_3 = 2500, \xi_4 = 3200\}$. Therefore, this method is a non-equal interval interpolation method. The problem in the above method is easy to spot. Firstly, the method does not consider the law of recurrence of the price index (that is, the more frequency a price index appears in the past, the more likely it is to appear in the future). Such as the value 2200 appeared twice and is more like appear in the future, but not be selected to the contour. Secondly, the prediction result is greatly affected by the number of contour lines initially determined (if s assign different values, the contour values of the determined contour lines are different). In order to overcome the shortcomings of above method, we propose a contour selection method. Our method selects the contour line with twice, according to the frequency rule of the indexes. At the first interpolation, the number of contour lines is given externally, and the second interpolation is determined by the frequency which the indexes appear, so the prediction is less affected by the number of contours given initially. In addition, there is another difference between our model and above method. In above model the contour selection method is based on the order of the price indexes sequences, however, we directly divide the indexes value. Doing so, it will be more reasonable to select a higher frequency index as a contour to predict. In this paper, the predicted object is the stock indexes time sequences. In the next section we mainly introduce the details of contour line selection method.

3 Index Frequency-Based Contour Selection

In the first step, we interpolate the contours for the first time by using the equally spaced distances of the sequence values as follows:

Let the indexes time sequences to be $X = (x(1), x(2), \ldots, x(n))$, and $\sigma_{\max} = \max_{1 \le k \le n}\{x(k)\}$, $\sigma_{\min} = \min_{1 \le k \le n}\{x(k)\}$, $\forall \xi_i \notin [\sigma_{\min}, \sigma_{\max}]$, $X = \xi_i$ is called contour lines, or contours. Let $\xi_i = \sigma_{\min} + \frac{i}{s}(\sigma_{\max} - \sigma_{\min})$, where $\xi_s = \sigma_{\max}$, $\xi_0 = \sigma_{\min}$, $i = 2, 3, \ldots, s - 1$, then ξ_i is called contours line.

The s in above represents the number of contour lines to be inserted, which is given externally in the first interpolation. Since the given value of s is arbitrary, it is scarcely possible to ensure that these s contours contain indexes with high frequency using contour interpolation by only once. Therefore, the second contour interpolation is needed, in order to find the contour lines containing the high frequency indexes. Based on the first interpolated contour (old contour), the new contour lines which interpolates in the second time are inserted between old adjacent contour lines using the similar interpolation method, but the number of new contours is not arbitrary in the second interpolation. We call the old adjacent contour lines a interval. Before the second contour interpolation, we need statistics the indexes numbers in each interval and total number of indexes. Then calculate the ratio between the indexes number of each interval and the indexes number of total indexes. When the number of contour lines in the maximal interval (which means the maximal ratio) to be inserted is determined, then the number of contour line to be inserted in the other intervals is determined. The specific calculation is as follows:

Definition 1 Let the original time sequences to be $X = (x(1), x(2), \ldots, x(n))$, $\zeta = (\xi_0, \xi_1, \ldots, \xi_s)$ is frequency-based contours lines which define in Definition 1. Then indexes time sequences in interval w (that is $[\xi_{w-1}, \xi_w]$, $w = 1, 2, \ldots, s$) are $X_w^{(0)} = (x_w^{(0)}(1), x_w^{(0)}(2), \ldots, x_w^{(0)}(m))$, $m < n$. The ratio $\Pr o_w$ can be defined as $\Pr o_w = \frac{m}{n}$.

Then find an interval which contains the largest number of indexes and determine the number of contour lines to be added to the maximum interval. Contour lines in other intervals are added by the following method.

Definition 2 Let $\Pr o_j = \max\{\Pr o_1, \Pr o_2, \ldots, \Pr o_w\}$, $[\xi_{j-1}, \xi_j]$ is corresponding interval, where $\Pr o_w$ is defined in Definition 2. If number of contour lines l_j inserted in interval j, then the number of contour lines insert in interval $[\xi_{w-1}, \xi_w]$ can be calculated as $l_w = l_j \times \frac{\Pr o_w}{\Pr o_j}$, and the new added contours are defined as index frequency-based contours.

In the first step, we adopt twice equal interval interpolation which fully considers the recurrence law of the stock indexes (that is, the higher frequency one index appears, the more chance it occurs in the future) and ensures the initial s has less affection on the predicted value. The remaining calculation steps of the improved FWFM are similar to the classic model, see [9] for details. However, there are some calculation details that need attention, as follows:

The second step is to determine the contour time sequences (the time value of the intersection between contours and waveform curve), seeing [9, 12, 13] for details of this step. We use a simple method to find the contour time sequences. If two adjacent indexes (whose time point is adjacent) minus one contour' value is negative, then this contour is in the middle of the two adjacent indexes and one of the contour' time sequence is calculated by the time of this two adjacent indexes' time using method in [9].

The third step is to perform GM (1, 1) model to forecast the obtained time sequences of each contour, seeing [9] for details of this step. The GM (1, 1) model must satisfy the smoothness which is the important base and premise of gray system model so that we should set a threshold Pw externally and only predict values above the threshold. So how to deal with time series of contour lines that do not conform to gray predicting model? These data are simulated but not predict new values in this paper.

The last step is to rank the predicted indexes in chronological order. In this step, some issues should be addressed. Firstly, How to judge whether the two predictions are equal since the predicted values have many decimal digits? Here gives a criterion that the absolute value of the difference between the two predictions is less than 1. Secondly, how to rectify the forecast time, since sequence time must be an integer? A wise method is taking the smallest integer greater than the predicted value. Thirdly, how to evaluate the missed predicted values? Linear interpolation is adopted in this research.

4 Empirical Studies

4.1 Data and Forecasting Result

In this paper, we use weekly closing price data of Shanghai Composite Index. Using weekly closing prices, from January 1, 2015, to December 31, 2016, simulate the real data. The first week price of January 2015 is defined the first sequence point. The weekly closing price from January 1, 2017, to September 30, 2017, is used to forecast and compare with the actual closing price. The comparison results are shown in Fig. 1. The calculation parameters are given:

$$\xi_0 = 2700, \xi_s = 5400, s = 10, Pw = 0.1, l_{\max} = 3$$

In Fig. 1, the blue line is the actual closing prices, and the red line is simulating and predicting values. The red line in Fig. 1 can be divided into two sections. The red

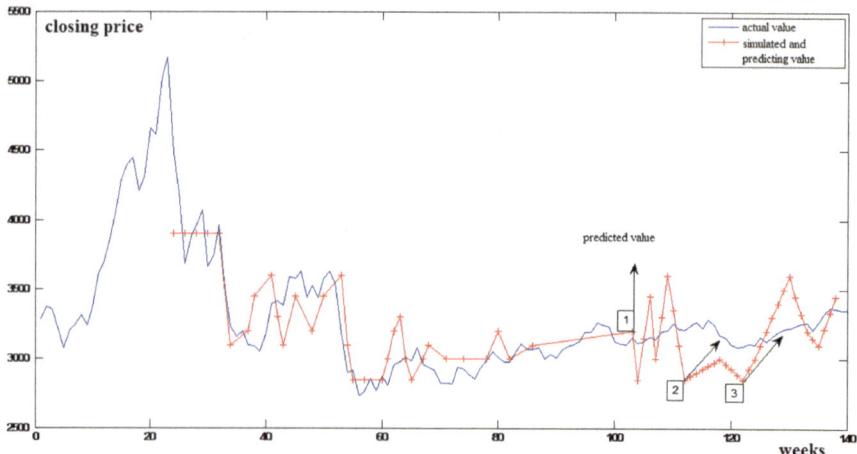

Fig. 1 Comparison chart between forecasting and actual closing price using improved GWFM

line before arrow 1 is simulating prices using the actual closing price, and the rest is the predicted closing prices. The more similar the shape of the red line segment in front of arrow 1 is to the shape of the blue line segment, the more reliable the prediction value of our model is. From Fig. 1, we can find that the trend of the red line before arrow 1 is quite like the blue line. The trend of red line after the arrow 1 has such characteristics: the red curve rises first, then falls then rises, just same as the trend of the blue line after the arrow 1. But there are still some differences. The red curve is peaked at 109-th week, while the blue curve is peaked at 116-th week. Simultaneously, the red curve reaches the trough in 113th week, but the blue curve reaches the trough in 118-th week. The predicted peak is five weeks earlier than the actual peak, and the predicted valley is two weeks earlier than the actual valley. So the model established in this paper can predict the peak and valley of the stock index in advance. However, there still is a time difference between the predictions and the actual price. This may be that the factors affecting the price of security are different at different times.

4.2   Model Comparison

Using the same data, we calculated it with the classical GWFM. The results are shown in Fig. 2. In Fig. 2, the blue line is the predicted closing prices, and the red line with notion "+" is the actual closing prices, which contrary to Fig. 1. In Fig. 2, the blue line (the predicted closing prices) is similar to serrated shape. The possible explanation is that in the classical GWFM the contour line is equidistant which can't contain data near mean prices. Therefore, the traditional GWFM is not effective for the data which is not sinusoidal waveform. This finding is similar to Chen [12].

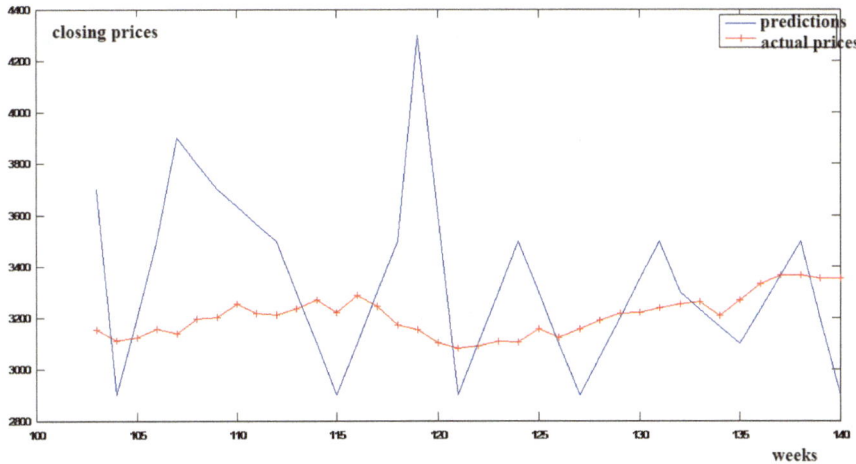

Fig. 2 Comparison chart between forecasting and actual closing price using classical GWFM

5 Conclusion and Discussion

The classical GWFM is often used in predicting model especially when data is scarce and data sequence is similar to the sine wave. However, stock market indexes are always not satisfied such characteristics. This paper proposes an improved GWFM with a kind of nonuniform-spaced contour line, based on the data density. Then collecting the data of Shanghai stock indexes from January 1, 2015, to September 30, 2017, and using the improved GWFM to analyze the predicted results. This paper finds that the fluctuation trend of Shanghai stock indexes is well forecasted, which demonstrated that the improved model has a certain advantage over the classical GWFM in forecasting stock market indexes.

However, this article also has some deficiencies. Some equal height time sequences data do not meet the Gray Model so these sequences data has to be discarded. Moreover, although the predicted closing prices trend is similar to the real one the predicting accuracy should need to improve in future research.

Acknowledgements The authors gratefully acknowledge financial support from the Science and Technology Commission of Shanghai Municipality (No.17DZ1101005).

References

1. H. Yao H, D. Li, J. Li, Stock market volatility forecast based on calculation of characteristic hysteresis. J. Comput. Appl. **35**(07), 2077–2082 (2015)
2. J. Lee, The evolving linkage between China and U.S. stock markets. J. Pediatr. Surg. **5**(4), 468–470 (2012)
3. L. Chijie, Hybridizing nonlinear independent component analysis and support vector regression with particle swarm optimization for stock index forecasting. Neural Comput. Appl. **23**(7–8), 2417–2427 (2013)
4. L.J. Kao, C.C. Chiu, C.H. Lu et al., Integration of nonlinear independent component analysis and support vector regression for stock price forecasting. Neurocomput. **99**(1), 534–542 (2013)
5. T. Xiong, Y. Bao, Z. Hu et al., Multiple-output support vector regression with a firefly algorithm for interval-valued stock price index forecasting. Knowl. Based Syst. 5587–55100 (2014)
6. G.S. Atsalakis, E.M. Dimitrakakis, D. Constantinos et al., Elliott wave theory and neuro-fuzzy systems, in stock market prediction: the WASP system. Expert Syst. Appl. **38**(8), 9196–9206 (2011)
7. M.Y. Chen, M.H. Fan, Y.L. Chen et al., Design of experiments on neural network's parameters optimization for time series forecasting in stock markets. Neural Netw. World J. **23**(4), 369–390 (2013)
8. P. Singh, B. Borah, Forecasting stock index price based on M-factors fuzzy time series and particle swarm optimization. Int. J. Approx. Reason. **55**(3), 812–833 (2014)
9. L. Sifeng, L. Yi, *Grey Systems: Theory and Applications* (Springer Science & Business Media, Berlin, Germany, 2010)
10. J. Wei, L. Zhou, F. Wang et al., Work safety evaluation in mainland China using grey theory. Appl. Math. Modell. **39**(2), 924–933 (2015)
11. Q. Wan, Y. Wei, X. Yang, Research on grey wave forecasting model. Adv. Grey Syst. Res. (2010)

12. Y. Chen, K. He, C. Zhang, A novel grey wave forecasting method for predicting metal prices. Resour. Policy **49**, 323–331 (2016)
13. Y. Chen, B. Liu, Forecasting port cargo throughput based on grey wave forecasting model with generalized contour lines. J. Grey Syst. **29**(1), 51–63 (2017)

Research on Information Dissemination Model in WeChat-Based Brand Community

Huijie Peng, Xingyuan Li, Yisong Zhang, Qifeng Tang and Liwei Zheng

Abstract The recent emergence of WeChat-based brand communities in China is regarded as an effective channel for user-centric service marketing. However, Brand-related information dissemination and acceptance are the main hindering force in their sustainability. Despite the growing popularity of these brand communities, there has been only limited research modeling the brand-related information dissemination rule and process. The present study examines the factors affecting brand-related information dissemination and models the process of brand-related information dissemination in WeChat-based brand communities. Based on the characteristics of WeChat, this paper presents a modified information dissemination model. It quantifies the brand information dissemination process through MATLAB simulation and gets the law of information propagation. The influence of user acceptance threshold and social motivation in information dissemination is examined. The study findings suggest that the user acceptance threshold and relative motivation have obvious positive effects on the width and the speed of brand information dissemination. The findings have implications for organizations intending to use WeChat-based brand communities to practice user-centric service marketing.

Keywords WeChat-based brand community · User acceptance threshold · Social motivation · Information dissemination · SIR model

H. Peng (✉) · X. Li
School of Business, East China University of Science and Technology, Shanghai, Xuhui 200237, China
e-mail: penghuij123@163.com

Y. Zhang
Business School, University of Shanghai for Science and Technology, Shanghai, Yangpu 200093, China

Q. Tang · L. Zheng
Shanghai Zamplus Technology Co., Ltd, Shanghai 200072, China

© Springer Nature Switzerland AG 2020
H. Yang et al. (eds.), *Smart Service Systems, Operations Management, and Analytics*, Springer Proceedings in Business and Economics, https://doi.org/10.1007/978-3-030-30967-1_27

1 Introduction

Social media, e.g., WeChat, has changed the way users communicate with and remain in contact with their social circles [1]. This has begun to affect the operation of the organizations. Organizations build online brand communities system based on the WeChat platform to realize the seamless connection with the users and promote brand information sharing. The brand community is a group of emotional exchange, word-of-mouth communication, enterprise value, and culture information sharing between users [2]. With WeChat becoming the main mobile social method, WeChat-based brand community has increasingly become an important service marketing channel for organizations to develop the brand building, attract new users, and retain old users.

Previous literature has focused on social media-based brand communities from the perspective of characteristics, antecedents and outcomes, and business and consumers [3]. However, WeChat-based brand community has a unique communication mechanism and characteristics. In WeChat networks, juxtaposed among the undirected communication are often directed consumer-to-consumer and marketer-to-consumer communication [4]. For example, consumers and marketers can identify each other's comments and respond to each party's content in a targeted manner. Gan and Wang [5] collected the empirical data among WeChat users and found that social motivation is the most important factor for driving users to adopt WeChat [5]. Chen et al. (2014) found that entertainment, sociality, information, and trust have a positive impact on WeChat users' attitudes and user trust [6]. Brand community target a wider range of corporate customers, not just fans. Therefore, Marketers in the brand community expect that marketing related information can be diffused as far as possible. Chen et al. [7] explored the motivations of people sharing crisis information through WeChat [7]. However, most extant research has been focused on empirical research, while brand information dissemination process has received less quantified attention. This may be difficult in revealing the mechanism and predicting the brand-related information diffusion process.

To fill this gap, we introduce "user acceptance threshold" and "social motivation" in this study as two key factors that influence the dissemination process. Our research differs from the extant studies by modeling the mechanism and process of marketing related information and quantifying the extent to which different aspects of consumers behavior drive information dissemination. We propose the brand-related information diffusion model for WeChat-based community based on the SIR (Susceptible, Infective, and Removal) model frequently used in previous research to analyze disease outbreaks and knowledge diffusion.

Our study makes the following contributions. First, our study unveils the intricate roles of users and marketers on WeChat-based brand community and provides a rigorous quantification of the impact of a WeChat brand community's information on consumers' information dissemination of an apparel brand. Second, our research serves as the first attempt to measure the impact of "user acceptance threshold" and "social motivation" on marketers' proactive marketing activities in WeChat-based brand community at the individual consumer level.

2 Modeling

WeChat-based brand community is an interconnected service marketing system [8]. A user participates in multiple communities and a community includes multiple users. With the help of graph theory in complex networks, this paper defines users on the social network as nodes, and the friendship between users is expressed as the edge between nodes. By social networking as investigated by ourselves, we describe the network topology shown in Fig. 1.

As shown in Fig. 1, user 1 share information with the circle of his friends, only some of the friends (users 2, 4, and 5) propagate the information. Other friends (users 3, 6, 15) haven't shared this information. Through the investigation, we obtain a specific relationship. The user 1 has a common interest with the users 2 and 5, and the user 1 and the user 4 are family members. Different users will generate information dissemination behaviors for different reasons. We classify the influencing factors that affect the user's information dissemination behavior into the user's acceptance threshold and social motivation.

2.1 Analysis of Influencing Factors

We introduce "user acceptance threshold" and "social motivation" as two main influence factors.

Firstly, after receiving the information, the user will influence the formation of information based on user cognitive needs, emotional needs, social capital maintenance, self-presentation, and other psychological motives to make a decision whether to accept the information and turn the information into an acceptable state [9]. The threshold is regarded as the user's acceptance threshold. Also, every user is heterogeneous.

Fig. 1 Online community network topology

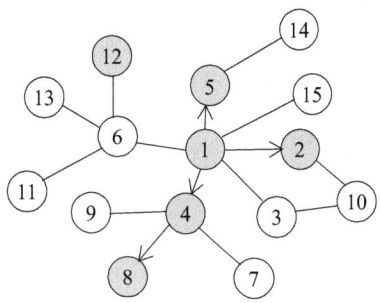

Secondly, we introduce "social motivation". Social motivation means that the user is more willing to receive and accept the information transmitted by close friends and information sharing. WeChat is a platform based on close social networks. At present, with the development of WeChat technology and user needs, there are more and more weak relationships entering into WeChat network. The research on the impact of user relationships on users' information dissemination is important.

In the traditional SIR model, there is only Infective who is the first one to share or post information in the brand community initially. The others are Susceptible. Then the first infective individual will disseminate the information with other users on their social network. Susceptible users are simultaneously affected by individual attributes and social attributes and will become infective with certain probability. In the process of dissemination, the information will reach a steady state at certain speed and all users will no longer share this information and become Removals [10]. We propose an improved SIR model based on the characteristics of the WeChat community based on the traditional SIR model.

2.2 Improved SIR Model

If a node j is susceptible at time t, it may be converted to an infective or removal state at time $t + \Delta t$. As is shown in Fig. 2, the nodes can be changeable between different states.

Here, take S as the Susceptible who does not know the information without sharing the information. Take I_a as the Infective who shares the information due to the user's psychological recognition information. Take I_b as the Infective who shares the information due to the user's acquaintance relationship. Take R as the Removal who will no longer share the information. In the social network, changes in the number and relationship of friends are not taken into account All individuals N in the social network are divided into three categories in *SIR* model: S (susceptible), I (Infective), and R (Removal). Among them, S will be turned into a Removal with a probability of $\beta < k >$. This type of user in the brand community is not interested in the information. Even if the information appears multiple times in the community. He still cannot be persuaded to pay attention to the information. However, some users will have an interest and resonate with the brand information. When the value of the information exceeds the user's acceptance threshold, They will communicate the information with a probability of $\alpha_1 < k >$. There is another possibility. Users not

Fig. 2 Flow chart of node
state transition

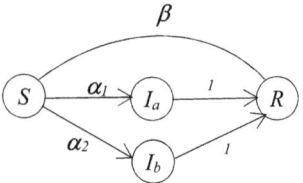

interested in the information itself will share the specific information with a probability of $\alpha_2 < k >$ because of the "acquaintance relationship". If the information is
shared by one user, the same information will no longer be shared. The Infective will
become the Removal with probability 1, and the information will stop spreading.
Therefore, our study uses the SIR model to simulate the information transmission
process in WeChat-based brand community. It is expressed by the dynamic differential equations as shown in Eqs. (1)–(4):

$$\frac{dS(k,t)}{dt} = -(\alpha_1 < k > +\alpha_2 < k > +\beta < k >)S(k,t) \tag{1}$$

$$\frac{dI_a(k,t)}{dt} = \alpha_1 < k > S(k,t) - I_a(k,t) \tag{2}$$

$$\frac{dI_b(k,t)}{dt} = \alpha_2 < k > S(k,t) - I_b(k,t) \tag{3}$$

$$\frac{dR(k,t)}{dt} = \alpha_1 < k > I_a(k,t) + \alpha_2 < k > I_b(k,t) + \beta < k > S(k,t) \tag{4}$$

Here, k represents the degree in the network, which is assumed to remain
unchanged in this model. N denotes the number of users in the brand community network $N = S(k,t) + I_a(k,t) + I_b(k,t) + R(k,t)$. In the process of node state transition,
users take the initiative to choose whether to disseminate information with a certain
probability. When users are exposed to information, the probability of disseminating
behavior due to the same factor is uniform and remains unchanged. Therefore, the
formula of propagating information probability under the psychological acceptance
of the user is shown in Eq. (5).

$$\alpha_1 = p < D_1(x_1, x_2, \ldots, x_n) > \tag{5}$$

$D_1(x_1, x_2, \ldots, x_n)$ denotes users acceptance threshold. It indicates the critical
value of the user's dissemination of the information after being influenced by his
own knowledge level, emotional appeal, and social capital maintenance. The value
of α_1 can be changeable with user's psychological characteristics. Secondly, since
the "acquaintance relationship" has a relatively larger impact on information dissemination, this paper defines the social motivation of users to share information as
shown in the formulas (6)–(8):

$$\alpha_2 = p < w_{ij}(\alpha, t) > \tag{6}$$

$$w_{ij}(\alpha, t) = \frac{\omega_{ij}^{\alpha}}{\sum_{m=1}^{k_1} \omega_{mi}^{\alpha}} \tag{7}$$

$$\omega_{ij} = \frac{\delta_{ij}}{k_i - 1 + k_j - 1 - \delta_{ij}} \qquad (8)$$

$\omega_{ij}(\alpha, t)$ indicates the strength of the social relationship between node i and node j. It represents the number of common friends. The more the number of common friends, the stronger the relationship between users. α denotes adjusting parameter. k_i denotes the degree of node i. k_j denotes the degree of node j. δ_{ij} denotes the number of common friends between i and j. In summary, the above two propagation probabilities depend on two influencing factors, indicating that the transition from a Susceptible to an Infective is a state transition that occurs under the influence of different factors.

3   Simulation

This article uses MATLABR 2014a for simulation experiment. We simulate the variation of the density of Susceptible, Infective, and Removal in WeChat community network structure with time. According to the characteristics of the model and the relevant data Settings in [11], we assume that: $S(0) = 199$, $I(0) = 1$, $R(0) = 0$, $N = 200$, $\alpha_1 = 0.5$, $\alpha_2 = 0.1$, $\beta = 0.4$. The number of iterations $T = 50$. Then, the results of the value on $S(k, t)/N$, $I_a(k, t)/N$, $I_b(k, t)/N$ and $R(k, t)/N$ are shown in Fig. 3.

It can be seen from Fig. 3 that density of the susceptible node $S(t)$ shows a sharp decrease at the initial stage, and when $t = 5$, it is close to zero. The information

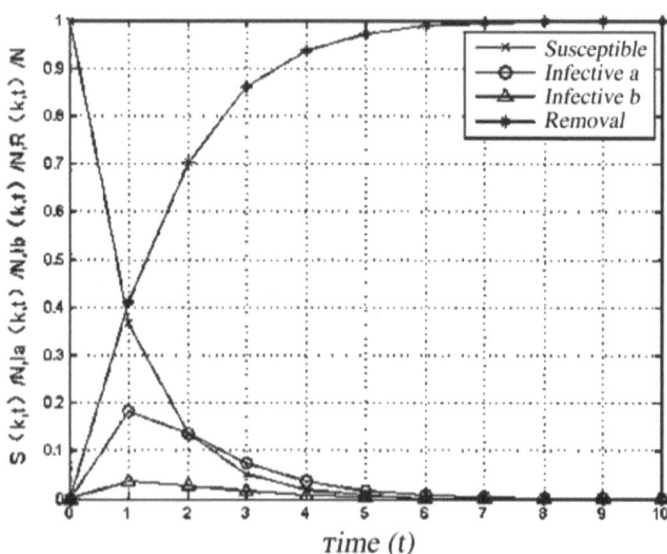

Fig. 3 Trend graph of node density with time t

spreads in the social network. The infective nodes show a relatively fast upward trend in the initial stage. When $t = 1$, both types of infective nodes reach the highest point at the same time, and at this point, the susceptible node intersects with the removal node. Then both infective nodes show a slow downward trend and return to zero. As we can see from this result, users who browse information will make quick choices in the early stage of information diffusion. Then, the Infective will share information simultaneously. After the number of communication reaches the highest level, it will slowly cool down and the information dissemination scope will be smaller. The removal node rapidly increases in the initial phase and continues to rise until it reaches a density of 1, indicating that all users no longer share information.

3.1 Sensitivity Analysis of User Acceptance Threshold

In the brand community, users are heterogeneous on user acceptance thresholds. When the value of brand information to users exceeds the acceptance threshold of each user, the user is willing to accept and share the brand information. The information will be spread and the user will turn into the Infective. If the value of information does not reach the critical point, the node will exit the transmission chain without the influence of other factors. In the simulation, α_1 is equal to 0.2, 0.5, 0.7, and 1, respectively. As the user acceptance threshold continues to decrease, the corresponding propagation probability gradually increases. The information propagation state in the scenario is shown in Fig. 4.

It can be seen from Fig. 4 that when α_1 takes different values, the most obvious change for node density is the Infective a. As time increases, the number of susceptible nodes decreases rapidly and then slowly approaches zero. However, the density of the infectious node a increases rapidly, reaching the highest point at $t = 1$. Then the number of Infective become less and less, and eventually, no user share the information. And as the value of α_1 increases, the density of the Infective a changes significantly, but even if $\alpha_1 = 0.9$, the density of the infectious node a is only close to 0.4 which is less than half of the total user. It can be seen that even with high quality information there is a limit to the speed and breadth of the information in the WeChat network. At a certain time t, if nearly one-half of the users share a piece of information the piece of information has a higher value content. On the one hand, this type of information is worth learning and learning from all organizations; on the other hand, the wide dissemination of such information indicates that this type of information is popular with consumers.

3.2 Sensitivity Analysis of Social Motivation

The WeChat brand community is a relatively private space, embodying a comprehensive new virtual community which has a dominant strong relationship and subsidiary

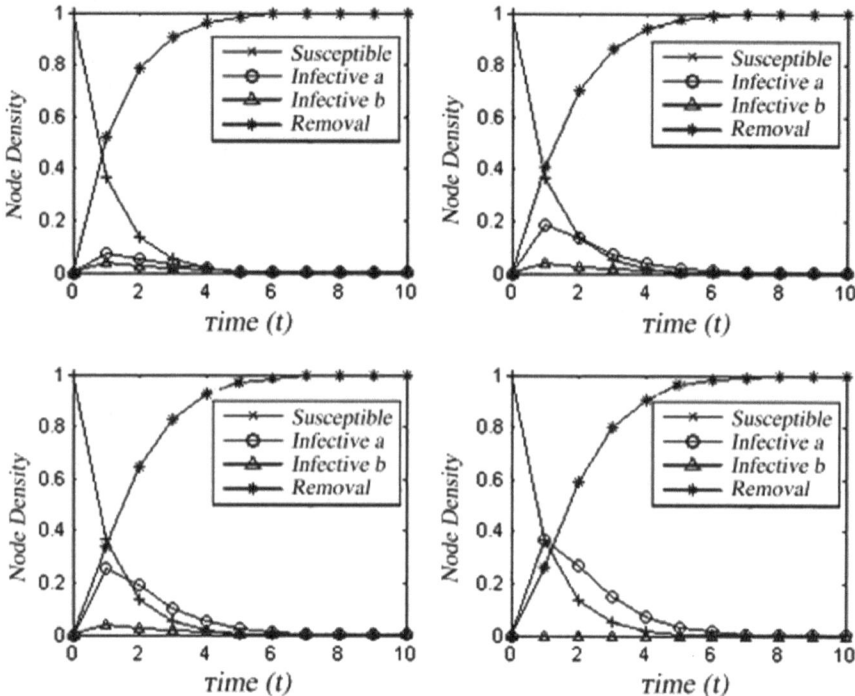

Fig. 4 The impact of α_1 on nodes density

weak connection. Previous studies have shown that information from "acquaintance relationships" is much more trustworthy. The "acquaintance relationship" mentioned in this paper not only refers to relatives with blood relationship but also includes the close friend relationship formed based on interaction and communication. The more frequent the communication between users, the closer the relationship will be, and the information is more likely to be transmitted by receivers. After receiving the information, although the user is not interested, he still accepts and shares the information. We assume that the α_2 is equal to 0.2, 0.5, 0.7, and 0.1 respectively. α_1 is equal to 0.1. The simulation results are shown in Fig. 5.

As is shown in Fig. 5, the magnitude of the propagation probability has a large influence on the node density. Here, $\alpha_1 + \alpha_2 = 1 - \beta$. When α_1 is equal to 1, the impact of α_2 on the Infective b is similar to the impact of α_1 on the Infective a. It can be seen that the spread of information in a strong relationship community is much broader. In the WeChat brand community, the higher the interpersonal relationship, the higher

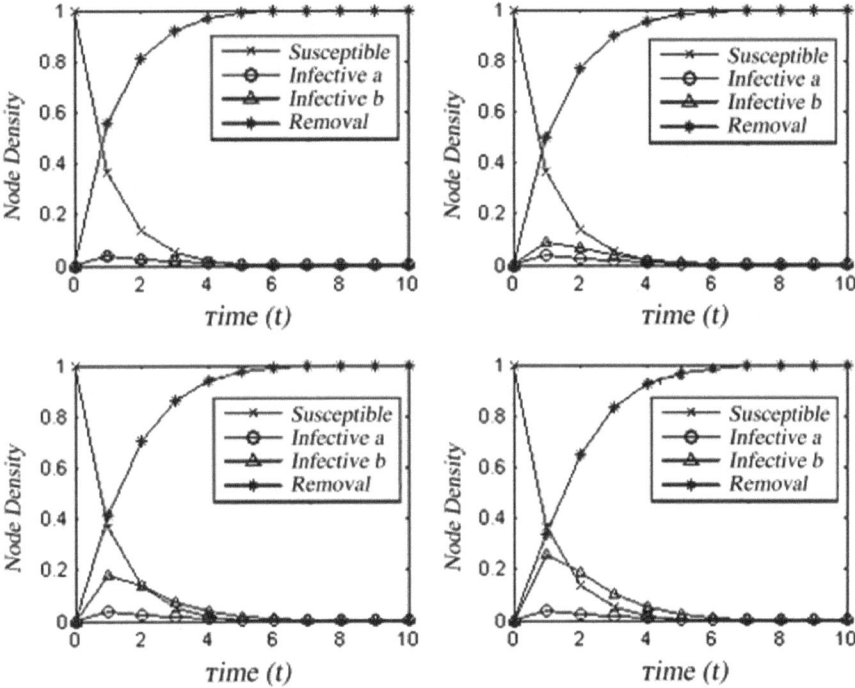

Fig. 5 The impact of α_2 on nodes density

the user's trust, and the user is willing to participate and share information. Therefore, in the WeChat brand community affected by strong social motivation, information has a strong communication effect in a short period of time.

4 Conclusion

This paper explores the degree of influence of user acceptance threshold and social motivation on brand information dissemination in the WeChat brand community. This study uses a simulation experiment to quantify the dynamic process of the actual communication process and provides a theoretical basis for effectively managing information dissemination on WeChat community marketing.

Our study has several important practical implications for WeChat marketers. Firstly, reducing user acceptance thresholds can promote information dissemination. Users hope that the information they receive to be beneficial to themselves and help their friends. Based on the user's psychological mechanism, users will measure the advantages and disadvantages of sharing information. Therefore, WeChat brand community managers or marketers should be user-centered to cater to or stimulate the psychology of the subject's cognitive and emotional needs. Adding knowledge, emotions, entertainment, and other elements to the brand information will make it easier for users to accept and share information with their friends around them. Secondly, the brand community should take advantage of the strong relationships in the WeChat network. From the results of this study, social motivation has a significant effect. As a network service system with strong relationship, WeChat has great value in its network structure, transforms a weak relationship into a strong relationship, and continuously enhances the strong relationship. Therefore, the managers of the brand communities of each organization should make full use of social values, motivate users to establish a strong network of relationships, and continuously expand and deepen strong relationships which is an important means for organizations to consider.

Acknowledgements The final draft of this paper was assisted by Zhang Yisong who comes from the University of Shanghai for Science and Technology. He collected a large number of social relationship data about WeChat community network and made a preliminary analysis. Moreover, he put forward constructive suggestions on the model construction of this paper. This paper was supported by the Science and Technology Commission of Shanghai Municipality (No.17DZ1101005).

References

1. P. Kaur, Underpinnings of user participation in service provider-hosted online communities. Serv. Sci. **8**(3), 249–262 (2016)
2. B. Rosenthal, E.P.Z. Brito, How virtual brand community traces may increase fan engagement in brand pages. Bus. Horiz. **60**(3), 375–384 (2017)
3. H.P. Zhang, Z.H. Wang, S.J. Chen et al., Product recommendation in online social networking communities: an empirical study of antecedents and a mediator. Inf. Manag. **56**(2), 185–195 (2019)
4. A.C. High, E.M. Buehler, Receiving supportive communication from facebook friends: a model of social ties and supportive communication in social network sites. J. Soc. Pers. Relationsh. **36**(3), 719–740 (2019)
5. C. Gan, W. Wang, Uses and gratifications of social media: a comparison of microblog and WeChat. J. Syst. Inf. Technol. **17**(4), 351–363 (2015)
6. Y.H. Chen, Lu, Y.B. Cao, B. Wang et al., How do product recommendations affect impulse buying? an empirical study on WeChat social commerce. Inf. Manag. **56**(2), 236–248 (2019)
7. Y. Chen, C. Liang, D. Cai, Understanding WeChat users' behavior of sharing social crisis information. Int. J. Hum.-Comput. Interact. **34**(4), 356–366 (2018)
8. Y.R. Chen, Y.R. Chen, Perceived values of branded mobile media, consumer engagement, business-consumer relationship quality and purchase intention: a study of WeChat in China. Public Relat. Rev. **43**(5), 945–954 (2017)
9. C.H. Yan, L.B. Yao, W. Bin et al., How do product recommendations affect impulse buying? an empirical study on WeChat social commerce. Inf. Manag. **56**(2), 236–248 (2019)
10. Y.P. Xiao, C.G. Song, Y.B. Liu, Social hotspot propagation dynamics model based on multi-dimensional attributes and evolutionary games. Commun. Nonlinear Sci. Numer. Simul. **67**, 13–25 (2019)
11. J.L. Wang, A.F. Liu, Z.F. Zhu, An information spreading model based on relative weight in social network. Acta Phys. Sinica **64**(5), 71–81 (2015)

Structure Evolvement and Equilibrium Analysis of International Credit Rating Market

Huang Xuewei and Wang Jianling

Abstract The purposes of this paper are (1) to introduce the process of structure evolvement of international credit rating market in chronological order, (2) to analyze S and P, Moody's, and Fitch's rating patterns and rating services, and (3) based on the content stated before, to conclude the characteristics of the evolvement of international credit rating market and analyze the cause of them. By using correlation analysis and statistical analysis we emphasized on doing research on the ratings released by S and P, Moody's, and Fitch (the Big Three) in 5 rating categories, financial institutions, insurance companies, corporate bonds, asset-backed securities, and government securities and gave our own opinion for causes of the current monopoly by the Big Three in credit rating market.7.

Keywords Credit rating · The Big Three · Monopoly

1 Introduction

Credit rating was derived in the United States in the early 20th century in order to address the problem of information asymmetry in the capital market. In 1909, the founder of Moody's, John Moody, began to rate the railroad bonds issued at that time, and later extended to various financial products and rating objects. After more than a hundred years of development, credit rating industry has gradually turned into a monopolistic industry from a free competitive one which are monopolized by three major international credit rating agencies, Standard and Poor's, Moody's Investors Services, and Fitch Ratings (also called the Big Three) [1]. In the second part, we are going to introduce the process of structure evolvement of international credit rating market.

H. Xuewei (✉) · W. Jianling
College of Economic and Management, Nanjing University of Aeronautics and Astronautics, Nanjing 210016, Jiangsu, China
e-mail: hxw_0806@126.com

© Springer Nature Switzerland AG 2020   303
H. Yang et al. (eds.), *Smart Service Systems, Operations Management, and Analytics*, Springer Proceedings in Business and Economics,
https://doi.org/10.1007/978-3-030-30967-1_28

In 2007, a large number of subprime mortgage defaults occurred in the United States, the personal housing mortgage market was in chaos, and the financial crisis was imminent. In this process, whether credit rating is the inducement of the financial crisis has been controversial. It is argued that the failure of the quantitative model used by rating agencies, the inherent conflict of interest in the payment model, and the monopoly situation of rating agencies are one of the factors leading to the financial crisis [2]. Therefore, in the second part of this article, we are also going to analyze the Big Three's rating patterns and rating services.

Facing with the current situation of credit rating industry, both of the U.S. and E.U. government enacted a series of laws and regulations to improve the quality of ratings by enhancing competition, reducing the dependence of external rating, and strengthening the responsibility of disclosure, etc. [3] In the third part, we will discuss the characteristics of the evolvement of international credit rating market and analyze the cause of them after the enactment of those regulations.

With China Financial Market developing swiftly, in order to figure out the present situation of the international credit rating market and keep pace with it, it is important to understand how the structure of international credit rating market evolved and analyze the characteristics and the trend of the international credit rating market in the future.

2 The Analysis of Evolvement of International Credit Rating Market

2.1 The Process of Structure Evolvement of International Credit Rating Market

(1) Free competition

Credit rating was derived from the United States in the early 20th century. In 1890, John Moody, the founder of Moody's Investors Service began to compile a financial information manual of American companies and gradually established an evaluation system to measure their default risk. In 1909, Moody first assessed the bonds of American Railway Corporation which showed the credit worthiness of the bonds issued by the railway companies then offered them to investors and in return obtained information usage fees. His rating manuals were widely welcomed by investors and he became the pioneer of the credit rating market in the United States.

After World War I, with American capital market boomed, investors' demand for credit rating soared correspondingly. Other credit rating agencies appeared in the market. In 1923, Standard and Poor's started its service of credit rating. Credit rating brought about a virtuous circle for the bond market and nourished the credit rating agencies at the same time.

During the Depression in the 1930s, investors suffered a huge loss due to the frequent bond defaults resulted from plenty of bankruptcies. That made investors realize the significance of credit rating. After that, the US Securities and Exchange Commission (here forth, SEC) strengthened their regulation on bond through dividing them into investment grade and speculative-grade. This reform facilitated the development of credit rating agencies. During that time, credit rating industry had a very low access threshold and no special business license restrictions. Credit rating industry is a free competition industry with no entry barriers at all.

(2) Monopoly by the Big Three

Penn Central Transportation Corporation, the largest debt servicing event in American financial history in 1970, went bankrupt with $1.6 billion in bonds and $125 million in commercial paper outstanding. The bankruptcy of Penn Central Transportation Corporation suggested the beginning of the modern credit rating industry. In 1975, SEC recognized Standard and Poor's, Moody's, and Fitch Rating Companies as the first "nationally recognized rating organizations" (NRSRO).

After 1975, SEC recognized four more credit rating agencies into NRSRO: Duff and Phelps in 1982, McCarthy Crisanti and Maffei in 1983, IBCA in 1991 and Thomson Bankwatch in 1992. Since then, companies in credit rating industry continued to merge and reorganize. IBCA was frustrated that the SEC did not allow them to carry out ratings other than bank ratings. Then it merged with Fitch IBCA in 1997 and added established its new headquarters in London. Fitch's impact on banks, financial institutions, and sovereign ratings has been significantly enhanced after the acquisition. In 2000, Fitch acquired Duff and Phelps, the fourth largest rating agency based in Chicago. Later in the same year, Fitch further acquired Thomson Bankwatch, the world's largest bank rating agency. These two acquisitions further expanded Fitch's service scope and its number of international branches. Fitch's merger and reorganization brought the international credit rating market back to the tripartite monopoly in 1997.

In 1997, SEC clarified the basic principles of "NRSROs" and established a formal procedure for the accreditation of qualifications. At that time, the power of market discipline had faded away by degrees because the original credit rating market with free competition and low entry barrier had transformed to a monopoly market dominated by the Big Three.

(3) NRSRO-oriented monopolistic competition

After the Crisis, regulatory agencies in Europe, America, and other countries actively committed to facilitating competition in the rating industry, speeding up the approval of small and medium-sized agencies to obtain rating qualifications, and intending to break the monopoly position of the Big Three. From 2003 to 2008, SEC recognized seven NRSRO institutions. There are currently ten NRSRO certificated credit rating agencies.

In the second half of 2006, the United States Congress passed the Credit Rating Agency Reform Act of 2006. The promulgation of this regulatory act strengthened the supervision of credit evaluation institutions and will have a far-reaching impact on the monopoly pattern of the credit rating market in the United States and even the whole international market for decades. However, at present, loose market access does not break the inherent advantages of the three oligarchs.

As shown in Fig. 1, there are mainly five types the international credit rating service: financial institutions, insurance companies, corporate bonds, asset-backed securities, and government securities. Among them, the government securities service accounts for 79.6%, and the insurance company service accounts for 0.9%.

As shown in Table 1, considering the market share of S and P, Moody's and Fitch in the five categories of service in 2017, the Big Three still have a monopoly in credit

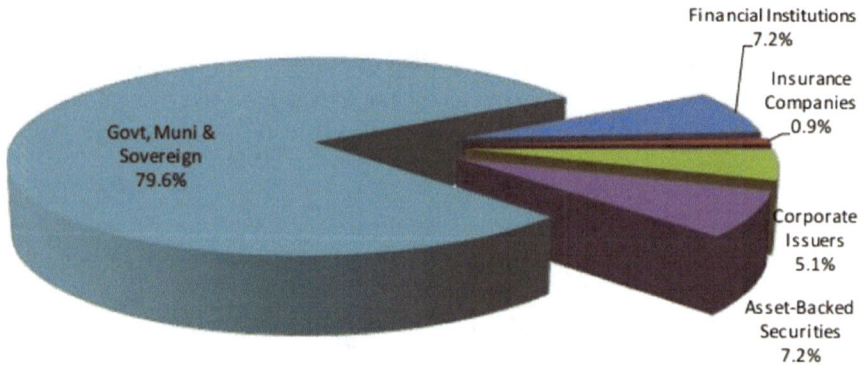

Fig. 1 Percentage by five international credit rating Category in 2017 (The pie chart was reorganized according to the Annual Report on Nationally Recognized Statistical Rating Organizations released by SEC, 2017)

Table 1 Number of outstanding credit ratings by rating category

NKSKO	Financial institutions	Insurance companies	Corporate issuers	Asset-backed securities	Government securities	Total ratings
A.M. Best	N/R	7,537	1,359	18	N/R	8,914
DBRS	7,969	158	3,037	12,757	16,784	40,705
EJR	11,112	837	6,480	N/R	N/R	18,429
Fitch	44,965	3,188	17,848	39,981	197,543	303,525
IIR Ratings	547	N/R	140	N/R	352	1.039
JCR	787	65	2,356	N/R	486	3,694
KBRA	705	5	1	5,561	63	6,335
Moody's	49,472	3,230	44,676	64,188	619,478	781,044
Morningstar	35	N/R	308	3,591	N/R	3,934
S&P	58,582	6,859	50,672	49,162	952,910	1,118,185
Total	174,174	21,879	126.877	175,258	1,787,616	2,285,804

The table was reorganized according to the Annual Report on Nationally Recognized Statistical Rating Organizations released by SEC, 2017

rating industry: their share of financial institution service is 84.7% in total, the share of insurance company service is 59.6% in total, the share of corporate bond service is 88.1% in total, the share of asset-backed securities service is 83.7% in total, and the share of government securities service 98.9% in total.

We can conclude that rising credit rating agencies have made little progress in the government securities service which is the largest part of credit rating service while in the insurance company service which is the smallest, their share accounts for nearly half of the total.

2.2 The Analysis of the Big Three's Credit Rating Patterns

(1) A brief introduction of the Big Three

Moody's was originally established by John Moody in 1900 and its headquarter is in Manhattan, New York. Moody's started its sovereign rating in 1909. It rated and issued its credit ratings of around 100 sovereigns. In January 2008, Moody's business was divided into two parts: Moody's Investor Services and Moody's Analytics. Moody's Investor Services provides credit rating reports of companies, government agencies and global structured financial securities, and Moody's Analysis provide risk management.

Standard and Poor's history can be traced back to 1860. It was combined by Poor's Publishing Company and Standard Statistics Company and its headquarter is located in New York. S and P began to rate national sovereignty since 1916 and up to now it has more than 1 million credit ratings outstanding on government, corporate, financial sector, and structured finance entities and securities.

Fitch rating was set up by John K. Fitch in 1913 with two headquarters in New York and London. In Apr. 2018, Hearst, an American media company purchased 20% stake of Fitch from a French venture capital company called Firmalac and make Fitch its wholly owned subsidiary. Fitch Rating is the only European-funded international rating agency among the three major international rating agencies in the world and the smallest among the Big Three.

(2) The comparison of the Big Three's rating level

CRAs rank securities and issuers on a relative scale. For example, Fitch and S and P use AAA, AA, A, BBB, and BB… while Moody's goes with Aaa, Aa, A, Baa, Ba… CRAs also use modifiers (such as "+" or "−"; "1, 2, or 3") appended to the rating level in order to denote relative status within the major rating levels. The combination of a level (for example "A") and a modifier (such as "+") constitutes a "notch" in the rating scale (for instance "A +") [2]. A change of a rating means assigning a credit rating a different notch in the rating scale. Levels do not reflect absolute measures of risk, only that securities classified in a higher level have higher quality than those in the levels below (Garcia 2012). S and P and Fitch's rating symbols have high similarity. Moody's uses different classification of symbols. There is incomparability among them but there is a certain mapping relationship between them. We sort out the mapping relationship of the rating levels of the Big Three by correlation analysis, and convert the rating letter grade into the corresponding digital grade so as to facilitate intuitive recognition and comparative analysis (Table 2).

(3)   The comparison of rating service of the Big Three

These three rating agencies all have the qualifications to issue ratings of financial institutions, insurance companies, corporate bonds, asset-backed securities, and government securities business but each has its own emphasis. Taking the NRSROs agency report issued by SEC in 2017 as an example, S and P is far ahead in government bond rating with a market share of 52.80% higher than the company's overall market share of 49.20%. It also has an obvious advantage in corporate bond rating service with a market share of 45.70%. Moody's focuses on asset-backed securities and a market share of 34.30% in government securities rating service are also very compelling. Meanwhile, Fitch is more focusing on financial institution rating service.

Table 2   The mapping relationship between the Big Three's rating levels and the digital grades

Credit rating	Moody's	Standard and poor's	Fitch	Numerical code
Highest grade	Aaa	AAA	AAA	1
	Aa1	AA+	AA+	2
High grade	Aa2	AA	AA	3
	Aa3	AA−	AA−	4
	A1	A+	A+	5
Upper medium grade	A2	A	A	6
	A3	A−	A−	7
	Baa1	BBB+	BBB+	8
	Baa2	BBB	BBB	9
	Baa3	BBB−	BBB−	10
Non-investment grade	Ba1	BB+	BB+	11
	Ba2	BB	BB	12
	Ba3	BB−	BB−	13
	B1	B+	B+	14
Low grade	B2	B	B	15
	B3	B−	B−	16
	Caa1	CCC+	CCC+	17
	Caa2	CCC	CCC	18
	Caa3	CCC−	CCC−	19
	Ca	CC	CC	20
	C	C	C	21
Default	N/A	D	DDD/DD/D	22

Fitch's ratings exceed Moody's with 24.90% of the market share in the business. Generally speaking, Fitch's rating performance is acceptable in the non-governmental bond rating service. Fitch's size is smaller than the other two rating companies, but in the global market, especially in emerging markets, Fitch's sensitivity is higher and its vision is more international.

3   The Characteristics of Evolvement of International Credit Rating Industry

(1)   The overall number of credit rating items declined steadily after the Crisis

According to the Annual Report on Nationally Recognized Statistical Rating Organizations from 2010 to 2017, we analyzed the number of rating items of the following five rating service statistically during that time: financial institutions, insurance companies, corporate issuers, asset-back securities, and government securities. Because of the total share of the former four rating service is far less than the share of government securities rating service, we called all of those four rating services together "non-government securities" and we found out that the overall number of credit rating items declined steadily after the Crisis. In 2010, the rating items of government securities is 2,185,726 while the number of non-government securities is 630,873. There were 2,816,599 ratings in total and non-government securities accounts for 22.40% of the total share. While in 2017, overall 2,192,211 ratings were issued and the number of government and non-government securities is 1,744,345 and 447,866, respectively. The share of non-government securities rating accounts for 20.43%. Compared with 2010, the average annual decline rate of total rating is 3.17%, the average annual decline rate of government rating is 2.89%, and that of non-governmental rating is the largest up to 4.14% (Fig. 2).

(2)   The newly recognized credit rating agencies have the benefit to activate the market

When we analyze the four types of non-governmental securities business rating it shows that the second largest asset-backed securities market has declined significantly with the number of rating items reaching 302,461 in 2010 and 157,784 in 2017, with the largest annual average decline of 6.83%. The number of rating items of financial institutions with the third largest business volume in 2010 is 209,338, and the number of rating items in 2017 is 157,368. The average annual decline was 3.55%, which exceeded the overall market level by 3.17%. Contrary to the overall market downturn, the trend of rating in small market corporate bonds and insurance companies is increasing. In 2010, the number of corporate bond ratings was 99,286 and in 2017, the number of corporate bond ratings was 112,163 with an average annual increase of 1.85%. In 2010, the number of insurance company ratings was 19,788 and in 2017, the number of corporate bond ratings was 20,551, with an average annual increase of 0.55% (Fig. 3).

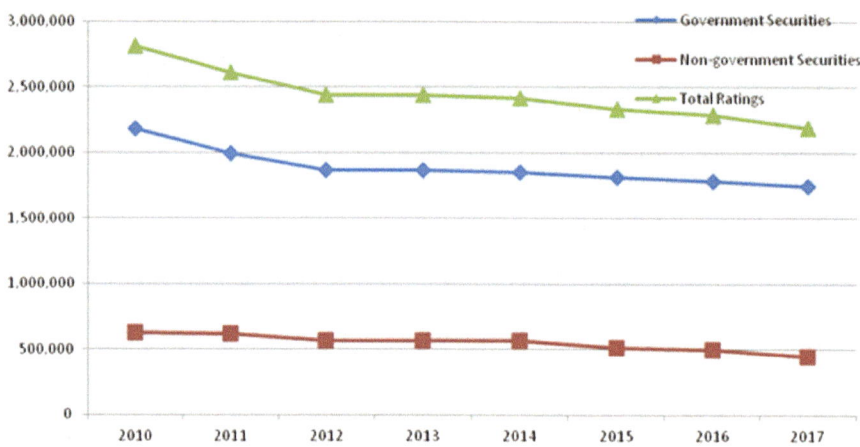

Fig. 2 The number of ratings released by NRSROs

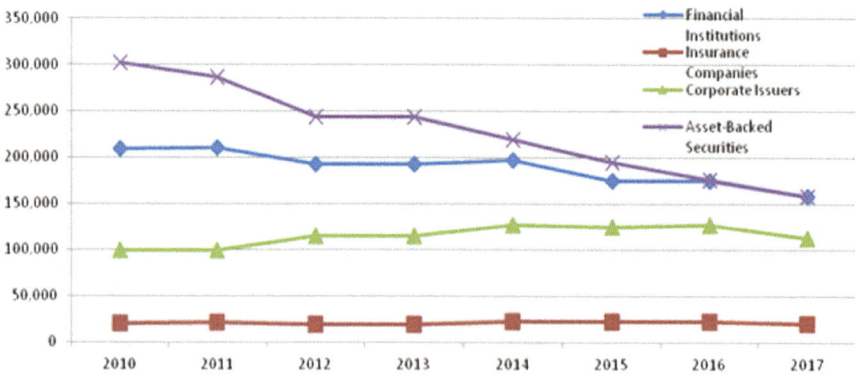

Fig. 3 The number of non-governmental rating service (The chart was reorganized according to the Annual Report on Nationally Recognized Statistical Rating Organizations released by SEC, 2017)

We further analyze the market share of NRSRO 10 certification agencies in corporate bonds and insurance companies in 2017. JCR and A.M. Best entered corporate bond rating business with 5.7%, 2.6%, 2.2%, and 1% share, respectively. The strength of emerging institutions is still weak. A.M. Best entered insurance company rating business with 35.0% market share and broke the market pattern of this business. Unfortunately, these two rating services account for very little in the international credit rating market. In 2017, insurance companies rating service only accounts for 0.94% of the whole market. and neither HR Ratings nor Morningstar is engaged in this business. These two small markets are growing against the background of the decline of the overall credit rating market which suggested that emerging institutions are beneficial to activate the market and promote the development of the market.

(3) The degree of monopoly in the credit rating industry is intensified instead of being alleviated.

The Huffindar-Herchmann Index (HHI) is a comprehensive index to measure industrial concentration. The measurement method of HHI is the quadratic sum of the market share of each major market competitor ranging from 0 to 10,000. When there are a large number of similar enterprises in the industry the index is close to 0. When one enterprise monopolizes the whole industry, the index is 10,000. According to the U.S. Department of Justice, the HHI index between 1,000 and 1,800 is considered appropriate concentration and more than 1,800 is a sign of overconcentration.

Besides, HHI Inverses can also indicate the degree of the market concentration. The lower the HHI Inverses is, the more centralized a market is. It can be calculated through dividing 10,000, the maximum concentration, by HHI index. It is used to indicate that the market of a specific industry is concentrated in several enterprises of similar size. According to the annual report of NRSROs issued by SEC in 2010, the HHI inverse index of the overall market concentration is from 2.88 in 2010 to 2.70 in 2017. Which means the monopoly degree of credit evaluation market has not alleviated but intensified after new institutions' entrance.

The market concentration of government securities business is higher than the overall market concentration. Since 2011, it has been distributed in 2.45 enterprises of similar scale with the highest barriers to entry. In contrast, the entry barriers of the other four types of business are slightly lower, and the HHI inverse index is higher than the overall market situation, especially in the financial institutions and insurance companies. In 2017, the financial institutions' credit rating market is equivalent to be distributed in 3.88 enterprises of similar scale and the insurance companies' is equivalent to be distributed in 3.79 enterprises of similar scale (Fig. 4).

The market share of the Big Three in the past few years suggests that the total market share of the Big Three has been 95% during 2010–2017, firmly monopolizing the credit rating market. In addition, S and P's market share is the largest and continues to expand from 42.27% in 2010 to 49.20% in 2017. The runner-up is Moody's market

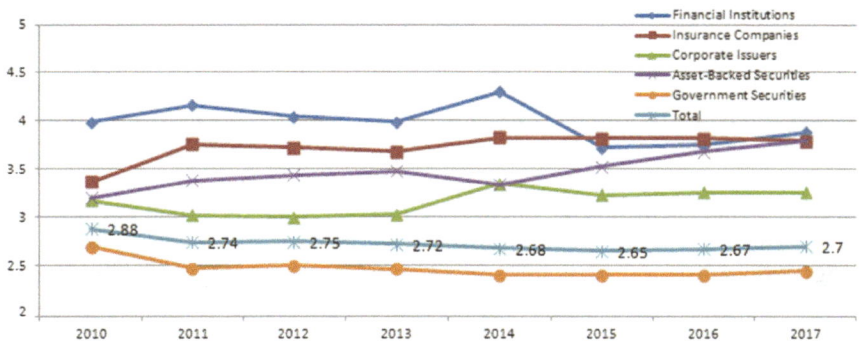

Fig. 4 2010–2017 HHI inverse index (The chart was reorganized according to the Annual Report on Nationally Recognized Statistical Rating Organizations released by SEC, 2017)

share which has a trend of decreasing from 36.90% in 2010 to 33.10% in 2017. Fitch's market has basically maintained at this level after a sharp decline to 13.35% since 2011.

4 Equilibrium Analysis on the Causes of Monopoly in International Credit Rating Market

At present, the credit rating market is still an oligopoly market which includes almost all the traditional characteristics of oligopoly market, including entry barriers to the market, a small number of market participants, interdependence, pricing power, and high profitability. There are two opposite views on which market structure is better, monopolistic or competitive: the competition theory and monopoly theory. The representative of competition theory study is Partnoy [4]. He believes that competitive market structure will make credit rating agencies pay more attention to their reputation then publish high-quality ratings. The monopoly theory holds that the competitive market structure of rating industry is likely to lead to the phenomenon of "rating shopping" and overrating. Market structure with stronger monopoly is conducive to the improvement of credit rating quality. The representative studies are Becker et al. (2011) [5] and Bolton et al. [6].

Currently, the credit rating industry has fallen into the paradox of free competition. According to Becker and Milbourn [7] research, reputation plays at least a relatively positive role in improving the quality of credit ratings. Empirical studies show that, after increasing competition, rating agencies become more friendly to issuers, less information is disclosed, and competition hinders reputation mechanism functioning in the credit rating market. As a result, the rating quality declined instead. The highly centralized market structure makes the Big Three lack of pressure to improve their rating methods and quality. However, the increase of participants has not played its due role and the market is still controlled by the Big Three.

(1) The recognition of NRSRO

For the international credit rating market, SEC only provides certification to credit rating agencies who are recognized by NRSRO. That forms a very obvious institutional barrier for rising credit rating agencies to entry. From 1975 to 2002, there were only three accredited institutions, which laid a solid foundation for the Big Three over the past two decades. After the Crisis, regulatory agencies in Europe, America, and other countries are actively committed to promoting competition in the rating industry, speeding up the approval of small and medium-sized agencies to obtain rating qualifications, and intending to break the monopoly position of the Big Three.

However, whether rising credit rating agencies can really be recognized by the international market requires rich experience and credibility and the qualification of rating is only one of the necessary conditions. The Big Three do not only have a long history but also benefit from the scale economy in the area of infrastructure costs, such as abundant human capital, fund, and IT support as well as a fixed customer base. Compared with the Big Three, the newly qualified agencies are still lack of credentials and unable to compete with them in the international market in the short term.

(2) Hegemonic Monopoly by the U.S.

As the origin of the credit rating industry, the U.S. owns the largest financial capital and financing instrument along with the most advanced capital market. It is the main channel of international open market financing. However, to enter this market, issues and issuers must accept the grade rated by American rating agencies. Rating opinions determine the level of financing costs, which gives the rating industry in the United States the edge when it competes with other countries'. It also has an absolute voice and monopoly status in the international capital market. So far, 9 of 10 NRSRO institutions which are headquartered in the United States. Taking Europe as an example, the credit rating business in Europe is mainly monopolized by the Big Three. Since the European Debt Crisis, the Big Three have repeatedly downgraded the sovereign debt rating of the relevant countries. The rating agencies have been accused of contributing to the occurrence of the European Debt Crisis. European Union, Japan, China, and other countries have realized the importance of establishing local international credit rating agencies. But it is undoubted for other countries credit rating agencies that to join the SEC-certified NRSRO group and occupy a seat have to face the oppression of American hegemony.

(3) Matthew Effect of credit rating market

Under the market structure of oligopoly, the Big Three have become market standard makers, and their credit rating standards are practiced in the worldwide. In order to seek market recognition, emerging agencies must use the same rating methods and standards. They either follow the existing methods and standards or innovate on the basis of them. However, no matter which way they adopt they cannot break the overwhelming advantages of the Big Three. In addition, the credit ratings of the Big Three cover a considerably wide range of areas because their accumulated reputation precedes them. Investors and issuers are generally eager to obtain ratings from credit rating agencies with abundant experience and solid reputation. As a result, the cycle of discrimination against emerging rating agencies will continue. Oligopoly institutions will still have a large market share.

(4) The characteristics of credit rating industry

The reputation mechanism plays a prominent role in the rating market. The maintenance of rating agencies' own reputation capital can regulate the healthy development of the rating market and effectively reduce the probability of rating agencies releasing false ratings. The reputation mechanism holds that because the rating results are open to the market, the public can continuously test the rating quality of a rating agency. Ratings which are independent, fair, and reliable will accumulate reputation capital for the rating agency, and vice versa, it will bring about loss of reputation capital. Rating agencies with more reputable capital are more likely to be trusted by investors so debt issuers will choose rating agencies with higher reputations to obtain investors' recognition.

Moderate competition will make a reputation mechanism play a more effective role. When investors and issuers have enough choices and alternatives, credit rating agencies which provide high quality ratings can have a foothold in the market. However, excessive competition will make the rating agencies choose between reputation and current earnings. The pressure of shrinking market share drives the rating agencies to reduce the expected market share returns from maintaining reputation and the whole financial market will suffer losses because of the "rating inflation" of the rating agencies. The industry characteristics of credit rating market also restrict the free competition of the market.

5 Summary

This article gives a summary of the evolvement of the international credit rating market and compares the organization information and rating patterns of the Big Three. According to the 2010–2017 Annual Report on Nationally Recognized Statistical Rating Organizations, we conducted a statistical analysis on the number of five types services offered by the Big Three which are financial institutions, insurance companies, corporate bonds, asset-backed securities, and government securities. And we conclude the latest characteristics of evolvement in credit rating industry are listed below:

(1) The overall number of credit rating items declined steadily after the Crisis;
(2) The newly recognized credit rating agencies are beneficial to activate the market;
(3) The degree of monopoly in the credit rating industry is intensified instead of being alleviated.

At present, the credit rating market is still an oligopoly market which includes almost all the traditional characteristics of oligopoly market including entry barriers to the market, a small number of market participants, interdependence, pricing power, and high profitability. We believe the equilibrium causes of Monopoly in International Credit Rating Market are stated below:

(1) The recognition of NRSRO;
(2) The Hegemonic Monopoly by the U.S.;
(3) The Matthew Effect of credit rating market;
(4) The characteristics of the credit rating industry.

Acknowledgements This work was done with great help from The Ministry of Education Research of Youth fund Projects on Humanities and Social Sciences of China (NO. 15YJC630122) and NUAA Graduate Base Innovation Open-End Fund (KFJJ20180911).

References

1. R Zhang, *Functional Disorder and Reform of Credit Ratings Mechanism,* China Money, No. 7, pp. 20–25 (2010). 张锐: "国际信用评级机构的功能扭曲与改革前景", 中国货币市场, No. 7, pp. 20–25 (2010)
2. Y. Jin, *Break the Monopoly Pattern.* China Urban Finance, No. 10, pp. 14–17 (2010). 金言: "打破垄断格局创建信用评级新秩序——国际信用评级体系改革与中国评级业发展展望", 中国城市金融, No. 10, pp. 14–17 (2010)
3. K. Boehm, Credit rating agencies: The EU regulatory framework assessment. Bachelor of business administration international business and logistics thesis, Helsinki Metropolita University of Applied Science, May 2013
4. F. Partnoy, Why markets crash and what law can do about it. Soc. Sci. Electron. Publ. **61**(3), 741–817 (1999)
5. V. Ivashina, B. Becker, *Cyclicality of Credit Supply: Firm Level Evidence* (Harvard Business School Finance Working Paper, 2011), pp. 10–107
6. P. Bolton, X. Freixas, J. Shapiro, The credit ratings game. J. Financ. **67**(1), 85–111 (2012)
7. B. Becker, T. Milbourn, *Reputation and Competition: Evidence from the Credit Rating Industry* (Harvard Business School, 2008)

Teaching a Man to Fish: Teaching Cases of Business Analytics

Sung-Hee "Sunny" Park, Soohoon Park and Linda B. Oldham

Abstract This paper gives an overview of a somewhat unusual business analytics initiative at a southeastern public university in the USA. The massive influx of data and the accessibility of analytics tools have presented a provocative opportunity for businesses to improve decision-making and have also created a demand for talent with data analytics skills. Though many universities have stepped up to meet this growing need for data analytics talent, the vast majority have done so by offering specialized programs at the M.S. level. The school offers embedded undergraduate analytics program with experiential learning. This paper includes our motivations and reasons behind the analytics program as well as how we have implemented it.

Keywords Business analytics · Experiential learning · Data lab

1 Introduction

Entrepreneur, Joris Toonders, mentioned "DATA in the 21st Century is like Oil in the 18th Century: an immensely, untapped valuable asset" [1]. While we totally agree with his argument, we would like to add this: An oil reservoir will not last forever; however, a data reservoir will continuously grow even if we are drilling and pumping from it massively. Reference [2] indicates data-driven decision-making is increasingly impacting organizations. As business becomes more data-driven, the abilities to manipulate, analyze, and leverage data are table stakes for many high-value jobs. A poll, conducted by Gallup for the Business-Higher Education Forum, revealed that by 2021, 69% of employers expect candidates with data analytics skills to get preference for jobs in their organizations [3]. Yet only 23% of college and university leaders say their graduates will have those skills.

S.-H. "Sunny" Park (✉) · L. B. Oldham
Department of Management Science, University of South Carolina, Columbia, SC 20208, USA
e-mail: sunny.park@moore.sc.edu

S. Park
Department of Business Administration, Bemidji State University, Bemidji, MN 56601, USA

© Springer Nature Switzerland AG 2020
H. Yang et al. (eds.), *Smart Service Systems, Operations Management, and Analytics*, Springer Proceedings in Business and Economics,
https://doi.org/10.1007/978-3-030-30967-1_29

Business schools in U.S. have responded by offering business analytics programs. The vast majority of those programs are stand-alone degree programs at the graduate level or for small cohorts of undergraduates. Our business analytics program initiative at Darla Moore School of Business (DMSB) for competing with other programs is to create an embedded program for all undergraduate DMSB students. Our initiative is based on following building blocks: (1) building analytics capabilities in two stages with experiential learning, (2) building a multidisciplinary and highly collaborative data analytics infrastructure, and (3) offering a four-course Business Analytics Concentration.

The remainder of the article is organized as follows. First, the curriculum changes at DMSB is outlined as to how we structure a set of business analytics courses and how we provide opportunities for more advanced analytical work on topics germane to each major via experiential learning. Next, we describe our infrastructures such as the Center for Applied Business Analytics, DMSB Data Lab, and Data Science Corps at USC. A final section is offered that denotes some student analytics projects as examples.

2 Curriculum Changes and Building Analytics Capabilities

Our vision at DMSB for teaching business analytics is based on building analytics capabilities in two stages: first by ensuring students' overall data proficiency, and second by providing opportunities for more advanced analytical work on topics germane to each major via experiential learning. To acquire this advanced analytical capability, first, we changed the business core curriculum for the following two courses:

- MGSC 290—Computer Information Systems in Business: An introduction to the effective use of information systems tools in day to day business communications, analysis, and decision-making.
- MGSC 291—Statistics for Business and Economics: Descriptive statistics, topics in probability, statistical inference, and modeling. Emphasis on the collection, summarization, analysis, and reporting of numerical findings relevant to business decisions and economic analysis.

MGSC 290 was offered to the sophomores at DMSB before the curriculum change. While we changed its level of offering to the first year students at DMSB, we teach Excel, Access, and PowerBI. MGSC 291, which used to teach basic business statistics, now focus on analytics project involving real-world data applied to a complex business problem while all DMSB students learn basic statistics from a new additional courses (STAT 206 Elementary Statistics for Business).

For this to be possible, a Data Lab with both virtual/cloud-based and physical characteristics, staffed 5 days a week (by Ph.D. and Master Students as data mentors) was required. In the DL, students will be trained on basic coding/computer logic using select statistics and computer languages/tools, where the projects will test a student's ability to extract, clean, load, visualize, and analyze large structured and/or unstructured data sets. Student teams will define the problem at hand, analyze the data using appropriate analytics methods/software (hard skills), and report findings to a general business audience (soft skills).

To acquire advanced analytical capability for DMSB students, the Concentration in Business Analytics is offered to all DMSB students. This concentration requires four courses (12 h of course work), as detailed below.

- MGSC 394—Data Analytics for Business (Required course): This course helps develop two sets of skills in students. First, effectively using standard business software tools that are used in data analysis. Secondly, using multiple analytics software tools to gain insights to guide decision-making.
- Three elective courses from the following, each of which has significant analytics applied in the area of study:
  - ACCT 404: Accounting Information Systems I
  - ECON 436: Introductory Econometrics
  - FINA 444: Corporate Risk Management
  - FINA 469: Investment Analysis and Portfolio Management
  - FINA 472: Student-Managed Investments
  - MGMT 425: HR Analytics
  - MGSC 390: Business Information Systems
  - MGSC 486: Service Operations Management
  - MKTG 352: Principles o f Marketing Research
  - MKTG 447: Pricing Strategy and Analytics.

3 Business Analytics Infrastructure

3.1 Center for Applied Business Analytics

The DMSB Center for Applied Business Analytics (CABA) was created to enhance business analytics teaching and research. The CABA builds capacity in the field of business analytics across all majors in the Moore School by providing students at the undergraduate and graduate levels opportunities to better understand how to transform data into meaningful decisions through the use of analytics. The CABA also works with business partners to bring real-world data sets into the classroom to enrich teaching and research. Students use real company data and are trained to make practical data-driven decisions that prepare them to better serve the needs of employers (state, regional, and national). The Moore School offers an undergraduate concentration in business analytics that approximately 300 students currently pursue.

3.2   Data Lab

The DMSB Data Lab is a vital resource to support data-driven student projects for the 1,500 freshman students admitted to the DMSB each year who are required to complete projects using R to analyze real-world business data. The Data Lab staffs will function as mentors in the Data Lab assisting these 1,500 students with their data projects and promoting data literacy and providing basic training on the use of R to analyze data. DMSB students will gain experiential learning, often considered the most powerful type of academic learning, by engaging in data projects and mentoring other students in the Data Lab. This type of engagement will expand their data science skills and broaden their understanding that proficiency in data analysis and data-driven decision-making are critical skills for everyone.

The Lab staff will facilitate the development of all data project communities, solicitation of project sponsors, access to data and the organization of projects. DMSB students will undertake projects provided by collaborating community, governmental and industry organization. These projects will teach the students how to extract, clean, load, analyze, and visualize large real-world datasets. Staff in the collaborating organizations will also sharpen their skills in data science by working with the DMSB students. Figure 1 shows the system architecture of Data Lab with Amazon Data Lake (Object Storage—Amazon S3; Backup and Archive—Amazon Glacier; Data Catalog—AWS Glue; Data Warehousing—Amazon Redshift; Big Data Processing—Amazon EMR).

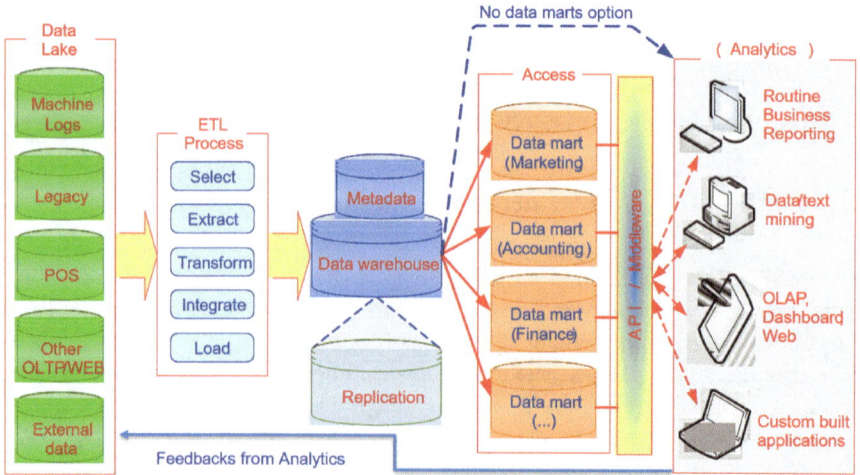

Fig. 1 DMSB data lab architecture

3.3 Data Science Corp

The University of South Carolina will be the coordinating organization for the creation of a Data Science Corp and the implementing organizations will be the Moore School of Business and the College of Engineering and Computing. The objective of our Data Science Corps is to build a strong, multidisciplinary, and multi-institutional partnership that develops data science professionals and data science students who will join the workforce with strong capabilities of not only unlocking the value of data but will also be able to apply these analyzes to create improvements and innovations. Figure 2 shows the relationship between the partners.

Multidisciplinary data science projects will be offered to business students by the College of Engineering and Computing. Select projects will be described and shared with business analytics faculty. Business faculty members will have the opportunity to select appropriate projects that match the curricular objectives in one of their courses. They will seek candidates in their courses and nominate appropriate students to join

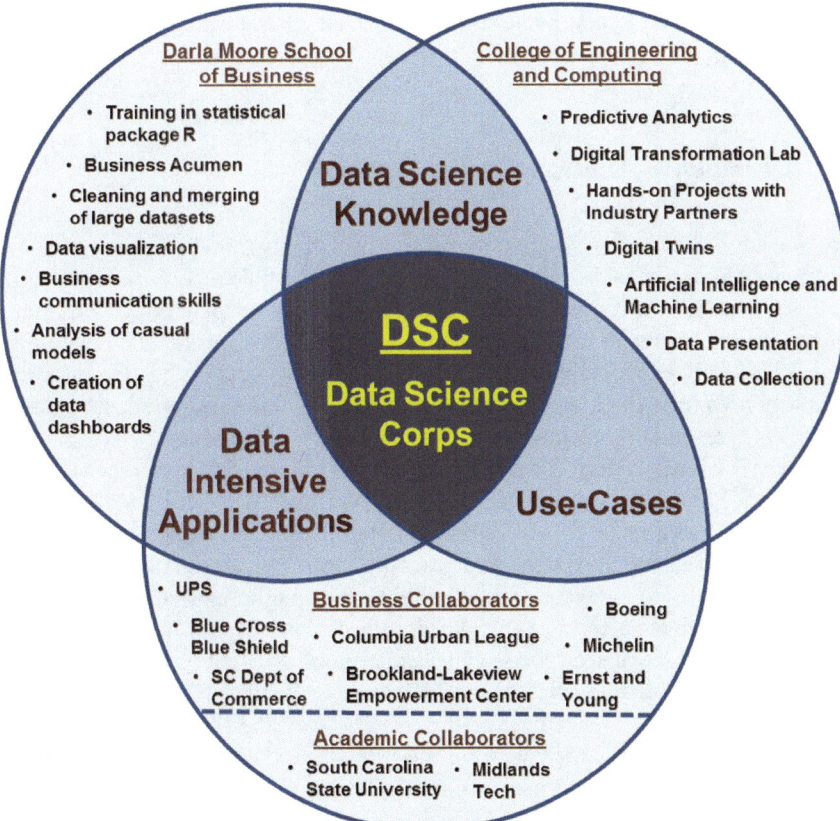

Fig. 2 Venn diagram for the DSC partners

an engineering data science project that would fulfill the requirements of a project in their business course. In addition, to the data science projects, participants will attend mandatory workshops to enhance their skills such as teamwork, entrepreneurship, critical thinking, communications, collaboration, creativity, and ethics. The Data Translation Skills Workshop is a 2-h workshop will teach students the skills that will help them translate data science knowledge into effective applications. Topics will include listening and communication skills, entrepreneurial concepts, teamwork, and collaboration techniques. Learning at the end of the workshop will be evaluated by appropriate responses to a data science case study that involves complex relationship and communication barriers. Another workshop will be the Ethics, Culture, and Diversity Workshop.

4 Business Analytics Projects

The following are student projects, which are provided as examples of business analytics projects. Note that some names and other identifications are masked in order to protect proprietary data and privacy.

4.1 A Predictive Analysis Project Sample

The company XXX tracks power outages to help in planning for future power outages. The key problem to be addressed is that there may be multiple records for any single event. It is implied the records of the event may cover more than 1 day, with more than one cause. The result is that the data gives an inaccurate appearance of multiple events where only one took place. In order to get any value out of these records, the multiple entries for a single event must be distilled, in some way, to reflect a unique event. This is essentially the transformation stage of an ETL process.

The team's primary efforts focused on SQL, using Microsoft SQL Server Management Studio. The data can be easily pulled into Microsoft SQL Server Management Studio (SSMS) as a table. SSMS offers an import/export function; however, care must be taken to ensure the proper data types are assigned to each column. To solve the problem of cleaning the data so that only the meaningful entries remain, Structured Query Language is recommended. Once the data is in SSMS, it must be sorted and filtered. Because we need to assign specific priorities based on event status, we can utilize a common table expression (CTE) to assign values to our priorities. The following SQL statement in Fig. 3 represents one way the required logic can be incorporated into a single statement to return the desired results from the dataset. Once a single data set was formed, we ran a Decision Tree to model a predictive analytics, where we can identify power outrages to be happened. We can sort all possible power outrages identified by the Decision Tree by the priorities that were set by the company.

```
WITH dataset AS
    (
    SELECT top 20000
        a.*,
        Priority =
            CASE
                    WHEN EVENT_STATUS = 'CLOSED' THEN '01'
                        …
                        …
                    WHEN EVENT_STATUS = 'REJECTED' THEN '10'
                    ELSE 99
            END
    FROM dbo.Import a
    WHERE a.[Outage Type] IN ('Major','Minor','MI','HGP','CI')
        AND a.[Serial Number] is not null
    ORDER BY [Serial Number] ASC, Priority ASC, [Actual Start Date] ASC,
        EVNT_ID DESC
    )
SELECT r.EVNT_ID, r.PROJ_ID, r.OTG_ID, r.[Outage Type],r.PROBABILITY,
r.EVENT_STATUS,
    r.[Planned Start Date], r.[Planned End Date], r.[Actual Start Date],
    r.[Actual End Date], r.[Serial Number], r.TECHNOLOGY_CD
FROM…
```

Fig. 3 Partial SQL codes

4.2 *A Prescriptive Analysis Project Sample*

Gamecock Drone Service, Inc. (GDS) is a drone delivery service operating in City of Columbia. In the city, GDS utilizes one drone to deliver packages to 125 customer locations each day. The drone starts at the GDS headquarters, makes the 125 stops, and then returns to GDS each day. In order to accomplish this in the shortest, most efficient time, GSD needs to develop an algorithm to identify each day's shortest route before the drone departs each morning. To develop the best algorithm, our team tested three approaches: (1) Excel delivery optimization model designed to solve for the shortest total distance given the coordinates of the day's stops, (2) Excel model similar to #1, but segmented into regional clusters (using Cluster Analysis) in the hope of finding optimal distance more quickly, and (3) Applying the existing traditional Traveling Salesman Problem models using R.

Approach #1 generated the lowest travel distance of 398.48 miles to cover the required 126 legs of the day's trip. We used our experience with excel modeling and the solver function to build a tool that can identify the optimal solution in a relatively short period of time. The first step was to assign a unique number (Assigned#) to each stop in the list. Headquarters (Home) was assigned number 1 and the rest followed up to 126. Next, we built a model that includes a row for each of the 126 required positions (Stop#) of the drone (from Home to 125 deliveries and back Home).

The model is driven by the number in the Assigned# column representing each of the stops in the list above. This number is used as the lookup value in a vlookup, using the table above as the array to bring in the LocationID, PackageID, x-Coordinate, and y-Coordinate for each location. Distance is calculated as $SQRT((x_2 - x_1)^2 + (y_2 - y_1)^2)$. In the assigned# column, 1 is hard coded for Home as the starting position (Stop# 0) and ending position (Stop# 126). The remaining 125 numbers, corresponding to Assigned# 2–126 are assigned using Excel's Solver function. The objective is the sum of all figures in the Distance column. The variable cells are the Solver column. The variable cells are constrained by >=1, <=125, Integer, and All Different. All Different requires that the values be 1 to n, n being the number of values being solved for. So the Solver column is a range if 1–125, one less than the 2–126 that was needed in the Assigned# column (see Fig. 4). The result is that solver assigns 125 numbers from 1 to 125 with no repetition in the sequence that minimizes the total travel time. When run on the Evolutionary solving method, the model produced an initial result of 721.75. The maximum time without improvement setting was extended to 1000 and the existing result was resolved three times. These iterations produced constantly improving results of 706.27, 701.95, and 698.48. The 698.48 was resolved three additional times with no further improvement.

To begin this final step of the strategy, an R package called "tspmeta" was down-loaded that provided R programs for distance optimization problems such as this one. The coordinates were translated into a distance matrix for R, and the "2-opt" method was utilized to formulate a delivery path with no intersections. A 2-opt method is an algorithm that plots a route so that there is never an overlap of the path. You never travel through the same place or spot multiple times.

This is visualized in Fig. 5, where the initial route is altered so that there is no intersection of the path (between b–e and c–f). The steps of the "tspmeta" package were as follows. First, we loaded in the data using a basic "read.csv" command in R. Then we created a distance matrix (a table of each of the 125 drop points

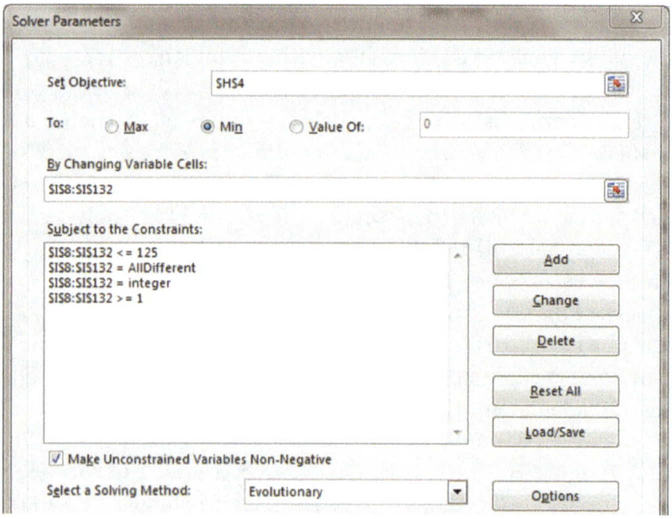

Fig. 4 Excel solver screenshot

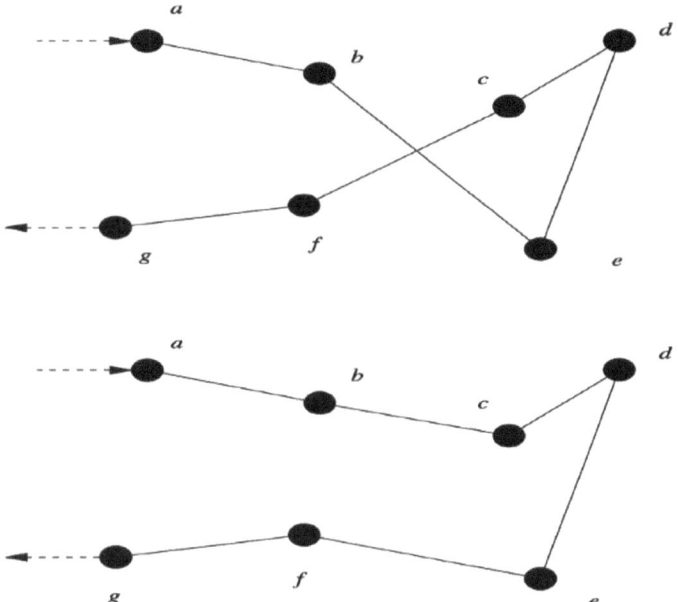

Fig. 5 An illustration that demonstrates the overlap issues and how the 2-opt algorithm fixes them

and their distances from each other) using a "dist.mx<-dist(coords.mx)" command. The distance matrix was made after making a coordinate matrix using "coords.df<-data.frame(x=xcoordinate,y=ycoordinate)" and "coords.mx<-as.matrix(coords.df)." The order and commands used can be seen in Fig. 6.

The final step was running the solver selecting the "2-opt" method. The command used, as shown below, was "tour<-run_solver (tsp.in,method="2-opt")". This used the previous steps and commands to create the path with no intersections. This "tour" output displayed the total distance traveled by the algorithm to the 125 drop points. However, the 2-opt method in R does not give a consistent value for distance traveled. Each time "run_solver" is used, a different distance traveled is given due to the countless paths that the algorithm could take without intersections. Each time the solver is run, a slightly different path is found by R and therefore a different distance value (tour length) is calculated. We knew because of this variation in the algorithm; we would need to find the optimal 2-opt path. The 2-opt method itself is the most efficient route for Gamecock Drone Service, but with infinite possibilities within the 2-opt method we created a system of picking the best path. We decided that we would run a 7-test check like Monte Carlo simulation. Once we found a distance traveled ("tour length") that was the lowest of all the others found, we would run the solver 7 more times. If the minimum value we found was still the lowest value after 7 more tests, we would stay with that value. As explained in the abstract, the most efficient path using the 2-opt method was 663.05 miles.

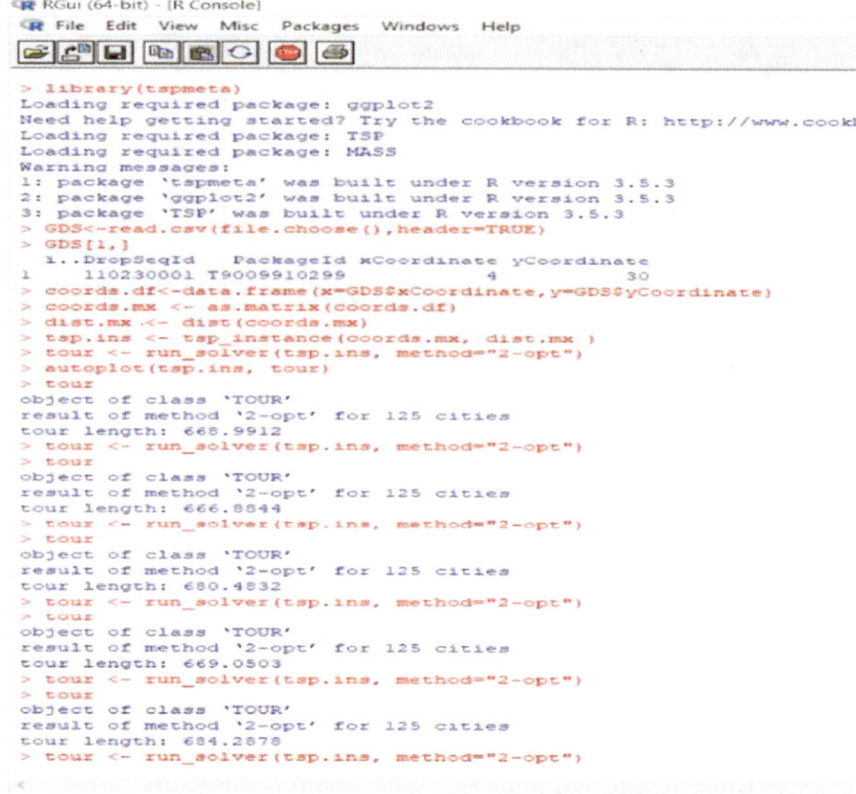

Fig. 6 Our R-code when beginning to use the 2-opt method

4.3 Battery Market Prediction Project

The automotive battery industry is expected to change in the next 10 years due to factors including the following:

- The rise in demand for electric vehicles
- The rise in rideshare apps like Uber and Lyft
- Decreasing prices of raw materials
- The rise in supportive regulation for renewable energy storage
- An increase in disposable income for Americans
- The availability of alternative products to batteries
- Technological development in the field.

This report explores these factors in relation to the automotive battery industry in the United States and predicts future trends using the relevant factors. First, we analyze the battery industry in the United States and found data to show those trends. Then, we chose the most relevant factors and its accompanying data to include in our prediction. Finally, we completed our prediction analysis using regression models for fields of data relevant to battery prediction.

We found that the automobile battery market will be growing rapidly over the next 10 years. We were able to prove this by collecting data for factors positively correlated to battery market growth over the past decade, exploring their positive linear relationships, determining the amount of statistical significance between the factors, and using our data to form predictive models. Please enjoy our comprehensive discussion of battery market growth over the next decade.

The purpose of this study is to predict the sales of automotive batteries in the United States for the next 10 years (2020–2030). In predicting this, there are major considerations in which one must account within the battery industry in the United States. The battery industry includes batteries made for automotives, communication equipment, lighting, audio and video equipment, as well as other portable equipment.

The regulation and policies in the battery industry are heavy in the US. Most of the regulations are related to the environment; specifically, companies need to meet EPA standards ISO 9000, QS 9000 or equivalent (IBISWorld 2019). In addition, factories need to be especially conscious about lead toxicity and there are many regulations set to ensure worker safety. Since the regulations are high in the United States and there are lower wages in these countries, batteries are largely imported from China and South Korea. On the other hand, the United States does not export many batteries making this industry have a trade deficit. This trade deficit is not ideal for the country as a whole because the US is giving more currency away than it is receiving. In terms of the battery industry, a trade deficit usually lower the cost of the product and simultaneously lowers the job market in that industry. Although policies are strict, many countries are encouraging and favoring renewable energy storage. These supportive policies are expected to increase domestic battery production and the domestic battery demand as a whole.

Testing and Evaluation: The first regression we ran compared the number of charging stations in the US for the past 10 years compared to the demand for electric vehicles overall in the US. The results listed below show there is a little variance in our model, with a high r-squared value of 0.988. It seems likely that as charging stations become more prominent and ubiquitous in the United States that people will begin to buy more electric vehicles as charging their vehicles becomes more convenient. According to our ERD, this implies that battery sales will increase in the future, as stronger EV sales imply more battery sales (Fig. 7).

SUMMARY OUTPUT

Regression Statistics	
Multiple R	0.994189583
R Square	0.988412928
Adjusted R Square	0.986964544
Standard Error	2437.620052
Observations	10

ANOVA

	df	SS	MS	F	Significance F
Regression	1	4054961274	4054961274	682.4246151	4.95195E-09
Residual	8	47535932.14	5941991.518		
Total	9	4102497207			

	Coefficients	Standard Error	t Stat	P-value	Lower 95%	Upper 95%	Lower 95.0%	Upper 95.0%
Intercept	-1663.258996	1272.172284	-1.30741647	0.227393149	-4596.893543	1270.375551	-4596.893543	1270.375551
X Variable 1	0.261870702	0.010024427	26.12325812	4.95195E-09	0.238754331	0.284987073	0.238754331	0.284987073

Fig. 7 Our regression analysis results

References

1. https://www.wired.com/insights/2014/07/data-new-oil-digital-economy/
2. J. Manyika et al., *Big Data: The Next Frontier for Innovation, Competition, and Productivity* (McKinsey Global Institute, 2011)
3. PwC and BHEF, *Investing in America's Data Science and Analytics Talent. The Case for Action* (2017)

The Study of Fresh Products Supplier's Comprehensive Evaluation Based on Balanced Scorecard

Xinyu Ma and Qing Zhang

Abstract Fresh supplier plays a vital role in the whole supply chain. How to evaluate the comprehensive ability of suppliers in a more scientific way so as to assist decision-making and establish long-term cooperative relationship is a problem worth thinking about by the enterprise members of supply chain. At present, there are few discussions targeting at the fresh industry in supplier capacity assessment researches, and most of them take KPI as the assessment standard and only focus on the current business ability of the enterprises while ignoring their long-term development. Even though some researches consider the multidimensional performance of the enterprise, the AHP method is often used to determine the weights, which lacks objectivity. Based on the characteristics of fresh products, this paper discusses the supplier capability evaluation in the field of fresh products. The BSC method is used to divide the four dimensions of evaluation indicators, comprehensively considering the financial and nonfinancial information, short-term performance and future development space of suppliers. This research collects data of the four representative suppliers and divides the weight of the subdivision index by the coefficient of variation method, so as to form the fresh supplier evaluation system. Combing the balanced scorecard and coefficient of variation method to establish an evaluation system, which can not only avoid the disadvantages of one-sidedness and lack of pertinence but also better assist the decision-making implementation in the supply chain, provides a new idea for supplier evaluation.

Keywords BSC · FAHP · Variation coefficient method · Fresh products supplier evaluation

X. Ma · Q. Zhang (✉)
College of Economics and Management, Nanjing University of Aeronautics and Astronautics, Nanjing 211106, Jiangsu, China
e-mail: karlnuaa@nuaa.edu.cn

© Springer Nature Switzerland AG 2020
H. Yang et al. (eds.), *Smart Service Systems, Operations Management, and Analytics*, Springer Proceedings in Business and Economics,
https://doi.org/10.1007/978-3-030-30967-1_30

329

1 Introduction

Fresh product as a kind of specific daily necessity is kept being paid close attention to for its characteristics such as easy to spoil, hard to transport, and preserve. With the current fresh products supply chain developing toward the multilevel and multichannel trend, fresh products supplying market competition is becoming more and more intense. Many disadvantages of fresh products supply chain like multiple links, low efficiency, and serious waste make members of supply chain realize the significance of cooperation and only by transforming the 'single-win' aimed at maximizing self-interests into the 'win-win' or even 'multi-win' for optimal collective interests can they finally gain a place in today's competition. Supplier, as the role of the source of the entire supply chain, takes the inevitable responsibility for product quality and is the most important factor determining customer satisfaction. Also, the cooperation among suppliers, retailers, and wholesalers affects the subsequent market reaction speed, service efficiency, and so on. The delivery capacity, price, logistics efficiency, and other factors of different suppliers will cause different fluctuation effects on market demand. A good supplier source is able to accurately position the market demand, respond to the cooperation plan of the seller, provide competitive price, quickly provide high-quality products, and finally, strongly ensuring sellers' market exploitation. Therefore, it is of great significance to conduct effective supplier evaluation and scientific selection.

At present, scholars at home and abroad have made a series of studies about scientific supplier evaluation method and system construction. Govindan [1] reviewed researches from 1997 to 2011 which talked about evaluating green supplier performance, finding that the applied approaches were mostly fuzzy-based single models and the most common criterion considered for green supplier selection was 'environmental management systems'. Luthra [2] used analytic hierarchy process (hereinafter the 'AHP'), vlseKriterijumska optimizacija i kompromisno resenje (hereinafter the 'VIKOR'), a multi-criteria optimization and compromise solution approach to establish a systematic and sustainability-focused evaluation system for suppliers and took an example of an automobile company in India to demonstrate the applicability of the frame. Beikkhakhian [3] applied hierarchical analysis into the evaluation of agile suppliers to determine the weight of data to increase the efficiency of the results of fuzzy technique for order preference by similarity to an ideal solution (hereinafter the 'TOPSIS') and employed interpretive structural model to interpret the effects of the criteria on suppliers. Banaeian [4] thoroughly discussed the incorporation of fuzzy set theory into TOPSIS, VIKOR, and grey relationship analysis (hereinafter the 'GRA') and utilized them to complete a green supplier evaluation and selection study for an actual company from the agri-food industry.

After reading a mass of relevant papers, we find that the major methods applied to evaluate fresh products suppliers' abilities focus on AHP, fuzzy AHP (hereinafter the 'FAHP'), TOPSIS, data envelopment analysis (hereinafter the 'DEA') and VIKOR, which can effectively carry out the ranking and evaluation of the scheme's advantages and disadvantages, providing auxiliary reference and theoretical support for the final decision. However, the key affecting the final results of evaluation are how to effectively select a proper evaluation criterion. There are few scholars who set indexes based on the characteristics of fresh products to assess suppliers but mostly expand by referring to existing experience and previous researches. The commonly used indicator determination methods are mainly economic value added (hereinafter the 'EVA') method and key performance indicator (hereinafter the 'KPI') method, but both of them have their own defects. EVA method mainly evaluates short-term financial performance while ignoring the long-term development potential and non-financial indicators in the assessment. The quantitative indicators formulated by KPI are difficult to define, and their real impacts on the final performance are immeasurable. There are also a series of problems in the implementation process. Therefore, this paper introduces the indicator determination method of Balanced Scorecard (hereinafter the 'BSC') in the evaluation of fresh suppliers to avoid the defects of the two methods above. The application of BSC takes both financial and nonfinancial information of suppliers into account, considers their future development space as well as integrates the strategic objectives of the organization into the short-term evaluation criteria, so that the target of more scientific and effective decision-making can be accomplished.

2 The Establishment of Fresh Products Supplier Evaluation Index System

Taking previous researches [5–7] and relevant traits of fresh products into account, this paper establishes the fresh products supplier evaluation index system from four dimensions o f BSC [8] (Table 1).

(1) *Finance*

Profit maximization tends to be the ultimate strategic goal of enterprises. Suppliers' financial performance evaluation indicators are mainly used to show their current and future profitability. The investing yield reflects the profitability of investment in a normal operating year, thus reflecting the investment effect of suppliers. Market share reflects the operation scale of supplier enterprises, symbolizing the corresponding market competitiveness, which is an important indicator to judge the profitability. Revenue growth rate reflects the future development trend. Under the traditional evaluation criteria, successful fresh food supplier enterprises often have three advantages, namely, short payback period (high return on investment), strong market competitiveness, and rapid income growth.

(2) *Customer*

With the rapid development of new market modes such as "new retail", the demand and experience of customers are increasingly valued by enterprises. The gradual improvement of people's living standards makes competitive price no longer the only factor affecting customer satisfaction. The quality (freshness) of products, the

Table 1 Fresh products supplier evaluation index system using BSC

Dimension	Index
Finance	Investing yield
	Revenue growth
	Market share
Customer	Price competitiveness
	Quality(freshness)
	Product traceability
	Customer responsiveness
Internal processes	Standardization level
	Informationization degree
	Transportation loss
	Inventory turn
Learning and growth	Informatization and mechanization construction
	Environmental protection and social benefits
	Brand building

traceability of product information, and the timely response of suppliers to customer needs will all affect customers' sense of service experience, thus affecting the operation and sustainable development of supplier enterprises [9].

(3) *Internal processes*

As the source of supply chain, the normal operation of supplier enterprises plays a crucial role in the operation of the whole supply chain. Therefore, a series of process evaluation within the supplier enterprise cannot be missed. In order to ensure the normal supply of supplier enterprises, storage, transportation, and packaging must achieve standardized management. At the same time, for improving the traceability of product information to optimize users' experience, the informatization degree of supplier enterprises should also be included in the evaluation indexes. In addition, storage and transportation efficiency indexes such as inventory turnover rate and transportation loss rate should also be considered.

(4) *Learning and growth*

Balanced scorecard is an indicator determination method that attaches importance to the strategic goals and long-term development of an enterprise. Therefore, in addition to the index dimension to assess whether an enterprise can operate normally, the efforts made by an enterprise for future improvement must also be included in the evaluation system. Under the general trend of technological development booming, the continuous improvement of mechanization and informatization level of supplier enterprises is the prerequisite for them to survive in the trend of the times and obtain competitive advantages. Besides, in order to expand the market share of suppliers, they are bound to build their own brands to achieve uniqueness in the market. Undertaking a series of social responsibilities and adopting environmental protection measures are also powerful means for enterprises to build their own brand power and generate social benefits.

3 Evaluation and Decision-Making of Fresh Suppliers Based on FAHP and CV Methods

Since the analytic hierarchy process (AHP) does not take the ambiguity of the evaluator's judgment into account, in order to evaluate the comprehensive ability of suppliers more accurately and practically, the fuzzy AHP is introduced to evaluate fresh suppliers for a better decision.

(1) *Construct the hierarchical structure model*

Under the premise of the balanced scorecard division index system, the comprehensive ability evaluation system of fresh suppliers is established as a two-level model. $A = \{B_1, B_2, B_3, B_4\}$, $B_1 = \{B_{11}, B_{12}, B_{13}\}$, $B_2 = \{B_{21}, B_{22}, B_{23}, B_{24}\}$, $B_3 = \{B_{31}, B_{32}, B_{33}, B_{34}\}$, $B_4 = \{B_{41}, B_{42}, B_{43}\}$ (Fig. 1).

(2) *Establish the fuzzy complementary judgment matrix and calculate the weights of each dimension*

The importance degree of each index is divided by 0.1–0.9 scale, satisfying
$$\begin{cases} a_{ij} = 0.5, i = j \\ a_{ji} = 1 - a_{ij}, i \neq j \end{cases} \text{(Table 2).}$$

Considering the strategic goal and long-term development of fresh suppliers, combined with the characteristics of fresh products and related literature, three experts were invited to evaluate the importance of fresh suppliers' four dimensions as follows, so as to reduce subjectivity and increase scientificity of the evaluation (Table 3):

The experts' evaluation values of dimensional significance obtained from the above table are averaged to obtain the final evaluation matrix of the first-level index significance.

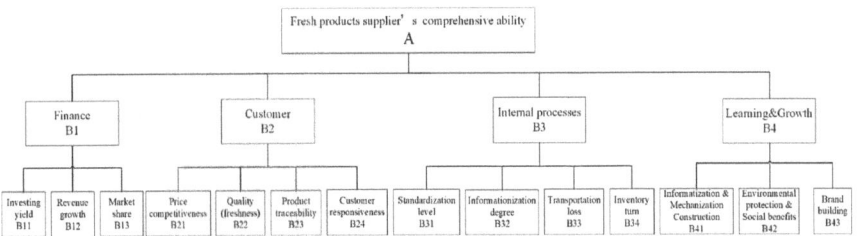

Fig. 1 Supplier comprehensive ability evaluation index system

Table 2 The corresponding meaning of 0.1–0.9 scale

Scale	Meaning
0.5	A and B have equal importance
0.6	A is slightly more important than B
0.7	A is obviously more important than B
0.8	A is much more important than B
0.9	A is extremely more important than B
0.1, 0.2, 0.3, 0.4	Compare from the contrary point of view

Table 3 The importance matrix of fresh suppliers' four dimensions

A	Finance (B1)	Customer (B2)	Internal processes (B3)	Learning and growth (B4)
Finance (B1)	0.5	0.3, 0.4, 0.3	0.6, 0.5, 0.4	0.3, 0.4, 0.2
Customer (B2)	0.7, 0.6, 0.7	0.5	0.7, 0.6, 0.5	0.6, 0.7, 0.6
Internal processes (B3)	0.4, 0.5, 0.6	0.3, 0.4, 0.5	0.5	0.4, 0.3, 0.4
Learning and growth (B4)	0.7, 0.6, 0.8	0.4, 0.3, 0.4	0.6, 0.7, 0.6	0.5

$$
D_0 = \begin{pmatrix} 0.5 & 0.333 & 0.5 & 0.3 \\ 0.667 & 0.5 & 0.6 & 0.633 \\ 0.5 & 0.4 & 0.5 & 0.367 \\ 0.7 & 0.367 & 0.633 & 0.5 \end{pmatrix}
$$

According to the formula of fuzzy complementary judgment matrix weight $W_i = \frac{\sum_{j=1}^{n} a_{ij} + \frac{n}{2} - 1}{n(n-1)}$, obtain the weight of the fuzzy complementary judgment matrix $W_{Bi} = (0.219, 0.283, 0.231, 0.267)^T$.

(3) *Divide specific subdivision indexes' weight*

Since AHP is subjective to some extent, if the weight determination of subdivision index continues the method of expert scoring, the final result will inevitably lack scientificity and objectivity. So when determining the weight of subdivision index, the coefficient of variation method is used for objective weighting.

We collect the relevant quantitative information of four representative fresh food suppliers in China in 2018, and the qualitative indicators are quantified by the percentage system score through the comprehensive evaluation of the information, so as to obtain the following data (Table 4):

Respectively get the standard deviation and mean values of each subdivision index under the classification of the four dimensions (Table 5).

According to the formula $v_i = \frac{\sigma_i}{\overline{x_i}}$ (v_i is the coefficient of variation, σ_i is the standard deviation, $\overline{x_i}$ is mean value) we can get the variation coefficient of the

Table 4 Data of four suppliers

	B11 (%)	B12 (%)	B13	B21	B22	B23	B24	B31	B32	B33 (%)	B34	B41	B42	B43
D	23.06	16.9199	0.060846	80	0.95	0.99	0.9	0.95	0.98	10	9.6789	16,000,000	85	90
E	37.87	−3.3301	0.186590	70	0.92	0.98	0.88	0.93	0.95	12	10.71	10,000,000	80	90
F	4.33	18.2398	0.007490	85	0.8	0.75	0.8	0.9	0.8	15	3.4632	7,000,000	70	80
G	4.7	4.6795	0.009944	85	0.86	0.7	0.78	0.8	0.7	18	9.6874	3,000,000	60	70

Table 5 Data processing

		Min	Max	Mean value	Standard deviations
B1	B11 (%)	4.33	37.87	17.49	16.16
	B12 (%)	−3.33	18.24	9.13	10.31
	B13	0.00749	0.18659	0.066217343	0.083932506
B2	B21	70	85	80	7.071
	B22	0.8	0.95	0.8825	0.06652
	B23	0.7	0.99	0.855	0.15155
	B24 (%)	78.00	90.00	84.00	5.89
B3	B31	0.8	0.95	0.895	0.06658
	B32	0.7	0.98	0.8575	0.13124
	B33 (%)	10.00	18.00	13.75	3.50
	B34	3.4632	10.71	8.384875	3.316632989
B4	B41	3,000,000	16,000,000	9,000,000	5477225.575
	B42	60	85	73.75	11.087
	B43	70	90	82.5	9.574

index. On the basis of the formula $w_i = \frac{v_i}{\sum\limits_{i=1}^{n} v}$, we can then obtain the weight division of each subdivision index in each dimension. Finally, use the weight of the fuzzy complementary judgment matrix of the importance of the four dimensions set by experts scoring at the beginning $W_{Bi} = (0.219, 0.283, 0.231, 0.267)^T$ to substitute into the subdivision index weight in the above table, the final specific weight of each index α_i under the four dimensions of BSC classification is obtained (Table 6).

Therefore, when choosing suppliers to establish a long-term cooperative relationship, fresh food enterprises can not only make a judgment based on their financial situation and business ability but also learn their long-term development ability, so as to make a more scientific decision.

4   Analysis of Examples

Assuming that Supermarket A needs to evaluate D, E, F, G four fresh suppliers' comprehensive abilities in order to establish long-term cooperation relations, it acquired the relevant financial data of four suppliers in the past 2 years for the percentage processing, and invited a number of industry experts to score their qualitative indicators, getting weighted average value, and finally obtain the score data as follows (Table 7):

(1) In the traditional KPI assessment method, it is often inclined to compare the data of quantitative indicators of suppliers. In reality, it i s often the assessment of their financial indicators, so as to evaluate and make decisions. In this case, if

Table 6 The establishment of weight of the indicators

		v_i (%)	w_i (%)	α_i
B1 0.219	B11	92.378	27.820	0.060924914
	B12	112.930	34.009	0.074479311
	B13	126.753	38.172	0.083595775
B2 0.283	B21	8.839	21.500	0.060844347
	B22	7.538	18.335	0.051887997
	B23	17.725	43.115	0.122016681
	B24	7.009	17.050	0.048250975
B3 0.231	B31	7.439	8.477	0.019582494
	B32	15.305	17.441	0.040288338
	B33	25.455	29.007	0.067005831
	B34	39.555	45.075	0.104123337
B4 0.267	B41	60.858	69.555	0.185712241
	B42	15.033	17.182	0.045874827
	B43	11.605	13.263	0.035412932

Table 7 The scores of four suppliers

	Detailed indicators	D	E	F	G
Finance B1	Investing yield B11	80	70	90	95
	Revenue growth B12	75	79	84	92
	Market share B13	80	86	75	84
Customer B2	Price competitiveness B21	74	89	78	82
	Quality (freshness) B22	92	94	78	76
	Product traceability B23	84	86	89	91
	Customer responsiveness B24	90	92	86	74
Internal processes B3	Standardization level B31	90	84	82	88
	Informationization degree B32	88	85	83	90
	Transportation loss B33	86	84	90	84
	Inventory turn B34	78	80	76	76
Learning and growth B4	Informatization and mechanization Construction B41	80	90	92	76
	Environmental protection and social benefits B42	75	82	86	92
	Brand building B43	90	86	82	78

Table 8 Mean values of four suppliers' finance scores

D	E	F	G
78.33333	78.33333	83	90.33333

KPI performance assessment is applied, mean value comparison shall be made based on the financial performance of the four suppliers, and the following scores shall be obtained (Table 8).

Finally, supplier G is selected as the long-term strategic partner of supermarket A.

Quantitatively analyzing, in terms of the current operating status judged from KPI, supplier G's financial status is the best. But it is not enough to prove that it has a good development trend in the future, and the evaluation is not comprehensive and scientific.

(2) Qualitatively, the performance of four suppliers can be evaluated from the perspective of experts from four dimensions by AHP, and the following evaluation results are obtained (Table 9).

By calculating and passing the consistency test, the final scores obtained in the AHP analysis evaluation are as followed (Table 10):

From the results above, it can be seen that the optimal scheme obtained through experts' scoring is F. However, the AHP method cannot avoid the subjectivity of experts, so evaluation by experts is obviously defective. According to the performance of F, this supplier does not attach enough importance to customers compared with other suppliers. It is well known that customer satisfaction is a major determinant for the survival of an enterprise in today's market, judging from which, the pure qualitative scoring and evaluation method is not reasonable.

(3) According to the fresh food supplier evaluation system obtained based on BSC dimension division, the scores of each dimension index are calculated, and the final scores of four fresh food suppliers are obtained as follows (Table 11).

Table 9 Scores of suppliers using AHP

B1-B2-B3-B4	D	E	F	G
D	1	1, 3, 1/4, 1/4	3, 1/2, 1/5, 1/5	5, 1/3, 1/3, 1/2
E	1, 1/3, 4, 4	1	3, 1/5, 1/2, 1	5, 1/7, 2, 1/2
F	1/3, 2, 5, 5	1/3, 5, 2, 1	1	3, 1/2, 4, 1/2
G	1/5, 3, 3, 2	1/5, 7, 1/2, 2	1/3, 2, 1/4, 2	1

Table 10 Final scores in AHP

D	E	F	G
0.2075	0.2369	0.2935	0.2621

Table 11 Mean values of four suppliers' comprehensive scores	D	E	F	G
	81.69244496	85.11213371	84.63959549	83.25672

Through the comparison, we can know that the overall situation of suppliers can be understood by comprehensively considering the performance of each supplier, namely the internal operation, future development and the achievement of strategic goals of suppliers by considering the information input, brand building, standardization degree, and other indicators. Although the profitability of Supplier G is good, for its informationization construction is insufficient, it will be unable to keep up with the future time informationization trend, so it is difficult for it to cope with diversified and immediate consumer demands. Also, without a strong commitment to brand building, it is impossible for G to establish long-term public praise and produce brand effects, which will result in constraints on its future development in many aspects.

According to the multidimensional evaluation results, Supplier E is finally selected as the long-term strategic partner of Supermarket A.

Compared with Supplier G with the best financial capacity and F which experts choose, from the perspective of the other three dimensions, supplier E pays more attention to the demands and experience of customers, invests more in informatization, mechanization construction and brand building, integrating long-term corporate strategic goals into daily operations as well as reducing the one-sidedness brought by subjective judgment to some extent. While keenly aware of the trend of social development, E forms and accumulates its own competitive advantages and has greater development space, which is a better choice for A to establish long-term cooperation with.

References

1. K. Govindan, S. Rajendran, J. Sarkis, P. Murugesan, Multi criteria decision making approaches for green supplier evaluation and selection: a literature review. J. Clean. Prod. **98**, 66–83 (2015)
2. S. Luthra, K. Govindan, D. Kannan, S.K. Mangla, C.P. Garg, An integrated framework for sustainable supplier selection and evaluation in supply chains. J. Clean. Prod. **140**, 1686–1698 (2017)
3. Y. Beikkhakhian, M. Javanmardi, M. Karbasian, B. Khayambashi, The application of ISM model in evaluating agile suppliers selection criteria and ranking suppliers using fuzzy TOPSIS-AHP methods. Expert Syst. Appl. **42**(15–16), 6224–6236 (2015)
4. N. Banaeian, H. Mobli, B. Fahimnia, I.E. Nielsen, M. Omid, Green supplier selection using fuzzy group decision making methods: a case study from the agri-food industry. Comput. Oper. Res. **89**, 337–347 (2018)
5. A. Azar, L. Olfat, F. Khosravani, R. Jalali, A BSC method for supplier selection strategy using TOPSIS and VIKOR: a case study of part maker industry. Manag. Sci. Lett. **1**(4), 559–568 (2011)

6. M.S. Dizaji, M.M. Mazdeh, A. Makui, Performance evaluation and ranking of direct sales stores using BSC approach and fuzzy multiple attribute decision-making methods. Decis. Sci. Lett. **7**(2), 197–210 (2017)
7. Y.H. Lin, C.-C. Chen, C.F.M. Tsai, M.-L. Tseng, Balanced scorecard performance evaluation in a closed-loop hierarchical model under uncertainty. Appl. Soft Comput. J. **24**, 1022–1032 (2014)
8. Y. Wang, Fresh supplier selection and evaluation. Logist. Eng. Manag. **38**(07), 175–176 (2016)
9. J. Zhang, Researching on the evaluation and construction of cold-chain logistics mode of dairy products. J. Tech. Econ. Manag. **11**, 29–33 (2015)

Maintenance Architecture Optimization of a Distributed CubeSat Network Based on Parametric Model

Honglan Fu, Hao Zhang and Yang Gao

Abstract Due to shorter development cycle and lower cost, CubeSats have been widely used in space science, technology, and business missions. CubeSat usually forms a formation/constellation, which boosts the capability of implementing complex space missions. However, the distributed CubeSat network is prone to malfunction. This paper envisions a maintenance architecture that includes launching and replenishing spare CubeSats to replace the faulty one on a regular basis. The major effort is to optimize this architecture in terms of total cost by taking the stochastic failures into consideration. In particular, a parametric model fitted from practical data is used to represent the realistic CubeSat lifetime distribution. A CubeSat lifetime database of 111 CubeSats has been built. The parametric model is obtained via a Bayesian estimation scheme. A cost model that is composed of fixed cost, holding cost, and shortage cost have been proposed. Then, A Monte Carlo simulation-based approach has been adopted to evaluate the cost. Finally, the optimal arrival time and quantity of backup CubeSats corresponding to minimal cost have been obtained by examining all the feasible combinations of arrival time and quantity of backup CubeSats. Results show that the CubeSat network should be replenished in the early stage which agrees with the high infant mortality trend of CubeSats.

Keywords On-orbit service · distributed CubeSat network · Weibull distribution · CubeSat failure database

H. Fu (✉) · H. Zhang · Y. Gao
Key Laboratory of Space Utilization, Technology and Engineering Center for Space Utilization, Chinese Academy of Sciences, Beijing 100094, People's Republic of China
e-mail: fuhonglan17@csu.ac.cn

H. Fu
University of Chinese Academy of Sciences, Beijing 100094, People's Republic of China

© Springer Nature Switzerland AG 2020   341
H. Yang et al. (eds.), *Smart Service Systems, Operations Management,*
and Analytics, Springer Proceedings in Business and Economics,
https://doi.org/10.1007/978-3-030-30967-1_31

1 Introduction

A CubeSat is a type of miniaturized satellite for space research which can be extended with 1U ($10 \times 10 \times 10$ cm^3) as the basic unit. The launching of the first CubeSat in June 2003 marks the beginning of a new era in space research. Compared with the long development period and high budget of large satellites, CubeSat has a shorter development cycle and lower cost which certainly benefits from the standardized and modular design strategy. However, small size and lightweight—the main characteristics of CubeSat—make the function of a single CubeSat inferior to that of a large satellite and thus CubeSats are mostly used as a technical demonstration and educational platforms [1].

To better taking advantage of CubeSat's benefits it is also desirable to consider the use of a large number of on-orbit CubeSats to form a distributed space network at a lower cost to perform functions that rival those of large satellites [1]. For example, aiming to image the entire Earth's surface every day and make global change visible, accessible, and actionable Planet Labs built a constellation consisting of more than 120 Dove satellites that provide a versatile dataset for geospatial analysis of markets, environments, and global change [2]. Chen et al. have proposed to deploy four CubeSats along Earth–Moon L_2 point to provide positioning service for users on the far side of moon [3]. Crisp et al. has investigated the launch and deployment strategies of small satellite constellations [4].

However, CubeSats are usually launched into space by rockets as piggyback payloads, as there is no dedicated launch vehicle for CubeSats. After arriving at orbit, the CubeSats are often ejected by deployment systems with the help of a spring-loaded mechanism [5, 6]. Vibrations they experienced during the launch and ejection processes are likely to reduce the mission success rate. Conversely, CubeSats wrapped in a soft bag are transported to the international space station (ISS), waiting for deployment command. This approach can decrease the vibration level during the launch process and improve the success rate of the mission benefitting from ISS's role as a staging post [7].

On the other hand, CubeSats in the network malfunction randomly causing system performance degradation. Therefore, new CubeSats are to be supplemented so as to restore the system. Compared with new launches from Earth directly, it is also possible to store backup CubeSats on-orbit, for example, on ISS. This maintenance architecture has a faster time of response. Usually, the on-orbit maintenance architecture is envisioned to be made up of launch vehicles, an orbital depot, a service spacecraft, and spare CubeSats. However, because of the uncertainty of the demand for spare CubeSats used to maintain the distributed CubeSat network, it is difficult to determine how many spare CubeSats should be stored beforehand. In addition, the spare CubeSats should be constantly replenished from Earth. Therefore, an important issue arises that is to determine the launching frequency and the number of CubeSats replenished, so as to minimize the operational cost of the maintenance architecture while guaranteeing the system performance. This issue is challenging due to the malfunction's randomness. Gu et al. [8] proposed an ordering strategy based on the lifetime distribution of aerospace parts to deal with the randomness of demand for aerospace parts. In that paper, the lifetime of the parts is assumed to follow a normal distribution. Sarton et al. [9] described a robotic on-orbit servicing infrastructures comprised of a launch vehicle, a robotic servicer, and an orbital depot and analyzed its responsiveness to random failure of space platforms. That paper assumed that failure advent follows a Poisson process. However, the spacecraft fault behavior does not necessarily follow a normal distribution or a Poisson process [10]. The realistic model should come from real practical data.

The current work aims to provide a more realistic treatment of optimizing maintenance architecture. The launch time of the launch vehicle and the amount of replenishment of the spare CubeSats are to be optimized to reduce the operational cost of the maintenance architecture by considering a random failure behavior. This current work is inspired by the spare parts management strategy in the aviation industry. The time of placing an order and the order quantity are optimized to minimize the total cost according to the lifetime distribution of the part. However, aviation is likely different from space in the sense of components fault characteristics.

To that end, a survey of CubeSats failure is first performed. Then, the Weibull distribution is selected to model the lifetime distribution of CubeSat due to its great usage in failure analysis and reliability research [10–12]. The two parameters of Weibull distribution are fitted using the obtained actual fault data of CubeSat and is implemented in computing the operational total cost. The total cost is composed of three parts: fixed cost, holding cost, and shortage cost. Finally, A Monte Carlo simulation-based approach is used to obtain the preliminary results. The arrival time and quantity of backup CubeSats corresponding to the minimal cost in the preliminary results are the optimal replenishing strategies.

This paper is organized as follows: Sect. 2 introduces the distributed CubeSat network maintenance architecture. Section 3 gives the CubeSat lifetime distribution parametric model. Section 4 presents the cost model of the maintenance architecture mentioned above. Section 5 describes the simulation methods and results. Section 6 provides the conclusion.

2 The Distributed CubeSat Network Maintenance Architecture

CubeSat has a short life span but a low-cost of manufacturing. Therefore, it is desirable to replace the faulty CubeSat by alternate CubeSats at an appropriate time in order to maintain the distributed space network. Some of the CubeSats packed in soft bag are launched to the ISS by cargo ship which safeguards them from the effect of vibrations and then deployed on the orbit from the ISS. To some extent, the ISS is equivalent to a staging post or an on-orbit CubeSat depot.

Based on the aforementioned background, this paper considers a CubeSat network maintenance architecture consisting of (1) launch vehicles, (2) an on-orbit CubeSat

Fig. 1 The distributed
CubeSat network
maintenance architecture

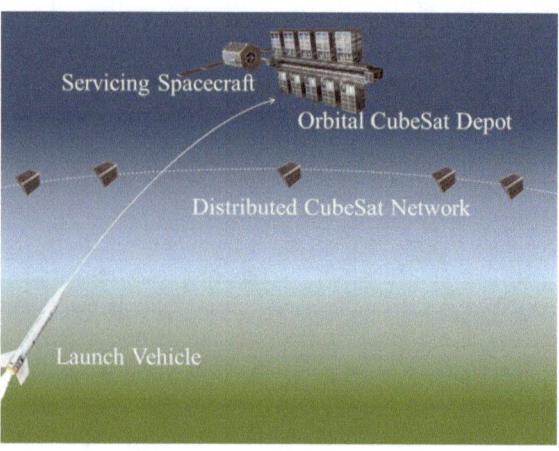

Fig. 1 The distributed CubeSat network maintenance architecture

depot, (3) a service spacecraft, and (4) spare CubeSats, as shown in Fig. 1. The launch
vehicle sends certain number of spare CubeSats to the orbital CubeSat depot at the
appropriate time. When the distributed CubeSat Network loses a CubeSat the service
spacecraft parked in the orbital depot transports a spare one to the orbit of the faulty
CubeSat and then the service spacecraft returns to the orbital depot waiting for the
next mission.

There are two reasons for using the on-orbit depot to temporarily store the spare
CubeSats: firstly, when the constellation loses a CubeSat, a spare one can be immediately released from the on-orbit depot, which improves the responsiveness to the
failure; and secondly, the orbital depot provides a relatively stable storage environment which can reduce the influence of harsh space environmental on CubeSat and
improve its lifetime.

3 Parametric Model

To minimize the operational cost of the maintenance architecture failure behavior analysis of CubeSat and cost model establishment are desired. For this reason, Weibull distribution is selected and its two parameters are estimated using the
Bayesian method in this section. Finally, the obtained parametric model is used to
generate fault time sequence to embed in the cost model. The cost model establishment will be detailed in Sect. 4.

3.1 CubeSat Lifetime Database

For the purpose of obtaining a meaningful parametric model for CubeSat's lifetime distribution, realistic CubeSat lifetime data are surveyed which contains launch dates, failure dates, project end dates, etc. Most of the data used in this paper come from the public nanosatellites database [13] which includes more than 1000 CubeSats launched as of January 2019. It should be noted that the obtained database is incomplete because some CubeSat operators have not published the specific operational status of their missions. However, we consider the resulting 111 data is sufficient for statistical analysis and for fitting parametric models.

Considering the limited observed failure data, this paper does not distinguish specific faults and all faults are treated equally, that is, once a CubeSat fails its life is considered to be finished. If a CubeSat has been working normally until the mission is completed, i.e., no failure occurred, it is called right censored data.

3.2 Parameters Estimation

Reliability analysis can be categorized as nonparametric analysis and parametric analysis. Nonparametric analysis does not rely on assumed distribution making it suitable for different types of failure data. It is also used for verifying the results of parametric analysis. Obviously, maintenance architecture optimization of distributed CubeSat network requires a parametric model. Therefore, in this paper, parametric analysis is applied. The main effort is to first find an appropriate model and then implement a parameter estimation algorithm. The parameter analysis method used in this paper is the Bayesian theory in combination with Markov Chain Monte Carlo simulations (Bayesian/MCMC) adopted by Guo et al. [10]. This method will be briefly introduced below and the specific description can refer to the literature [10, 11].

3.2.1 Weibull Distribution

The Weibull distribution widely used in reliability analysis can flexibly describe different failure behaviors. The Weibull distribution has two parameters: the shape parameter β and the scale parameter η. The probability density function of a Weibull random variable is given by Eq. (1):

$$f(t) = \begin{cases} \left(\frac{\beta}{\eta}\right)\left(\frac{t}{\eta}\right)^{(\beta-1)} \exp\left[-(t/\eta)^{\beta}\right], & t \geq 0 \\ 0, & t < 0 \end{cases} \tag{1}$$

where t presents the time before the malfunction.

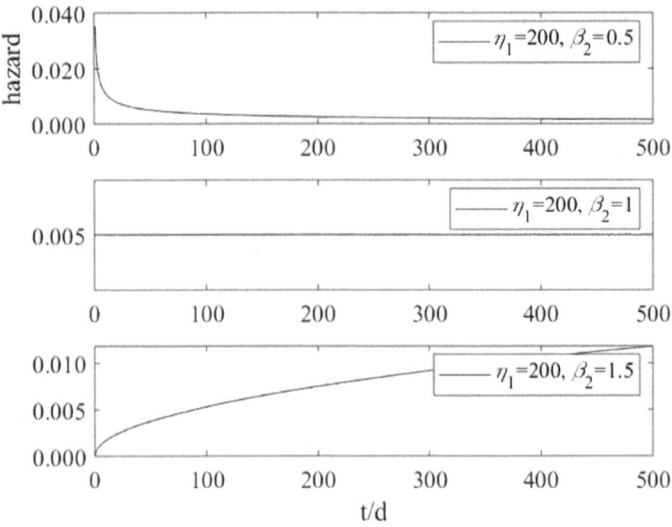

Fig. 2 Hazard function corresponding to different Weibull distribution

The resulted reliability function of the Weibull distribution is:

$$R(t) = \exp\left[-(t/\eta)^\beta\right] \tag{2}$$

The hazard function gives the instantaneous failure rate of an individual, which is related to the probability density function, $f(t)$ and cumulative distribution function, $F(t)$, as follows:

$$h(t) = \frac{f(t)}{1 - F(t)} \tag{3}$$

The hazard function corresponding to different Weibull distributions is shown in Fig. 2, indicating that Weibull distribution with different shape parameter can represent different fault behavior: (1) $0 < \beta < 1$, the hazard rate decreases with time which can be used to capture the infant mortality trend [10], (2) $\beta = 1$, the hazard rate is constant over time, and (3) $\beta > 1$, the hazard rate increases with time.

3.2.2 Bayesian/MCMC Method

First, the full conditional distribution of the parameter β and the parameter η is derived (see the literature [10] for details) as follows:

$$P(\beta | \eta, t_f, t_c) = \frac{\beta^{n-1}}{\eta^{(\beta-1)n}} \prod_{i=1}^{n} \left(t_{f_i}^{\beta-1} \right) \exp\left[-\sum_{i=1}^{n} \left(t_{f_i}/\eta \right)^{\beta} \right] \exp\left[-\sum_{j=1}^{m} \left(t_{c_i}/\eta \right)^{\beta} \right]$$

(4)

and

$$P(\eta | \beta, t_f, t_c) = \frac{1}{\eta^{\beta n+1}} \exp\left[-\sum_{i=1}^{n} \left(t_{f_i}/\eta \right)^{\beta} \right] \exp\left[-\sum_{j=1}^{m} \left(t_{c_i}/\eta \right)^{\beta} \right]$$

(5)

where t_f is the observed malfunction time, and t_c is the censoring time.

Then the Gibbs sampler in combination with Parsimonious Adaptive Rejection Sampling [14] is constructed:

1. initial value $\beta^{(0)}$
2. for $i = 1$ to N
    sample $\eta^{(i)} \sim P(\eta \mid \beta^{(i-1)}, t_f, t_c)$
    sample $\beta^{(i)} \sim P(\beta \mid \eta^{(i)}, t_f, t_c)$
    end for
3. outputs: the N samples, $(\eta^{(1)}, \beta^{(1)}), (\eta^{(1)}, \beta^{(1)}), ..., (\eta^{(N)}, \beta^{(N)})$

Here N equals 15,000. In order to avoid the influence of the initial value, the mean of the last 5000 samples out of 15,000 samples is taken as the final result.

For the collected CubeSat lifetime data in this work, the numerical results are shown as follows: The shape parameter β is 0.5467 that is less than 1, which reveals the infant mortality trend of CubeSat. The size parameter η is 385.5321 (day). The coefficient of determination (or R^2 value) is 0.9674. The closer the value of R^2 is to 1, the more accurate the parameter estimation result is. Here the R^2 is very close to 1, implying that this parametric model is a strong fit of the CubeSat lifetime data.

4   Cost Model

The cost model is established in this section. The obtained parametric model in Sect. 3 will be embedded in the cost model to optimize the maintenance architecture in the next section.

It is assumed that the total operational cost of the maintenance architecture consists of three parts: fixed cost, holding cost, and shortage cost. Fixed cost refers to the total cost of designing and building a CubeSat depot. Holding cost refers to the total cost of providing a stable storage environment for the spare CubeSats in the orbital depot. Shortage cost refers to the loss caused by the lack of spare CubeSats in the orbital warehouse. In order to simplify the analysis the time spent in transferring the spare CubeSat to the target orbit by servicing spacecraft is not considered in this paper.

It is further assumed that backup CubeSats are replenished only once during one planning period.

To model this cost formula, we define the following notations:

• s: shortage cost per unit time	• n: total number of failures
• h: holding cost per unit time	• j: total number of faults that occurred before
• c: fixed cost per spare CubeSat	the spare CubeSats arrive
• T: planning period	• t_j: the time when the jth fault occurs
• t: spare CubeSats arrival time	• t_k: the time when the kth fault occurs
• Q: total number of spare CubeSats replenished	

Let t be the arrival time of new CubeSats and we assume $t_j \leq t \leq t_{j+1}$, where t_j is the occurring time of the jth fault. We suppose that there are n faults in total during one planning horizon. The cost model can be categorized into five cases depending on the value of Q:

1. $Q < j$ (as shown in Fig. 3a)

In this case, Q spare CubeSats arrive at time t; and j malfunction events occur before time t. Therefore, we have Q damaged CubeSats being replaced at time t. The shortage time associated with these satellites is: $(t - t_k)$, $k = 1, 2, \ldots, Q$. The corresponding shortage cost is: $s \times (t - t_k)$, $k = 1, 2, \ldots, Q$. However, as $Q < j$, replenishment, in this case, is not enough to replace all the damaged CubeSats. There still exists $(j - Q)$ faulty CubeSats. In addition, the CubeSats that malfunction after time t also cannot be replaced. These two situations are collectively referred to as spares shortage that persists until the last day of the planning horizon. Thus, the shortage time is: $(T - t_k)$, $k = Q + 1, Q + 2, \ldots, n$. Accordingly, shortage cost is: $s \times (T - t_k)$, $k = Q + 1, Q + 2, \ldots, n$. The total cost, in this case, can be computed as:

$$R = s \left[\sum_{k=1}^{Q} (t - t_k) + \sum_{k=Q+1}^{n} (T - t_k) \right] + cQ \qquad (6)$$

2. $Q = j$ (as shown in Fig. 3b)

In this case, the new arrivals at time t are just enough to replace all the CubeSats that failed before time t. Their shortage time is: $(t - t_k)$, $k = 1, 2, \ldots, j$, and the shortage cost is: $s \times (t - t_k)$, $k = 1, 2, \ldots, j$. However, CubeSats that failed after time t cannot be replaced this results in a shortage situation that continues until the end of the planning period. Their shortage time is: $(T - t_k)$, $k = j + 1, j + 2, \ldots, n$. Shortage cost is: $s \times (T - t_k)$, $k = j + 1, j + 2, \ldots, n$. The total cost, in this case, can be calculated as:

$$R = s \left[\sum_{k=1}^{j} (t - t_k) + \sum_{k=j+1}^{n} (T - t_k) \right] + cQ \qquad (7)$$

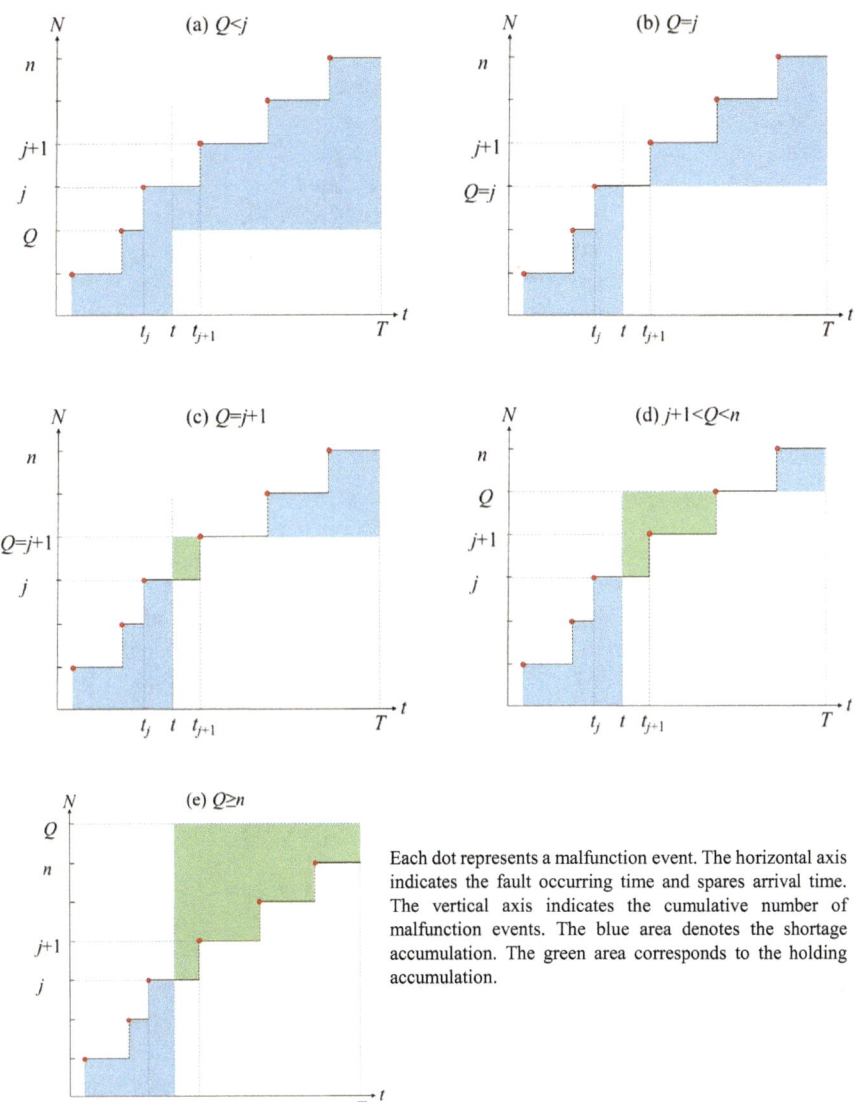

Each dot represents a malfunction event. The horizontal axis indicates the fault occurring time and spares arrival time. The vertical axis indicates the cumulative number of malfunction events. The blue area denotes the shortage accumulation. The green area corresponds to the holding accumulation.

Fig. 3 Cost calculation diagram

3. $Q = j + 1$ (as shown in Fig. 3c)

After the spare CubeSats arrive at the depot, j spares are used to replace the j faulty CubeSats immediately. One extra spare CubeSat will be stored from time t until the next fault occurs and then it is consumed at time t_{j+1}. The corresponding holding cost is: $h \times (t_{j+1} - t)$. All the CubeSats failed after time t_{j+1} cannot be replaced so shortage situation will continue until the last day of the planning period. The shortage time associated with these CubeSats is: $(T - t_k)$, $k = j + 2, j + 3, \ldots, n$. The corresponding shortage cost is: $s \times (T - t_k)$, $k = j + 2, j + 3, \ldots, n$. The total cost, in this case, can be computed as:

$$R = s \left[\sum_{k=1}^{j} (t - t_k) + \sum_{k=j+2}^{n} (T - t_k) \right] + h(t_{j+1} - t) + cQ \tag{8}$$

4. $j + 1 < Q < n$ (as shown in Fig. 3d)

After the spare CubeSats arrive, j spares are used to replace j faulty CubeSats instantly. The remaining $(Q - j)$ spares are stored in the orbital depot. Each time a new CubeSat malfunction occurs, one spare is consumed. This process continues until the Qth fault occurs when the last spare is consumed. The corresponding holding cost is: $(t_k - t)$, $k = j + 1, j + 2, \ldots, Q$. The $(n - Q)$ CubeSats failed after that could not be replaced due to the lack of spare CubeSats. The corresponding shortage cost is: $s \times (T - t_k)$, $k = Q + 1, Q + 2, \ldots, n$. The total cost, in this case, can be computed as:

$$R = s \left[\sum_{k=1}^{j} (t - t_k) + \sum_{k=Q+1}^{n} (T - t_k) \right] + h \sum_{k=j+1}^{Q} (t_k - t) + cQ \tag{9}$$

5. $Q \geq n$ (as shown in Fig. 3e)

Same as the fourth case, after the spare CubeSats arrive, j spares are consumed to replace j faulty CubeSats instantly. Of the remaining $(Q - j)$ spares, $(Q - n)$ spares will be stored from time t to the last day of the planning period. The holding cost associated with these CubeSats is: $h \times (T - t)$. The other $(n - j)$ parts are stored from time t, after that, each time a CubeSat malfunction, one spare is consumed until the $(n - j)$ spares are exhausted. The holding cost associated with these CubeSats is: $h \times (t_k - t)$, $k = j + 1, j + 2, \ldots, n$. The total cost, in this case, can be calculated as:

$$R = s \sum_{k=1}^{j} (t - t_k) + h \left(\sum_{j+1}^{n} (t_k - t) + \sum_{n+1}^{Q} (T - t) \right) + cQ \tag{10}$$

In summary, the cost model can be formulated as follows:

$$
R = \begin{cases}
s\left[\sum\limits_{k=1}^{Q}(t-t_k) + \sum\limits_{k=Q+1}^{n}(T-t_k)\right] + cQ, & Q \le j \\[3ex]
s\left[\sum\limits_{k=1}^{j}(t-t_k) + \sum\limits_{k=Q+1}^{n}(T-t_k)\right] + h\sum\limits_{k=j+1}^{\min(Q,n)}(t_k-t) + h(T-t)[\max(Q,n)-n] + cQ, & Q \ge j+1
\end{cases}
\tag{11}
$$

5   Optimization and Results

The operational cost of the maintenance architecture of distributed CubeSat network is related to the fault event of the CubeSat, the launch time, and quantity of the spares.

Firstly, the failure time sequence is generated. Setting a planning period to be two years, 100 random numbers are generated using the Weibull model obtained by Bayesian methods. These numbers are then sorted from low to high and the numbers that are less than planning period are retained. This number sequence is regarded as a fault time sequence which is shown in Fig. 4. Therefore, the total cost of a specified fault time sequence can be computed by substituting the arrival time and quantity of the spare CubeSats. Since that the fault time sequence is random, total cost of a single fault time sequence fluctuates sharply, while a more meaningful results comes the average of a Monte Carlo simulation. Therefore, 1000 fault time sequences are generated to calculate the mean cost. As shown in Fig. 5, mean cost curve of 1000 fault time sequences is relatively smooth.

Then, a total cost calculation is carried to find the optimal arrival time and quantity of the spare CubeSat. Cost model proposed in Sect. 4 contains three cost parameters: fixed cost, holding cost, and shortage cost. These three parameters used in the optimization are not from practical applications but assumed a ratio of 1:1:1. Substituting the 1000 fault time sequences into the Cost model, the average operational

Fig. 4 Time sequence of failure event

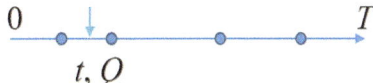

t, Q

- failure
- t : spare CubeSats arrival time
- Q: quantity of spare CubeSats replenished
- T : planning period

Fig. 5 Total cost
corresponding to different
arrival time

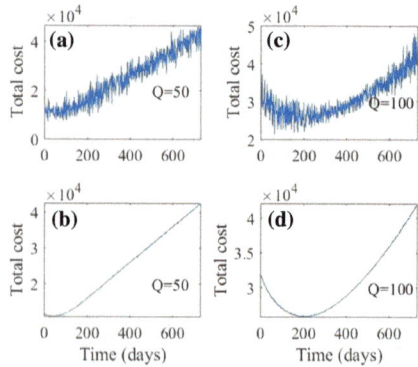

(a), (c): cost of a single fault time sequence;
(b), (d): mean cost of 1000 fault time sequences.

costs corresponding to different arrival time (i.e., 1–730) and different quantity of replenishment (i.e., 1–100) are calculated. Part of the results is shown in Fig. 6. Obviously the time and the amount corresponding to the lowest cost are the optimal results.

The results are that the optimal arrival time is the 58th day and the optimal total number of spare CubeSats is 63, corresponding the minimal cost of 8165.7. Whereas, the highest total cost is 42870.1 with terrible arrival time as 727th day and total number of spare CubeSats as 9, respectively. Compared with the worst result, the optimal total cost is only 19% of the highest total cost, saving around 81% of the

Fig. 6 Total cost
corresponding to different
arrival time and different
quantity of replenishment

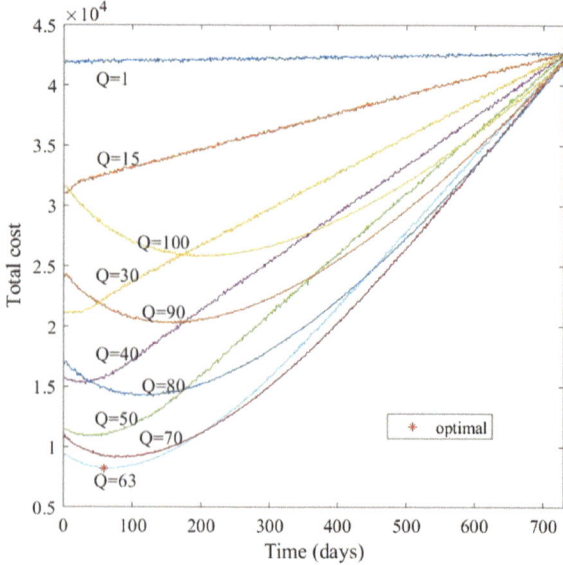

cost. The result shows that the optimal supply time is obtained at the initial phase of the planning period which is consistent with the infant mortality of CubeSats, that is, the CubeSat failure rate is higher in the early life and gradually decreases with time. Therefore, in the distributed CubeSat network there are more CubeSats that malfunction in the initial stage. If they are not repaired in time, users will suffer great loss because of significant shortage cost.

On the other hand, as shown in Fig. 6, total cost increases as the arrival time is delayed when the quantity of replenishment is small. This is because those spare CubeSats are almost always in a shortage state in this case, therefore, the influence of holding cost is small and the shortage cost increases as the arrival time is delayed resulting in an increase in the total cost. In contrast, although CubeSat suffers from infant mortality, total cost decreases first and then increases with time when the quantity of replenishment is large (about more than 50) for the following reasons. In such a situation, holding cost caused by redundant spare CubeSats will decrease with the delayed arrival time. The trends of these two costs over time are just the opposite and affect the total cost together. To summarize, in case of large replenishment, the total cost is mainly affected by holding cost when arrival time is early while it is mainly affected by shortage cost when arrival time is late.

6 Conclusion

In this paper, we aim to establish a reasonable analysis method to solve the spare CubeSats management problem in the maintenance architecture of a distributed CubeSats network. The objective is to find the optimal quantity and arrival time of spares. The procedure consists of two major parts: obtaining a meaningful failure event model and evaluating the operational cost. The parametric model of CubeSat obtained by Bayesian method using fault database of 111 different CubeSats reveals the infant mortality trend of CubeSats. The cost model has been developed, that includes three parts: fixed cost, holding cost, and shortage cost.

During the two-year planning period optimization results are 58th day and 63 spare CubeSats, respectively. Comparing the terrible results which are 727th day and 9 spare CubeSats, it indicates that 81% of the cost can be saved. The rationality of this analysis method is that the CubeSat lifetime distribution (i.e., Weibull distribution) is obtained by fitting the parameters using practical fault data which is more suitable to reflect the actual situation rather than assuming a certain distribution arbitrarily such as the normal distribution. In fact, the fitted shape parameter β of the Weibull distribution is 0.5467 that is less than 1, revealing the infant mortality trend of CubeSat. Therefore, the CubeSat network should be replenished as soon as possible which agrees with the result obtained from optimization. In future work, actual cost parameters in the cost model from practical contexts and their effects on results will be further investigated.

Acknowledgements The work was supported by the Key Research Program of the Chinese Academy of Sciences (CAS), Grant No. ZDRW-KT-2019-1-0102.

References

1. P. Armen, A. Golkar, CubeSat evolution: analyzing CubeSat capabilities for conducting science missions. Prog. Aerosp. Sci. **88**, 59–83 (2017)
2. https://www.planet.com/products/planet-imagery/
3. H. Chen, J. Liu, L. Long, Z. Xu, Y. Meng, H. Zhang, Lunar far side positioning enabled by a CubeSat system deployed in an earth-moon halo orbit. Adv. Sp. Res. (2019)
4. N.H. Crisp, K. Smith, P. Hollingsworth, Launch and deployment of distributed small satellite systems. Acta Astronaut. **114**, 65–78 (2015)
5. S. Bandyopadhyay, R. Foust, G.P. Subramanian, S.J. Chung, F.Y. Hadaegh, Review of formation flying and constellation missions using nanosatellites. J. Spacecr. Rockets **53**(3), 567–578 (2016)
6. C. Wen, H. Zhang, P. Gurfil, Orbit injection considerations for cluster flight of nanosatellites. J. Spacecr. Rockets **52**(1), 196–208 (2014)
7. https://www.nasa.gov/mission_pages/station/research/benefits/cubesat/
8. J. Gu, G. Zhang, K.W. Li, Efficient aircraft spare parts inventory management under demand uncertainty. J. Air Transp. Manag. **42**, 101–109 (2015)
9. T. Sarton du Jonchay, K. Ho, Impact evaluation of an orbital depot on on-orbit servicing infrastructures dedicated to modularized earth-orbiting platforms. Acta Astronaut. **132**, 192–203 (2017)
10. J. Guo, L. Monas, E. Gill, Statistical analysis and modelling of small satellite reliability. Acta Astronaut. **98**, 97–110 (2014)
11. J.F. Castet, J.H. Saleh, Satellite reliability: statistical data analysis and modeling. J. Spacecr. Rockets **46**(5), 1065–1076 (2009)
12. M.J. Crowder, *Statistical Analysis of Reliability Data* (Routledge, 2017)
13. https://www.nanosats.eu/
14. L. Martino, Parsimonious adaptive rejection sampling. Electron. Lett. **53**(16), 1115–1117 (2017)

Study on the Control Measures of MDRO Transmission in ICU Based on Markov Process

Zhu Min and Su Qiang

Abstract Intensive care unit (ICU) has the highest outbreak rate of Multidrug-Resistant Organisms (MDRO) infection. In this paper, the control measures of MDRO transmission in ICU were studied. Considering the incubation period and media transmission of MDRO, we combine the compartmental model and continuous time Markov chain (CTMC) t o build the stochastic model of MDRO transmission in ICU. The model was expanded into a bidimensional Markov transmission model based on the heterogeneity of population. By simulation, the key factors of the transmission model were quantitatively analyzed, and the state evolution rules of patients and medical staff were studied. Then, through the sensitivity analysis, we get some manage insights to provide control suggestions for MDRO transmission in each scenario.

Keywords Multidrug-resistant organisms · Control measures · Bidimensional Markov transmission model · State evolution rules · Sensitivity analysis

1 Introduction

A. *Background*

Nosocomial Infections (Nosocomial Infections (NI), Hospital Infections (HI), or Healthcare-Associated Infections (HAI)) are hospital infections acquired by patients or hospital staff. Nosocomial infection is a major cause of morbidity and mortality [1]. The infection seriously affects the treatment, extends the length of hospital stay, aggravates the economic burden of patients and consumes a lot of medical resources.

Nowadays, the main type of nosocomial infection is Multidrug-Resistant Organisms (MDRO) infection, which refers to the bacteria simultaneously presenting resistance to three or more types of antibiotics used in the clinic. Patients in ICU are characterized by severe illness, multiple underlying diseases, and poor collective immunity, as well as more invasive operations and extensive prophylactic use of

Z. Min (✉) · S. Qiang
School of Economics and Management, Tongji University, Shanghai 200082, China
e-mail: zhumin960906@163.com

© Springer Nature Switzerland AG 2020
H. Yang et al. (eds.), *Smart Service Systems, Operations Management, and Analytics*, Springer Proceedings in Business and Economics, https://doi.org/10.1007/978-3-030-30967-1_32

antibiotics, which greatly increase the risk of the patient's MDRO infection. According to the national nosocomial infection detection network, the percentage of hospital infections caused by MDRO in ICU patients is as high as 20–30%, and the rate of MDRO hospital infections in ICU patients is 5–10 times higher than that in general wards [2]. These seriously threatened the patient's life and increased the patient's economic burden, and meanwhile consumed and occupied extremely precious and scarce medical resources such as ICU [3]. Therefore, how to effectively prevent and control the MDRO infection in the ICU has become one of the most difficult and urgent problems in hospital quality management.

B. *Literature*

At present, using mathematical model to study the influencing factors, transmission mechanism, evolution rule, and control measures of nosocomial infection is a research hotspot. Through the control group experiment, many scholars (Kollef, Xie, Liu) introduce statistical model to analyze risk factors of MDRO infection and predict the occurrence of the infection [4–6]. This type of study is mainly retrospective, lacking a prospective study on MDRO infection and a quantitative description of key factors of the infection.

In recent years, the in-depth study of complex networks and the expansion of infectious disease dynamics methods have provided new ideas, theories and methods for the study of MDRO infection [7].

(Barnes S., Worby C. J., Robotham J. V.) expand and optimize the Susceptible-Infected-Recovered(SIR) model to construct the MORO infection transmission model of ICU department, based on the characteristics of MDRO transmission and the nursing strategy of ICU department, and finally, verify the effectiveness of different control strategies by simulation [8, 9]. Although this kind of research quantifies the key factors in the transmission path, it only considers static prevention in the control measures of nosocomial infection, failing to combine the actual situation of the ICU departments, which is not very operable.

The rest of this paper is organized as follows. In Sect. 2, according to characteristics of MDRO propagation in ICU, we build MDRO transmission stochastic model in ICU, combining Markov process with methods of dynamics of infectious diseases. Through simulation, we quantitatively describe the key factors in the chain of transmission and simulated ICU patients and medical staff in the state of evolution in Sect. 3. In Sect. 4 we draw the final conclusion.

2 Stochastic Model of MDRO Transmission in ICU

There are two main types of MDRO generation in the ICU: (1) patients admitted to the hospital carry MDRO (colonized). (2) inpatients generate MDRO. This paper only considers the cross-infection phenomenon caused by patient carrying and inappropriate care. According to the nature of MDRO transmission, poor hand hygiene and improper medical practice [10] by health care workers (HCWs) are the main

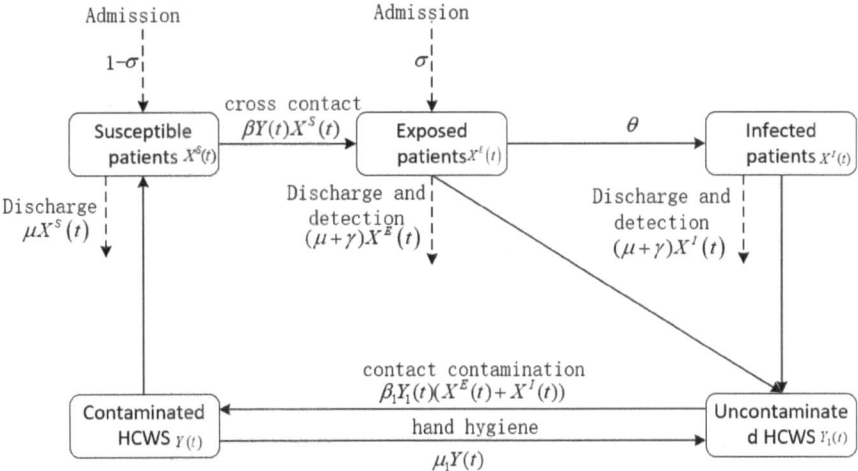

Fig. 1 The diagram of ICU transmission path

reasons for the transmission of MDRO. Therefore, HCWs play the role of the transient vectors for the transmission of MDRO in ICU. Figure 1 for the diagram of ICU transmission path.

2.1 Model Description

According to the characteristics of the MORO, due to incubation period, we introduce the Susceptible-Exposed-Infected-Susceptible (SEIS) model [11] to describe the patients' state transition. When patients admitted to hospital with σ probability of carrying the MDRO, patients are partitioned into three categories: susceptible, exposed, and infected, which exposed patients transform into infected with θ probability. Since HCWs become carriers due to improper practice in the process of nursing, we divide HCWs into two categories: contaminated and uncontaminated. See Table 1 for a summary of the parameters.

In order to facilitate model construction, the following assumptions are made:

(1) The number of patients and HCWs in the ICU are constant value (N and N'), and when there are unit patients discharged, there will be unit patients admitted. The patients are completely random and evenly mixed together.

(2) In infinitesimal time, the change in the number of infected patients is independent of the change of HCWs.

(3) The rate of MDRO transmission is equal between the HCW and the patient ($\beta = \beta_1$)

Table 1 Parameters for the model

Description	Symbol
Number of patients in ICU	N
Number of HCWs in ICU	N'
Number of susceptible patients in ICU at time t	$X^S(t)$
Number of exposed patients in ICU at time t	$X^E(t)$
Number of infected patients in ICU at time t	$X^I(t)$
Number of contaminated HCWs in ICU at time t	$Y^C(t)$
Number of uncontaminated HCWs in ICU at time t	$Y^U(t)$
The probability of carrying MDRO for patient admission	σ
HCW-patient transmission rate	β
Patient-HCW transmission rate	β_1
Detection rate of colonized (exposed and infected) patient	γ
Patient discharge rate	μ
Handwashing rate	μ_1

$$(X^S(t) + X^E(t) + X^I(t) = N, Y^C(t) + Y^U(t) = N')$$

(4) Once the patient is detected to be carrying or infected with the MDRO, the patient is immediately isolated (that is, the source of infection is cut off without any effect on the transmission of the MDRO system).

Build the SEIS-MDRO model: as shown below

$$
\begin{cases}
\frac{dX^S(t)}{dt} = (1 - \sigma)N_1 - \mu X^S(t) - \beta Y X^S(t) \\
\frac{dX^E(t)}{dt} = \sigma N_1 - (\mu + \gamma)X^E(t) - \theta X^E(t) + \beta Y(t)X^S(t) \\
\frac{dX^I(t)}{dt} = \theta X^E(t) - (\mu + \gamma)X^I(t) \\
\frac{dY^U(t)}{dt} = \mu_1 Y^C(t) - \beta_1 Y^C(t)\left[X^E(t) + X^I(t)\right] \\
\frac{dY^C(t)}{dt} = \beta_1 Y^c\left[X^E(t) + X^I(t)\right] - \mu_1 Y^C(t)
\end{cases}
\tag{1}
$$

$N_1 = \mu X^S(t) + (\mu + \gamma)X^E(t) + (\mu + \gamma)X^I(t)$, N_1 represents the number of patients who leave the ward due to discharge and detection at time t. Because the model is compartmental model, it cannot describe the change of individual state precisely. Considering the exponential distribution of admission and discharge and the independence of each event (detection, nursing, admission, decontamination, etc.), we introduce the continuous time Markov chain (CTMC) [12] to extend the model.

2.2　Stationary Distribution

The higher the dimension, the more complex the calculation is. We define $X^C(t)$ for the number of patients carrying the MDRO (colonized or infected), $X^C(t) = X^E(t) + X^I(t)$. The process $\{X^C(t), Y^C(t); t > 0\}$ result to be a bidi-mensional CTMC with state space $S = \{(i, j); 0 \le i \le N, 0 \le j \le N'\}$. $P_t(i, i')$ defines the probability X^C transfer from state i to state i'. $P_t(j, j')$ defines the probability Y^C transfer from state i to state i'. $P_t(i, i') = P(X^C(t) = i' | P(X^C(0) = i))$, $P_t(j, j') = P(Y^C(t) = j' | P(Y^C(0) = j))$. Because the change of X^C and Y^C state are independent of each other, we first calculated stationary distribution of X^C.

Basing on the Chapman–Kolmogorov equation [13]: $\sum_k P_s(i, k) P_t(k, i') = P_{s+t}(i, i')$.

$$\Delta P_h = P_{t+h}(i, i') - P_t(i.i') = \sum_k P_h(i, k) P_t(k, i') - P_t(i.i')$$

$$= \sum_{k \ne i} P_h(i, k) P_t(k, i') + [P_h(i, i) - 1] P_t(i.i')$$

$$= \sum_{k \ne i} P_h(i, k) P_t(k, i') - \sum_{k \ne i} P_h(i, k) P_t(i, i')$$

$$q(i, i') = P'_t(i, i') = \lim_{h \to 0} \frac{\Delta P_h}{h} = \sum_{k \ne i} q(i, k) P_t(k, i') - \sum_{k \ne i} q(i, k) P_t(i, i')$$

(2)

Introduce new matric $Q(i, i') = \begin{cases} q(i, i'), i' \ne i \\ -\sum_{k \ne i} q(i, k), i' = i \end{cases}$, $P'_t(i, i') = Q(i, i') P_t(i.i')$. As the same, $P'_t(j, j') = Q(j, j') P_t(j, j')$. Considering Eq. (1), there are four types of state transition processes, as shown in the Table 2.

As a result, it's stationary probability vector $\boldsymbol{\pi}$, where $\boldsymbol{\pi} = (\boldsymbol{\pi}_0, \boldsymbol{\pi}_1, \dots, \boldsymbol{\pi}_N)$, where $\boldsymbol{\pi}_0 = (\pi_{01}, \pi_{02}, \dots, \pi_{0N'})$ which defines the probability of stationary state $l_{ij} = \{(i, j); 0 \le i \le N, 0 \le j \le N'\}$. Considering Eq. (2) and characteristic of vector, $\boldsymbol{\pi}$ satisfies

$$\boldsymbol{\pi} Q = 0_f, \boldsymbol{\pi} e_f = 1$$

(3)

Table 2 Events and transition rate

	Events	Transition	Rate
1	$(X^E \text{ or } X^I)^-$ and $(X^S)^+$	$q(i, i-1)$	$(1-\sigma)(\mu+\gamma)i$
2	$(X^E \text{ or } X^I)^+$ and $(X^S)^-$	$q(i, i+1)$	$\sigma\mu(N-i) + \beta(N-i)j$
3	$(Y^C)^+$ and $(Y^U)^-$	$q(j, j+1)$	$\mu_1 j$
4	$(Y^U)^+$ and $(Y^C)^+$	$q(j, j-1)$	$\beta_1 i(N'-j)$

Section 3 simulates and analyzes two variates:

1. The expected proportion of colonized patients

$$X_p = \frac{E[X^C]}{N} = \frac{\sum_{i=0}^{N} \sum_{j=0}^{N'} i\pi_{ij}}{N}$$

2. The expected proportion of contaminated HCWs

$$Y_p = \frac{E[Y^C]}{N'} = \frac{\sum_{j=0}^{N'} \sum_{i=0}^{N} j\pi_{ij}}{N'}$$

3 Simulation and Analysis

Based on field research, we selected the ICU department of Shanghai first people's hospital, which has 40 beds (patients, $X^C + X^S$) and 6 HCWs ($Y^C + Y^U$). According to the data of 17–18 years, the proportion of newly admitted patients carrying MDRO was 0.02 ($\sigma = 0.020$, the detection rate (γ) is 0.2, and the discharge rate is 0.2 ($\mu = 0.2$). The other parameters were selected from [14], as shown in the Table 3.

In daily nursing, handwashing of HCWs is important. When the handwashing frequency raise from 0.2 to 0.6, the expected proportion of colonized patients, X_P, reduces from 0.5 to 0.1 (Fig. 2, $\sigma = 0.01$). Combine Fig. 2 with Fig. 5, when the probability of carrying MDRO, σ, is low, such as 0.01, these control measures is effective, where we should increase the handwashing frequency and discharge rate to reduce the probability of MODR transmission. However, when σ is higher (0.05 0.1), no matter how much we improve handwashing frequency and discharge rate, μ, the proportion of colonized patients can not decrease. Therefore, when patients were admitted to hospital, we should detect each patient thoroughly to lower X_P, as

Table 3 The default value of parameters

Description	Symbol	Default value
Number of patients in ICU	N	40
Number of HCWs in ICU	N'	6
The probability of carrying MDRO for patient admission	σ	0.02
HCW-patient transmission rate	β	1/6
Patient-HCW transmission rate	β_1	1/6
Detection rate of colonized (exposed and infected) patient	γ	0.2
Patient discharge rate	μ	0.2
Handwashing rate	μ_1	14

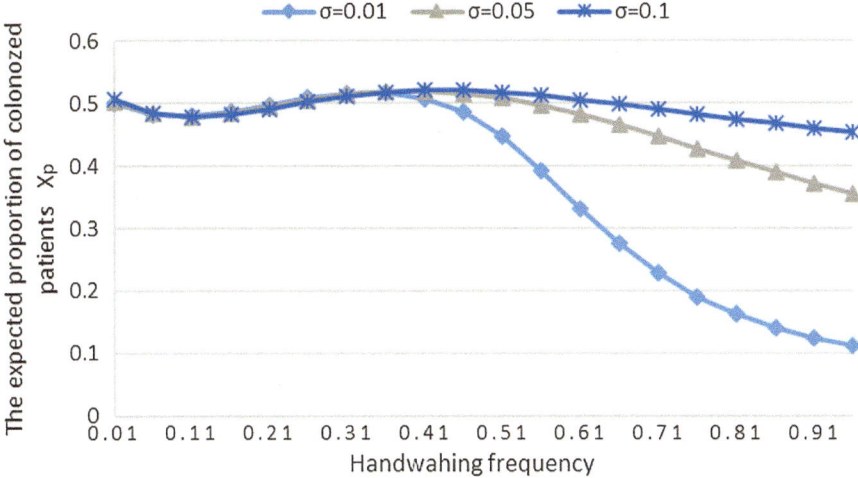

Fig. 2 X_p as a function of σ and handwashing frequency

shown in Fig. 6. We should keep the transmission rate between patients and HCWs at low level (β, $β_1$ = 1/6, 1/12) to lessen the proportion of colonized patients and contaminated HCWs (Figs. 3 and 4).

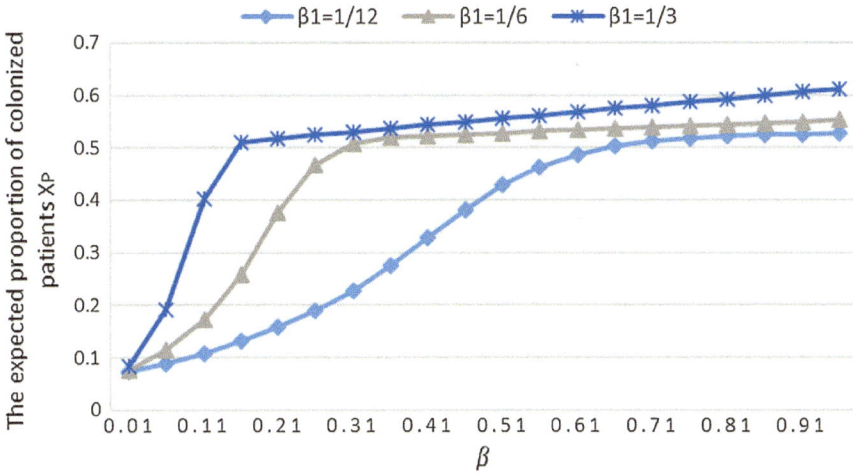

Fig. 3 X_p as a function of β and $β_1$

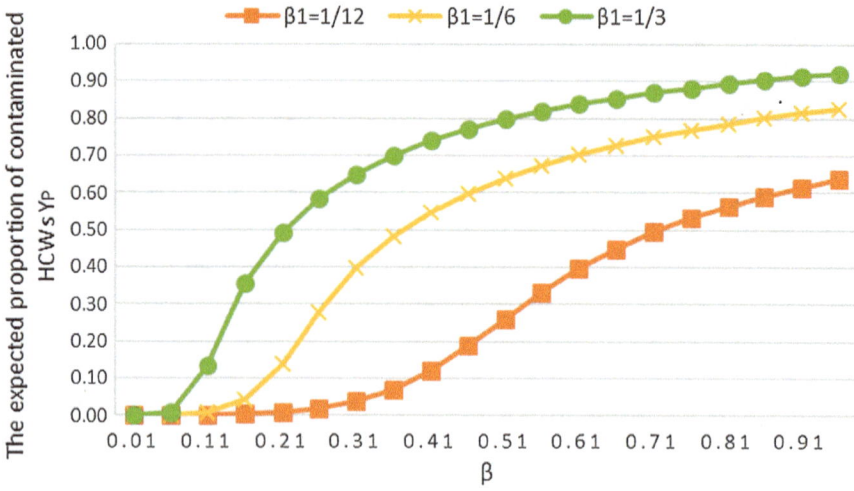

Fig. 4 Y_p as a function of β and β_1

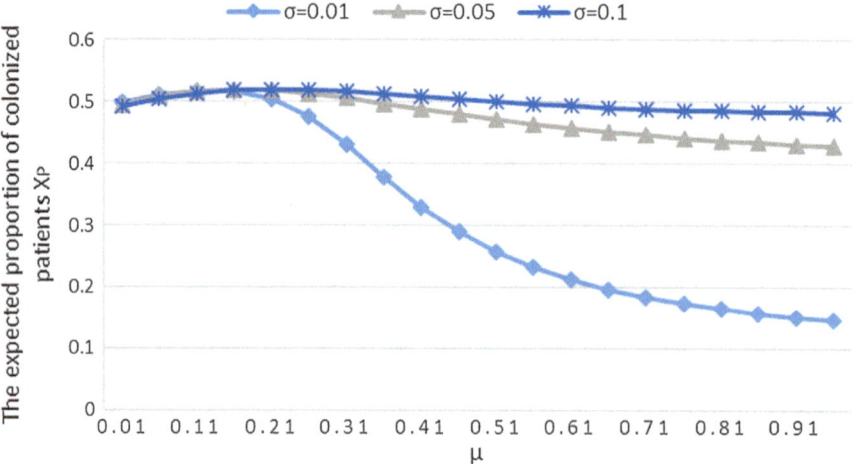

Fig. 5 X_P as a function of σ and μ

4   Conclusion

Markov stochastic process is more suitable for modeling the MDRO transmission in ICU. Considering the incubation of MDRO and HCWs as vector of infection, we combine the compartmental model (SEIS) and continuous time Markov chain (CTMC) to build the stochastic model of MDRO transmission in ICU. We introduce stationary probability vector π to calculate the expected proportion of colonized patients and contaminated HCWs, X_P *and* Y_P. Finally, by simulation and sensitivity

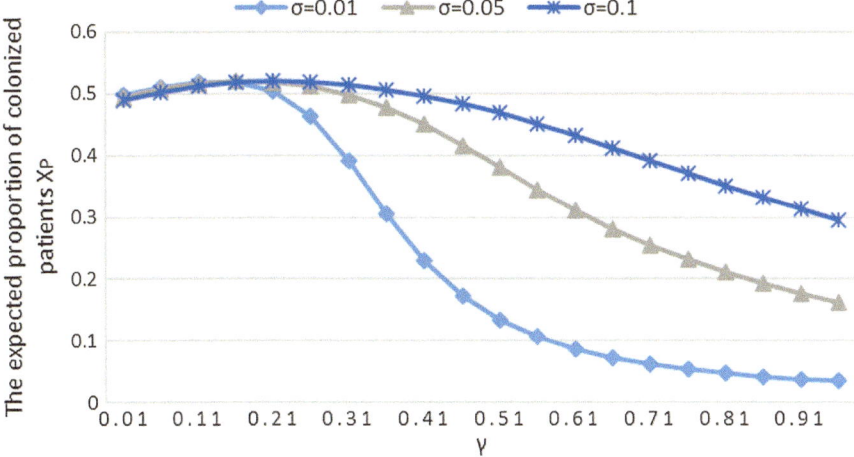

Fig. 6 X_P as a function of σ and γ

analysis, we assess the effect of each key factor to provide control suggestions for the transmission process in each scenario. Improving detection rate and reducing the probability of carrying germs, can reduce MODR transmission at the source. In daily nursing, handwashing of HCWs is important which is more in line with the actual situation.

References

1. P. Baylina, P. Moreira, Challenging healthcare-associated infections: a review of healthcare quality management issues. J. Manag. Mark. Healthc. **4**(4), 254–264 (2011)
2. E. van Kleef, J.V. Robotham, M. Jit, S.R. Deeny, W.J. Edmunds, Modelling the transmission of healthcare associated infections: a systematic review. BMC Infect. Dis. **13** (2013)
3. J.C. Lucet, D. Koulenti, J.R. Zahar, Persistence of colonisation with MDRO following discharge from the ICU. Intensive Care Med. **40**(4), 603–605 (2014)
4. M.H. Kollef, Prevention of hospital-associated pneumonia and ventilator-associated pneumonia. Crit. Care Med. **32**(6), 1396–1405 (2004)
5. D.S. Xie, W. Xiong, L.L. Xiang et al., Point prevalence surveys of healthcare-associated infection in 13 hospitals in Hubei Province, China, 2007–2008. J. Hosp. Infect. **76**(2), 150–155 (2010)
6. J.Y. Liu, Y.H. Wu, M. Cai et al., Point-prevalence survey of healthcare-associated infections in Beijing, China: a survey and analysis in 2014. J. Hosp. Infect. (2016). S019567011630007X
7. A. Dadlani, M.S. Kumar, S. Murugan et al., System dynamics of a refined epidemic model for infection propagation over complex networks. IEEE Syst. J. **10**(4), 1316–1325 (2016)
8. H. Laarabi, A. Abta, K. Hattaf, Optimal control of a delayed SIRS epidemic model with vaccination and treatment. Acta Biotheor. **63**(2), 87–97 (2015)
9. C.J. Worby, D. Jeyaratnam, J.V. Robotham et al., Editor's choice: Estimating the effectiveness of isolation and decolonization measures in reducing transmission of Methicillin-resistant Staphylococcus aureus in hospital general wards. China Health Stand. Manag. **177**(11), 1306–1313 (2013)

10. B. Allegranzi, D. Pittet, Role of hand hygiene in healthcare-associated infection prevention. J. Hosp. Infect. **73**(4), 305–315 (2009)
11. A. Korobeinikov, Lyapunov functions and global properties for SEIR and SEIS epidemic models. Math. Med. Biol. J. IMA **21**(2), 75 (2004)
12. J.R. Artalejo, On the Markovian approach for modeling the dynamics of nosocomial infections. Acta. Biotheor. **62**(1), 15–34 (2014)
13. R. Metzler, Generalized Chapman-Kolmogorov equation: a unifying approach to the description of anomalous transport in external fields. Phys. Rev. E Stat. Phys. Plasmas Fluids Relat. Interdiscip. Top. **62**(62), 6233–6245 (2000)
14. B.S. Cooper, G.F. Medley, G.M. Scott, Preliminary analysis of the transmission dynamics of nosocomial infections: stochastic and management effects. J. Hosp. Infect. **43**(2), 0–147 (1999)

What Makes a Helpful Online Review for Healthcare Services? An Empirical Analysis of Haodaifu Website

Ya Gao and Ling Ma

Abstract The online healthcare websites bring more healthcare resources to patients, reduce time cost, and break geographical restrictions. However, the information explosion brought by the healthcare website also increased the difficulty in information screening and trust establishment for risky healthcare service. The online review is an important resolution for information asymmetry. This paper explores the review significance of healthcare websites by examining the impact of review depth and valence on the review helpfulness, especially in the context of different risk level diseases. We employed a secondary data econometric analysis obtained as 44,938 pieces of reviews from Haodaifu.com. We found that both review depth and valence have a significant impact on review helpfulness. And review depth has a more significant impact when review valence is low. But the disease risk moderates the impact, that is, review depth is more useful for low-risk diseases than high-risk diseases. Also, the disease risk moderates the impact of review valence. For low-risk diseases, neutral reviews have a more positive impact on the review helpfulness. For high-risk diseases, extreme reviews have a greater impact on the review helpfulness. These findings can help to understand users' needs on healthcare websites and establish more effective ways to do communication.

Keywords Review depth · Review valence · Review helpfulness · Healthcare service

1 Introduction

The online healthcare websites, which are popular in recent years, have expanded the online service to the healthcare field. Traditional healthcare services is no longer limited to face-to-face treatment modes, breaking geographical and temporal limitations. This mechanism greatly enhanced healthcare information transparency and patient choice and saved searching and treatment costs. The patients can use the information

Y. Gao · L. Ma (✉)
School of Business, East China University of Science and Technology, Shanghai 200237, China
e-mail: maling@ecust.edu.cn

© Springer Nature Switzerland AG 2020   365
H. Yang et al. (eds.), *Smart Service Systems, Operations Management, and Analytics*, Springer Proceedings in Business and Economics,
https://doi.org/10.1007/978-3-030-30967-1_33

displayed on the website including doctors, hospitals, and historical healthcare cases for the relevant patients to choose from different regions and different levels. According to the data, 79% [1] of users have used such healthcare websites for consultation in the United States. However, the massive amount of information brought by Internet tools has greatly increased the difficulty of screening and selecting information. The information on healthcare websites is mostly professional so that users without experience or healthcare knowledge cannot quickly make decisions and build trust.

Online Word-of-Mouth (WOM) represented by user-generated online reviews [2], becomes a "translator" of information, an important piece of information that reveals the quality of doctors and a factor that influences consumer choices and trust [3]. Users are not only information seekers of healthcare websites but also information providers [4] which are called "online mutual assistance." Users can get similar disease descriptions and easier to understand information from patient reviews with service experience, as well as deeper information such as personal experience that the website's homepage cannot provide for information selection and trust building [5].

Some studies have explored the impact of user-generated online reviews about experience products [6] like the movie market [7–9] and book market [10]. There are little researches on online expert services (such as healthcare websites) that involve the "double risk" of the service itself and the Internet environment [11]. Based on the service situation of healthcare websites, patients often have different decision-making processes for diseases of different risk levels. The decision-making process for ordinary diseases is shorter, the degree of cautiousness is lower, and the medium rating reviews that objectively describe the facts are often preferred. The decision-making process of severe diseases with high risk is longer and more cautious. It is necessary to carefully search for famous doctors with an extremely good reputation and high evaluation. The research of Li et al. [12] also proved that health consultation market meets the Pareto principle and usually has been dominated by a few best famous doctors. However, current literature do not yet provide factual evidence to verify how the attributes of a review affect its helpfulness on a healthcare website environment. It is also necessary to further verify that this helpfulness differs between diseases of different risk levels and different psychological motivations.

2   Research Hypothesis

2.1   Theory of Rational Action

Rational Behavior Theory (TRA) [13] assumes that a person's behavior depends on his/her intention to perform a particular behavior. Behavioral intention is influenced by two key factors: an individual's trust and attitude toward social behavior and beliefs or what one thinks that others think he/she should do. Basically, the theory explains the factors that influence people's behavioral decisions. More specifically, it recognizes that when people make behavioral decisions, they are not only individual factors of beliefs and attitudes, but also social factors of social norms. Therefore, it is necessary for this study to consider personal and social factors when examining the behavior of a doctor. The use of TRA in this study is appropriate and could be used in the healthcare field to explain certain trusts and behaviors in the healthcare field. TRA explores the factors that influence a person's formation of trust and behavior and it will help explain the factors that influence the patient's use of online healthcare sites to build trust and then do action.

The review valence is a comprehensive evaluation of the quality or usage of the product or service by the reviewer. It is usually expressed in terms of star rating. One star (lower level) indicates the negative evaluation of the consumer and is not satisfied with the overall experience; Five stars (higher level) indicate positive feedback from consumers and are very satisfied with the overall quality; Three stars reflects a moderate view, off course. Usually, extreme evaluation refers to the evaluation of one star or five stars, and neutral evaluation refers to three-star evaluation. A large number of consistent valences show the attitude of the general users to the product or service indicating the consensus of the majority. Ghose and Ipeirotis's research [14] on film and music found that commentary titers tend to be bimodal with extreme ratings accounting for the majority. Forman's research on books shows that the impact of neutral evaluation on sales is not as extreme as the evaluation. In addition, Hao [4] found that positive sentiment orientation (extreme evaluation) has a significant positive correlation with consumer perceived film quality when studying online film reviews; but Eisend's research [15] on advertising shows that consumers trust with positive neutral evaluation more. Therefore, this paper proposes the following assumption:

H1   In the online healthcare website the review valence has a significant impact on the review helpfulness.

In the Internet healthcare platform, information depth is one of the factors that affect user decision-making. The individuals take their trust and attitude towards social behavior and beliefs as the basis for decision-making. It is an important factor in judging information reliability and increasing trust. When users read and compare reviews from peers interpreted the content, the judgment of information reliability is formed and the judgment result of trust or distrust is formed. Therefore, they use it to guide final decisions. At this point, the amount of information can be important and the depth of the information in the review content is useful for improving diagnostics and impacting perceived helpfulness. Especially when the review valence is getting lower which can be understood that users would like to read content of negative reviews. Salehan's [16] research solidly proved review depth positively influences review helpfulness in the context of online shopping websites. Karimi and Wang [17] also found mixed support to the effects of various review attributes to review usefulness at mobile gaming applications, including review depth. The increased depth of information can help decision-making process by increasing users confidence. Longer reviews usually include more service details, including service experience process, experience feeling, and more. Therefore, this paper proposes the following assumption:

H2a   In the online healthcare websites, the depth of reviews has a significant impact on the review helpfulness.

H2b   In the online healthcare websites, the depth of reviews has a more significant impact on the review helpfulness when the review valence is getting lower.

2.2   Product Involvement Theory

Product involvement refers to the level of care and attention that consumers pay to the products they are about to purchase. Product involvement is based on consumer self-perception and self-concept and is influenced by consumer values, needs, and interests. Product involvement can be divided into high product involvement and low product involvement [18]. High-involvement products are important to consumers and consumer perceived risk increases when purchasing highly involved products. Therefore, consumers with high product involvement have a more complex, extensive, and rational decision-making process. For low-involvement products, it often means that consumers have a wide set of considerations for certain types of products which are less likely to form preferences or loyalty than consumers with high product involvement.

Therefore, in the Internet healthcare website we believe the influence factors of the review helpfulness is not the same under different disease risk levels. For low-risk diseases, patients may not have specific preferences for the most famous and most highly regarded doctors as long as they are not too inexperienced. Concerns about the evaluation may focus on the pertinent objective description of the condition and service. However, in the case of serious diseases, the patients may have no significant effect after several rounds of treatment. There is a strong preference for most popular and famous doctors. It is urgent to understand all the information. The information demand is large, the eager to find positive evaluations to judge the quality of the doctor's services is strong, and pay much attention to extreme negative evaluations to test whether the doctor has irreversible and even life-threatening consequences. The process of trust and decision-making is more complicated.

Therefore, this paper proposes the following assumptions:

H3a   The disease risk moderates the impact of review depth on review helpfulness, the review depth has a greater impact on the review helpfulness of low-risk diseases than high-risk diseases.

H3b   The disease risk moderates the impact of review valence on review helpfulness. For low-risk disease, the neutral review has a greater impact on the review helpfulness than the extreme review.

H3c   The disease risk moderates the impact of review valence on review helpfulness. For high-risk disease, the extreme review has a greater impact on the review helpfulness than the neutral review.

3 Empirical Research

3.1 Data Collection

This paper uses the empirical research method of secondhand data to verify the hypothesis proposed above. The online crawler tool was used to collect indirect data from the Haodaifu.com which is one of the most competitive websites for online healthcare consultation websites in China and it is also one of the main ways for relevant patients in China to seek online healthcare services. The website provides each doctor with the functionality of a personal homepage. This study captured the data of doctors in Shanghai for the two diseases (pediatric pneumonia and pancreatic cancer) until the end of 2016, pediatric pneumonia is the representative of low-risk diseases and pancreatic cancer is the representative of high-risk diseases. The total number of reviews was 44,938, including 25,589 from pediatric pneumonia and 19,349 from pancreatic cancer.

3.2 Data Model

This article uses the variables of doctors' pages in the pediatric pneumonia and pancreatic cancer area on the website including the number of words in each review, review stars, useful votes, hospital levels, and doctor position. The measurement model constructed in this paper is shown in the following formula:

$$
\begin{aligned}
Votes = {} & beta\_0 + beta\_1 * ReviewDepth + beta\_2 * ReviewValence \\
& + beta\_3 * Disease\_risk + beta\_4 * Disease\_risk * ReviewDepth \\
& + beta\_5 * Disease\_risk * ReviewValence \\
& + beta\_6 * ReviewDepth * ReviewValence \\
& + beta\_7 * Position + beta\_8 * HospitalLevel + \alpha i + \varepsilon i
\end{aligned}
\tag{1}
$$

"Votes" indicates the websites users' opinions on that if this review is helpful for them. "ReviewDepth" indicates the number of words in each review, and "ReviewValence" indicates the rating of each review which is the average of the efficacy and attitude star rating. "Disease risk" indicates the risk of disease, 0 is pediatric pneumonia, 1 is pancreatic cancer. "Position" indicates the level of the doctor. In the doctor's hierarchy in China, we assigned chief physician = "5", deputy chief physician = "4", attending physician = "3", resident physician = "2", and others = "1". "HospitalLevel" indicates the hospital levels. The third-level hospital is the highest. Each level of hospitals is divided into A, B, and C. We divided the hospital equivalent into six grades, namely, third-level-A hospital = "6", third-level-BC hospital = "5", secondary-level-A hospital = "4", secondary-level-BC hospital = "3", first-level-A hospital = "2", and first-level-BC hospital = "1".

3.3 Evaluation Outcomes

Table 1 shows a descriptive analysis of the sample variables used and Table 2 shows the results of the correlation between the variables. In order to maintain magnitude consistency, all variables were standardized before performing regression. The regression results of the model are shown in Table 3. The model analysis results were significant, and most of the important variables had a highly significant p-value level ($P = 0.000$), indicating that the independent variables had a significant effect on the dependent variables.

Model 1 contains only control variables; Model 2 adds the risk level of the diseases, Model 3 focuses on the review depth and the interaction between review depth and disease risk, and the results show the coefficient of review depth ($0.217, P < 0.001$) which is positive and highly significant indicating that regardless of the disease risk the review depth has a positive impact on the review helpfulness, H2a is supported and the interaction between review depth and disease risk is also highly significant indicating that the disease risk moderated the impact of the review depth on review helpfulness; Model 4 focuses on the review valence and the interaction between review valence disease risk, and the results show the coefficient of reviews valence ($0.0644, P < 0.001$) is positive and highly significant indicating that regardless of the disease risk, the review valence has a positive impact on the review helpfulness, H1 is supported. In addition, the interaction between review valence and disease risk is also highly significant indicating that disease risk moderated the impact of the review valence on review helpfulness.

In addition, in Model 5, it can be seen the coefficient of interaction between review depth and review valence is negative (-0.0588) and highly significant ($P < 0.001$), indicating review depth is more helpful when the review valence is low, indicating that people like to read long negative reviews. H2b is supported. It can also be seen that the coefficient of interaction between disease risk and review depth is negative (-0.0447) and highly significant ($P < 0.001$) indicating the disease risk moderated the impact of the review depth on review helpfulness. As the low-risk disease was set to "0" and high-risk disease was "1", so it can be explained that when review depth has a more positive impact on low-risk diseases, H3a is supported.

Table 1 Descriptive statistical analysis of variables

Variables	Observations	Mean	Standard deviation	Minimum	Maximum
Votes	44,938	5.91	12.61	0	446
ReviewDepth	44,938	172.85	189.94	0	3928
ReviewValence	44,938	4.54	0.94	1	5
DiseaseRisk	44,938	0.43	0.50	0	1
Position	44,938	4.63	0.66	1	5
HospitalLevel	44,938	5.88	0.49	1	6

Table 2 Correlation coefficient matrix

	Votes	ReviewDepth	ReviewValence	DiseaseRisk	Position	HospitalLevel
Votes	1.0000					
ReviewDepth	0.1756	1.0000				
ReviewValence	0.0282	0.0666	1.0000			
DiseaseRisk	−0.0088	0.0946	0.2549	1.0000		
Position	0.1076	0.0453	0.0277	0.0475	1.0000	
HospitalLevel	−0.0456	0.0268	0.0374	0.1525	0.1038	1.0000

Table 3 Regression result 1

Variables	(1)	(2)	(3)	(4)	(5)
	stdvote	stdvote	stdvote	stdvote	stdvote
stdPosition	0.114***	0.114***	0.107***	0.112***	0.105***
	(0.00471)	(0.00471)	(0.00464)	(0.00470)	(0.00463)
stdHospitalLevel	−0.0574***	−0.0566***	−0.0587***	−0.0562***	−0.0590***
	(0.00471)	(0.00476)	(0.00469)	(0.00475)	(0.00467)
Diseaserisk		−0.0112	−0.0433***	−0.00655	−0.0326***
		(0.00957)	(0.00946)	(0.0102)	(0.0101)
stdReviewDepth			0.217***		0.212***
			(0.00715)		(0.00714)
stdDisk_Depth			−0.0741***		−0.0447***
			(0.00938)		(0.00962)
stdReviewValence				0.0644***	0.0481***
				(0.00540)	(0.00535)
stdRisk_Valence				−0.128***	−0.127***
				(0.0123)	(0.0121)
stdReviewDepth_Valence					−0.0588***
					(0.00571)
Constant	0	0.00482	0.0221***	0.0196***	0.0374***
	(0.00468)	(0.00624)	(0.00616)	(0.00635)	(0.00627)
Observations	44,938	44,938	44,938	44,938	44,938
R-squared	0.015	0.015	0.046	0.019	0.052

Note The number in parentheses are standard error. $+p < 0.1$; $*p < 0.05$; $**p < 0.01$; $***p < 0.001$

Table 4 Regression result 2

Variables	(Low Risk)	(High Risk)
	stdVote	stdVote
stdPosition	0.110***	0.101***
	(0.00613)	(0.00700)
stdhospitalLevel	−0.0838***	0.0229***
	(0.00613)	(0.00699)
stdReviwDepth	0.175***	0.185***
	(0.00622)	(0.00705)
stdReviewValence	0.273***	−0.319***
	(0.0376)	(0.0487)
stdReviewValence2	−0.329***	0.252***
	(0.0376)	(0.0487)
StdReviewDepth*ReviewValence	−0.0226***	−0.0947***
	(0.00657)	(0.00780)
Constant	0.00112	0.00443
	(0.00608)	(0.00700)
Observations	25,589	19,349
R-squared	0.056	0.056

Note The number in parentheses are standard error. $+p < 0.1$; $*p < 0.05$; $**p < 0.01$; $***p < 0.001$

To further examine this regulatory relationship data was divided into two samples, low-risk diseases and high-risk diseases. This is shown in Table 4. For low-risk diseases, the review valence (0.273, $P < 0.001$) and square of the review valence (−0.329, $P < 0.001$) were highly correlated with the review helpfulness. Moreover, the positive coefficient of the valence (0.273) and the negative coefficient of square of the valence (−0.329) indicate that data has an "inverted U" relationship, that is, for low-risk diseases, if the neutral review has more positive effect on helpfulness than the extreme evaluation, H3b is supported.

For high-risk diseases, the review valence (−0.319, $P < 0.001$) and the square of the review valence (0.252, $P < 0.001$) were also highly correlated with the review helpfulness. Moreover, the negative coefficient of the valence (−0.319) and the positive coefficient of square of the valence (0.252) indicate that under high-risk diseases, there is a "U-shaped" relationship between the effect of review valence on the review helpfulness, that is, if the impact of extreme evaluation on review helpfulness is greater than the medium rating, H3c is supported.

3.4 *Robustness Check*

In the robustness test, the number of review words is counted by the actual unequal number ladder. According to the word distribution in the Fig. 1, words of 0–100 are set to 1, 100–200 are set to 2, and 200–300 are set to 3, 300 or above is set to 4, and the variable is named "ReviewDepth2". The regression results are shown in Model 6 in Table 5.

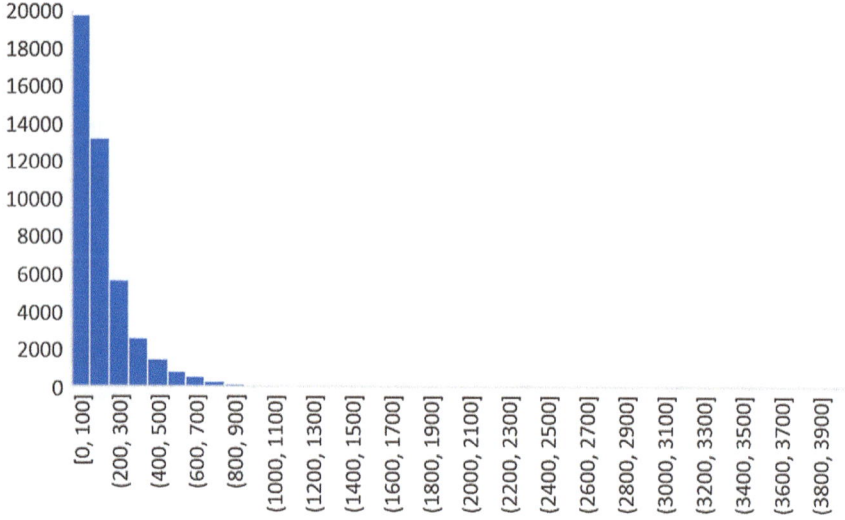

Fig. 1 Distribution of ReviewDepth

Table 5 Robustness check result

Variables	(1)	(2)	(3)
	stdvote	stdvote	stdvote
stdPosition	0.105***	0.105***	0.105***
	(0.00465)	(0.00463)	(0.00463)
stdHospitalLevel	−0.0590***	−0.0591***	−0.0591***
	(0.00469)	(0.00467)	(0.00467)
stdReviewDepth2	0.178***		
	(0.00642)		
stdReviewValence	0.0528***		
	(0.00537)		
Diseaserisk	−0.0251**	− 0.0368***	−0.0229**
	(0.0101)	(0.00989)	(0.0101)

(continued)

Table 5 (continued)

Variables	(1)	(2)	(3)
	stdvote	stdvote	stdvote
stdDisk_ReviewDepth2	−0.0444***		
	(0.00956)		
stdRisk_ReviewValence	−0.130***		
	(0.0122)		
stdReviewDepth_Valence	−0.0256***	−0.0604***	−0.0613***
	(0.00560)	(0.00570)	(0.00570)
stdReviewDepth		0.214***	0.213***
		(0.00714)	(0.00714)
StdReviewValence(lx)		0.0376***	
		(0.00537)	
stdRisk_ReviewDepth		−0.0471***	−0.0444***
		(0.00962)	(0.00962)
stdRisk_ ReviewValence(lx)		−0.0991***	
		(0.0115)	
StdReviewValence(td)			0.0405***
			(0.00520)
stdRisk_ ReviewValence(td)			−0.151***
			(0.0130)
Constant	0.0317***	0.0343***	0.0354***
	(0.00629)	(0.00625)	(0.00625)
Observations	44,938	44,938	44,938
R-squared	0.043	0.051	0.052

Note The number in parentheses are standard error. $+p < 0.1$; $^{*}p < 0.05$; $^{**}p < 0.01$; $^{***}p < 0.001$

In addition, in the main model, the review valence is the combination of the doctor's service attitude and efficacy rating. In the robustness test the attitude (Taidu) and the efficacy (Liaoxiao) are separated, which are the model 7 and model 8 in Table 5. The related variables are still significant indicating that the results are robust.

4   Discussion

4.1   Principal Findings

In this study, we found that review depth and valence of online healthcare websites have a significant impact on reviews helpfulness, regardless of low-risk or high-risk diseases. And the review depth has a more significant impact on the review helpfulness when the review valence is low indicating that people usually like to read negative reviews carefully. The disease risk moderates the impact of review depth on review helpfulness, that is, review depth is more useful for low-risk diseases than high-risk diseases. Also, the disease risk moderates the impact of review valence on the review helpfulness. For low-risk diseases, neutral reviews have a more positive impact on the review helpfulness. For high-risk diseases, extreme reviews have a greater impact on the review helpfulness.

In addition, from the perspective of life experience and patient psychology the above hypothesis verification results are also easy to explain. The review depth often represents the degree of information disclosure when users make healthcare consultations. The more information can be provided to users, the more valuable reviews are. However, in the case of high-risk diseases, the user's attention to the review depth is not that high, it is not required most to describe the condition and the treatment process in the review. It is understandable that the patients who have severe diseases for a long time are often familiar with the symptoms and treatment process, they pay more attention to the treatment effect and doctors' reputation. So the patients with severe diseases are sensitive to the evaluation of extreme reviews (such as great or no effect), which reflects the users want to find the best-performing doctors and try to avoid the less effective treatment. On the contrary, for low-risk diseases, users pay more attention to the review content, the description of the symptoms, and treatment process (such as the degree of pain). In terms of review valence, extreme derogatory reviews (such as abuse) or extreme compliments (such as exaggeration) are less useful than moderate descriptive reviews. The users build trust in reviews that objectively describe the healthcare treatment process and service.

4.2   Managerial Contributions

The results of this paper validated the influencing factors and impact effects of reviews helpfulness on online healthcare websites. The research results can increase the significance of online healthcare platforms when users make online healthcare decisions which is conducive to promote mutual interaction between doctors and patients and help to promote the prosperity of the healthcare platform.

Second, our findings provide platform with a more valuable online review guide. For example, our results suggest that online healthcare sites should consider the risk of diseases when applying different reviews guidelines. For low-risk diseases (such as pediatric pneumonia), patients who have already visited the hospital are encouraged to provide a more moderate evaluation and describe the process and experience of the treatment objectively and particularly. In the case of high-risk diseases, patients are encouraged to use emotional words such as "very good", "therapeutic effect", and give full marks. In addition, the healthcare platform can also setup some "excellent reviewers" selection and reward mechanisms.

4.3 Limitations

This study has several limitations. First, data comes only from one online healthcare website. The singularity of the research may affect the applicability of the research results. Subsequent research can use the data from multiple healthcare websites to verify the conclusions of this study. It is possible to collect data from different healthcare websites and find out the general rules that can be applied to the entire healthcare websites.

Second, the study in this paper has only selected two representative disease types. Although we have consulted the authoritative literature on the classification of diseases when selecting, this paper still needs to be improved in sample diversity. In the future, more researchers can combine and compare data of more different level diseases to further expand the research samples.

5 Conclusions

Our findings suggest that review depth and review valence on healthcare platform (such as Haodaifu.com) have a significant impact on review helpfulness. And people pay more attention to negative reviews. But under different risk level diseases, the review depth and valence reflect different nuances. That is, review depth is more useful for low-risk diseases than high-risk diseases, neutral reviews have a more positive impact on the review helpfulness for low-risk diseases, and extreme reviews have a greater impact on the review helpfulness for high-risk diseases.

In addition, this research result found the difference in users' psychology when they select doctors online and make healthcare decisions. That is, people with severe diseases pay more attention to the treatment effect and doctors' reputation than symptoms and treatment process. So they are sensitive to the evaluation of extreme reviews (such as great or no effect) in order to find best-performing doctors and avoid the less effective treatment. On the contrary, people with low-risk diseases (such as pediatric pneumonia) pay more attention to the description of the symptoms and treatment process (such as the degree of pain and convenience) in reviews than extreme derogatory reviews or extreme compliments reviews. This kind of difference is reflecting the different users' psychology of healthcare websites, which is worthy of distinction and must be identified. This result is of great significance for healthcare websites to do review management and user communication.

References

1. S.R. Mano, Social media and online health service: a health empowerment perspective to online health information. Comput. Hum. Behav. **39**, 404–412 (2014)
2. A. Ghose, P. Ipeirotis, The Econo Mining project at NYU: studying the economic value of user-generated content on the internet. J. Revenue Pricing Manag. **8**, 241–246 (2009)
3. F.M. Bass, Comments on a new product growth for model consumer durables the bass model. Manag. Sci. **50**(12_supplement), 1833–1840 (2004)
4. F. North, L.D. Uthke, S.M. Tulledge-Scheitel, Integration of e-consultations into the outpatient care process at a tertiary healthcare centre. J. Telemed. Telecare **20**(4), 221–229 (2014)
5. G.G. Gao, B.N. Greenwood, J. McCullough, R.A. Agarwal, Digital soapbox? the information value of online physician ratingsm, in *Conference on Information Systems and Technology* (2011)
6. S.M. Mudambi, D. Schuff, What makes a helpful online review? a study of Customer reviews on Amazon.com. MIS Q. **34**(1), 185–200 (2010)
7. Q. Ye, Y. Hao, Y. Li, Research on the influencing factors of online review helpfulness based on film critic data. J. Manag. Sci. **13**(8), 78–96 (2010)
8. W. Duan, B. Gu, A.B. Whinston, Do online reviews matter?—an empirical investigation of panel data. Decis. Support Syst. **45**, 1007–1016 (2008)
9. Y. Chen, J. Xie, Online consumer review: word-of-mouth as a new element of marketing communication mix. Manag. Sci. **54**, 477–491 (2008)
10. J.A. Chevalier, D. Mayzlin, The effect of word of mouth on sales: online book reviews. J. Mark. Res. **43**, 345–354 (2006)
11. C. Forman, A. Ghose, B. Wiesenfeld, Examining the relationship between reviews and sales: the role of reviewer identity disclosure in electronic markets. Inf. Syst. Res. **19**, 291–313 (2008)
12. J. Li, Y. Zhang, L. Ma, X. Liu, The impact of the internet on health consultation market concentration: an econometric analysis of secondary data. J. Med. Internet Res. **18**(10), e276 (2016)
13. N.J. Vriend, Rational behavior and economic theory. J. Econ. Behav. Organ. **29**(2): 263–285 (1996)
14. A. Ghose, P. Peirotis, Designing ranking systems for consumer reviews: the impact of review subjectivity on product sales and review quality, in *Proceedings of the 16 Annual Workshop on Information Technology and Systems* (2015), pp. 156–167
15. L. Eisend, Two sided advertising: a meta-analysis. Int. J. Res. Mark. **23**(2), 187–198 (2006)
16. M. Salehan, D.J. Kim, Predicting the performance of online consumer review: a sentiment mining approach to big data analytics. Decis. Support Syst. **81**, 30–40 (2016)
17. S. Karimi, F. Wang, Online review helpfulness: impact of reviewer profile image. Decis. Support Syst. **96**, 39–48 (2017)
18. W.-I. Lee, S.-Y. Cheng, Y.-T. Shin, effects among product attributes, involvement, word-of-mouth, and purchase intention in online shopping. Asia Pac. Manag. Rev. **22**, 223–229 (2017)

Analyzing WeChat Diffusion Cascade: Pattern Discovery and Prediction

Ruilin Lv, Chengxi Zang, Wai Kin (Victor) Chan and Wenwu Zhu

Abstract WeChat social network is one of the most popular social platforms in China, providing not only communication services but also enabling a number of service innovations. Understanding how information diffuses in an online social network such as WeChat is critical to the design and evaluation of existing or new services. This paper studies the diffusion pattern and predictability of WeChat cascade. We propose an analysis framework for WeChat cascade based on the characteristics of cross-scenario diffusion. By analyzing a real WeChat dataset, we reveal some typical diffusion patterns. We also obtain good prediction performance.

Keywords Social network · WeChat · Information diffusion · Prediction

1 Introduction

The boundary of service systems is disappearing with the emerging of various kinds of new services thanks to the advancement in information technology such as IoT, social networks, and so forth [1]. The increasing popularity of social network platforms such as Facebook and WeChat not only facilitates communication between people but also creates new service innovations. For example, WeChat is providing many mini-apps that provide all kinds of services to users, including payment, finance, food, entertainment, travel, education, and so forth.

In China, WeChat is probably the most popular platform for online social interactions currently. It has more than 1 billion monthly active users. Almost all Chinese net citizens use WeChat to communicate with friends and obtain multimedia information. WeChat offers multiple ways for information diffusion, including Chat and

R. Lv · W. K. (Victor) Chan (✉) · W. Zhu
Tsinghua-Berkeley Shenzhen Institute, Tsinghua University, Shenzhen 518071,
People's Republic of China
e-mail: chanw@sz.tsinghua.edu.cn

C. Zang
Department of Computer Science and Technology, Tsinghua University, Beijing 100084,
People's Republic of China

© Springer Nature Switzerland AG 2020
H. Yang et al. (eds.), *Smart Service Systems, Operations Management,
and Analytics*, Springer Proceedings in Business and Economics,
https://doi.org/10.1007/978-3-030-30967-1_34

Moments, which inherently leads to high complexity of information diffusion. Therefore, understanding the diffusion patterns of WeChat can help improve the design of online services.

Using real WeChat diffusion dataset, this paper analyzes the diffusion process of WeChat articles by incorporating diffusion cascade and user behavior record. First, we obtain the diffusion pattern of WeChat articles. Then, we use machine learning algorithm to predict future evolution of WeChat diffusion. Finally, using visualization, we perform case studies on cascades in different content topics.

There are existing studies on diffusion mechanism of social platforms like Facebook, Twitter, and Weibo [7, 10–14]. However, the study of diffusion mechanism in WeChat network is still a relatively new area. This paper tries to enrich this research area by analyzing the WeChat data. It was found that WeChat network is rich in diffusion patterns. WeChat exhibits a cross-scenario diffusion characteristic that makes its structural pattern of diffusion cascade full of complexity. This paper uses advanced machine learning algorithms to predict WeChat diffusion process. It was found that content topics might be an important factor that influences the evolution of the diffusion structure. Results of this paper add values to research on analyzing information diffusion and communication behavior in WeChat network. The results may be used to better design social networks to achieve beneficial goals such as blocking spreading of rumors, promoting positive emotions, and even facilitating new service innovations.

2   Related Works

This section reviews some important work in social network analysis and information diffusion process. In social network analysis, a well-known phenomenon: "Six-degree-of-Separation" was proposed and experimentally verified [2]. Under this phenomenon, a person could reach any other one in the world along the relationship chain within six people on average. The small world theory [3] introduced a model via rewiring to capture characteristics of real-world networks, such as local clustering and short distance.

For network structure, Barabasi and Albert proposed the scale-free network model in which degree distribution follows a power law. This model relies on the preferential attachment and growth process that generate a network with a few "hub" users, which also represents many real-world networks [4]. Community detection is another hot research topic that studies heterogeneous communities [5].

The process of information diffusion is one of the key research questions in social network analysis. In order to explain how information diffuses across social networks, traditional explanatory model can be used. One example is the Epidemic Model. In this model, each user can receive and distribute diseases/information, which is governed by a certain probabilistic infection mechanism [6]. Another explanatory model is the Social Influence model, which evaluates and maximizes the influence of individual (or community) in a social network [7, 8].

One major goal of the study of information diffusion is to predict the diffusion process in advance. Predictive models concentrate on improving the predictability of information cascade. Along this line of research is the independent cascade model, which regards the diffusion process as a tree structure and consider users who contribute to information spreading. Reference [9] claimed that the cascade predictability was determined by "skill" and "luck" and discussed possible sources of prediction error. Reference [10] examined the prediction error when adding different categories of features and explored the limitation of prediction accuracy. Reference [11] considered the diffusion process based on truth and rumor and compared the cascade structural pattern using a Twitter dataset. Reference [12] explicitly defined the cascade prediction problem to avoid the side effect of sample imbalance. Reference [13] proposed the metric of structural diversity and distinguished viral cascade from nonviral ones. Reference [14] presented other alternative structural metrics, such as the number of connected components within a user's friends network.

There are also other predictive models. For instance, Linear Threshold Model and Game Theory Model. The former model focuses on the study of the triggering threshold of diffusion [15], while the latter one concentrates on equilibrium analyses of among information spreaders [16].

In summary, existing works have built a basic research framework for the analysis and prediction of diffusion cascade in social networks. This paper contributes to this area by providing an in-depth study into the diffusion process of WeChat network.

3 Basis of WeChat Diffusion Research

3.1 Cross-Scenario Diffusion Characteristic

WeChat has three diffusion scenarios, which make the diffusion process different from those in other social platforms (such as Weibo, Facebook, and Twitter). These three diffusion scenarios are initial release scenario, chat scenario, and moments scenario (Fig. 1).

WeChat diffusion process cascades step by step as follows: an article is initially published by Official Account. The link of this article first appears in the Initial Release Scenario. Then subscribers are able to share this link to both Chat scenario and Moments scenario, thus cascading the diffusion of the article.

3.2 Measuring WeChat Cascade Structure

WeChat cascade is composed of "feeds". Each feed is a link to an article. The feeds can appear in any scenarios. In order to evaluate the influence and structural complexity

Fig. 1 Cross-scenario
diffusion characteristic of
WeChat

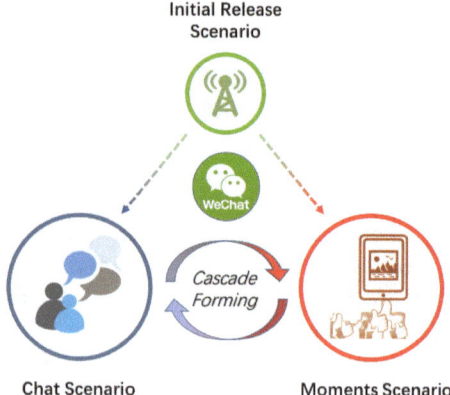

of WeChat cascade, four basic metrics are employed, including mass, breadth, depth, and wiener index (Fig. 2).

Mass is defined as the number of feeds in cascade. Mass value reflects the level of overall diffusion coverage. Therefore, the order of magnitude of mass is a main measure for the diffusion performance.

Breadth is defined as the maximum number of feeds that single cascade layer holds. Breadth value reflects the horizontal size of cascade. Larger breadth could imply the characteristic of "broadcast diffusion". The order of magnitude is also a main measure.

Depth is defined as the maximum distance between leaf nodes to root node. Depth value equals the number of layers, indicating the penetrability of information diffusion. Larger depth value suggests that the cascade have at least a few long chains that enrich its diffusion structure.

Wiener index is defined as the average distance of all node pairs. We need to obtain cascade wiener index by treating the cascade in undirected view, then construct the distance matrix and apply the division formula. Wiener index reflects the overall structural complexity and provides a quantitative measure of the structural pattern of the diffusion.

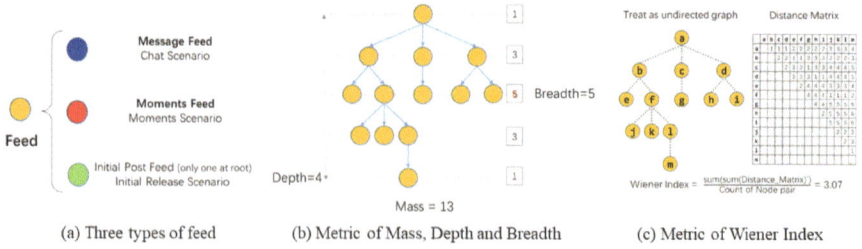

Fig. 2 Component feed and metrics of WeChat cascade

3.3 Dataset Description

Our dataset includes 7504 independent diffusion cascades. Each cascade corresponds to one article randomly sampled from overall WeChat Official Account published content. The corresponding articles were released between March 1 and March 7 in 2016.

The dataset is composed of five data fields, including sharing cascade record, article reading record, user relationship record, user attribute record, and Official Account subscribers record.

4 Discovering Diffusion Pattern in WeChat

4.1 Distribution of Structural Metrics on Final Cascade

We first examine the frequency distribution of WeChat cascade structural metrics. From the complementary cumulative distribution function curve in logarithmic coordinate system, it is observed that the distribution of these structural metrics is extremely uneven. The major cascades are in small mass and breadth, with low depth and Wiener Index. However, there exist some extremely huge cascades in terms of coverage size and structural complexity. There are nearly 1% of cascades that could be called "enormous" as their mass exceeded 10,000. This situation is similar to the distribution of breadth value. For metrics of depth and Wiener Index, the largest observed values are around 33 and 20, respectively, confirming the existence of a highly structural and complicated cascade in WeChat diffusion (Fig. 3).

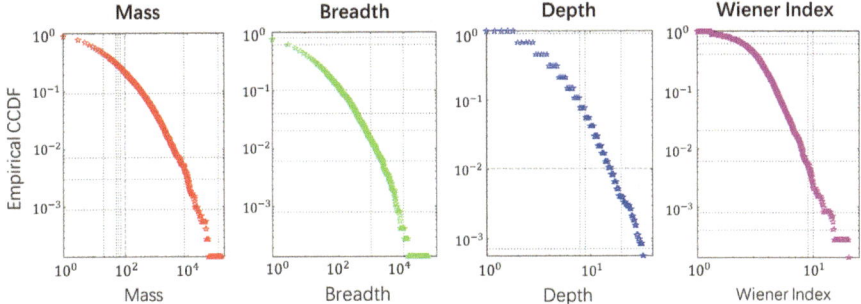

Fig. 3 Distribution of final cascade metrics value

4.2   Correlation Pattern of Structural Metrics

Next, we analyze the correlation between mass and other structural metrics. The heat
maps in Fig. 4 reveal several findings. First, breadth and mass values are positively
correlated. This means that if a WeChat cascade is large in mass, it has very high
probability to hold high breadth value at a similar magnitude. Mass is also found
to have positive correlation with depth and Wiener Index. The heat map presents a
diverging trend especially when mass is large. Cascades tend to hold lower depth and
smaller Wiener Index when their mass value is below medium. This phenomenon
may be explained by using the uncertainty theory. Once a WeChat cascade grows
larger in mass, it suffers from more uncertainty in structure evolution. The current
cascade could either stay in simple structure or transfer to complex structure.

In order to distinguish whether a WeChat cascade is structurally simple or complicated, we define the structural patterns as "star-like" and "chain-like". The numerical
indicator is Wiener Index. Our definition starts from the boundary condition: if the
mass is n, the Wiener Index value of "pure star" is $(2 - 2/n)$, and $((n + 1)/3)$ for "pure
chain" structure. From a mathematical point of view, "pure chain" is structurally more
complicated than "pure star" at any mass value n. From the empirical dataset, the
structural complexity of most WeChat cascades are somewhere in between "pure
star" and "pure chain", except for a few extremely small ones. The distribution area
is restricted between the asymptote of "pure star" and "pure chain" as shown in the
right heat map. It is observed that the major structural pattern of WeChat cascades is
closer to star pattern, especially for large cascades. In addition, with the increasing
uncertainty in diffusion process when cascade mass grows larger, the element of
chain is gradually added into the cascade structure, leading to a larger variance of
Wiener Index.

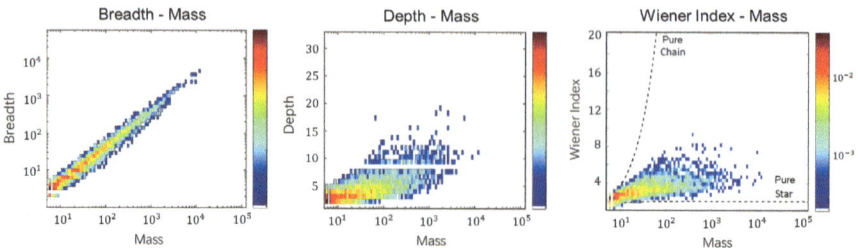

Fig. 4   Joint distribution of structural metrics to mass, plotted in heat map

5 Predicting Cascade Structural Evolution

5.1 Feature Extraction

Our empirical dataset provides information about cascade diffusion and related user behavior. After preprocessing and feature engineering, six categories of useful features are extracted for each cascade. They would serve as the input for prediction in the next subsections. All these features are captured in cascade midway at mass of k.

First category is *Content Feature*. This category includes three features as follows: article content length (in byte count), subscriber amount of Official Account, and post time indicator (to distinguish A.M. post or P.M. post).

Second category is *Cascade Structural Feature (at mass of k)*. This category includes three features as follows: cascade depth, cascade breadth, and cascade Wiener Index.

Third category is *User Relationship Graph Feature*. As the relation graph is very sparse, two extracted features are considered: number of connected components, and fraction of users belonging to the largest connected component

Fourth category is *Forwarding User Attribute Feature*. Features in this category are only about those users who have contributed to the cascade. This feature category provides user identity record, such as average age, sex ratio, average educational level, average account activation years, and average number of friends. In addition, this category also consists of features that reflect a user's activity level on WeChat in the recent month (i.e., March 2016), including average amount of chat messages sent and received by this user, and average amount of moments post and view by this user.

Fifth category is *Reading User Behavior Feature*. As some users may not share an article with others after reading it, this feature category evaluates the relationship between reading and forwarding: reading frequency of all reading users, reading frequency o f forwarding users, ratio of users who have read the article more than once, ratio of reading platform (mobile/PC), user conversion rate (from reading to forwarding).

Sixth category is *Temporal Feature*. This feature category embodies the effect of dynamics, including time elapsed (since original post), average time gap between two consecutive feeds (from the first feed to the $k/2$ feed, and from the $k/2$ feed to the k feed), and the variance of time gap.

5.2 Problem Definition

The objective of our prediction task is to achieve a high prediction accuracy and stability of the final diffusion coverage and structural complexity when a cascade is at its beginning stage. Specifically, the prediction problem is defined as a binary classification task: for a given cascade, predict whether the final metric value (mass, depth, breadth, Wiener Index) of this cascade would exceed the median value comparing with its peers that also currently hold a mass of k. Using the median value to distinguish positive and negative samples can balance the sample ratio, thus reducing the risk of sample imbalance.

5.3 Prediction Performance of Machine Learning Models

To perform the binary classification task, several machine learning algorithms are tested, including Logistic Regression, Naïve Bayes, Support Vector Machine, and Decision Tree. Furthermore, in order to improve the model performance, ensemble learning methods like bagging and boosting are applied to Decision Tree, obtaining the Adaboost and Random Forest model. The observation size is $k = 100$ and F1-score is used to evaluate the classification performance.

Figure 5 gives the prediction performance of single and ensemble models. For single models, Decision Tree and Logistic Regression generally perform better than the other two models. As for the predictability of the four metrics, breadth and mass are obviously more predictable than depth and Wiener Index. Decision Tree is generally a reliable model, which gives F1-scores of 0.85, 0.92, 0.68, 0.65 for the four metrics, respectively.

The prediction performance of ensemble models suggests that both Random Forest and Adaboost performs better than the single Decision Tree model, while the performance of Random Forest is slightly better. For instance, in the depth prediction task, Random Forest model could improve F1-score from 0.67 to 0.73, compared to the single Decision Tree model. In addition, based on bagging method, Random Forest model is efficient to deploy in application with the help of parallel computing. Therefore, Random Forest model is used in the following prediction task.

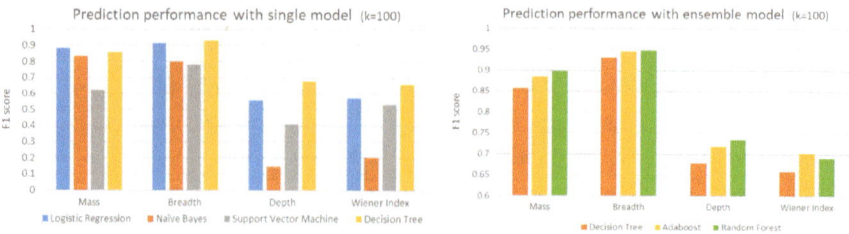

Fig. 5 Prediction accuracy with different machine learning algorithms, in single model and ensemble model

5.4   Prediction Performance with Feature Set and Observation Size

In real application, observation size k and selected feature set are two key factors affecting system performance. It is desired to obtain a certain level of prediction accuracy by consuming the least computational resources and detecting abnormal diffusions as early as possible. To obtain a certain level of prediction accuracy, smaller size of feature set and smaller observation size k are preferred. Therefore, the prediction performance is tested under different observation sizes and different single feature categories (Fig. 6).

In predicting mass and breadth, it is found that, the complete feature set performs the best, with a prediction accuracy around or higher than 0.9 at all observations. When single feature set is considered, structural features and temporal features give the top two performances. This indicates that the current structure and dynamics are intrinsically reflective to future coverage of cascade. For other feature categories, prediction accuracy varies from 0.6 to 0.75, due to large fluctuation.

For the prediction of depth and Wiener Index, the complete feature set performs the best, with a prediction accuracy around 0.8. The performance of single feature sets is generally poorer and has no essential difference among each other. For single feature categories, prediction accuracy ranges from 0.5 to 0.7, also suffering from large fluctuation as the observation size increases. In addition, an apparent trough at $k = 100$ is detected in almost all curves. This observation implies that $k = 100$

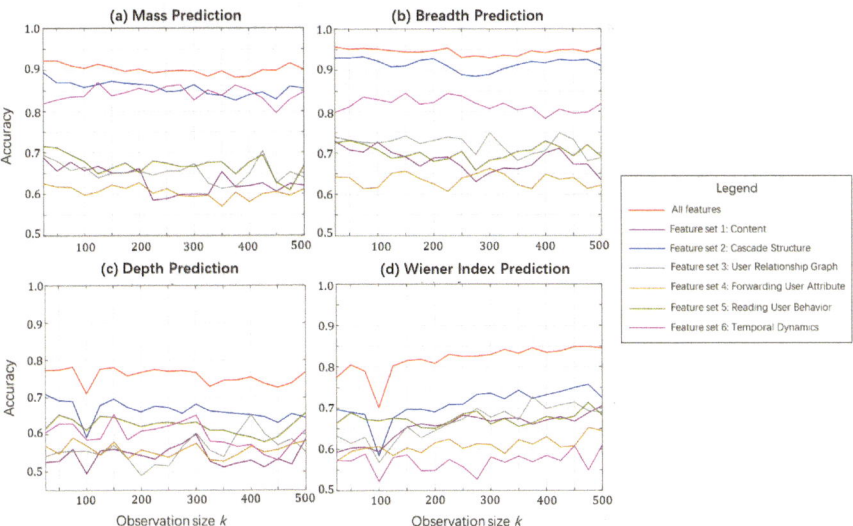

Fig. 6 Evolution curve of prediction accuracy with different input feature set and observation size

might be a key demarcation point of predictability for cascade structural complexity. Once cascade mass grows larger than 100, the structural predictability goes back to its early stage.

In summary, first, mass and breadth are more predictable than depth and Wiener Index. Second, the complete feature set leads to the best prediction performance, while the structural and dynamical features set seem to contribute a lot. Third, observation size k has little impact to prediction performance, confirming the robustness of our prediction model.

6 Case Study of WeChat Cascade

6.1 Case One: "Star-Like" Cascade

Case one examines the typical "star-like" pattern. The article content is public instruction about how to choose best insurance to buy. The content topic could be categorized as "Life Guidance".

In this case, the cascade structure presents a "star" like shape. From the visualized topology in Fig. 7, it is clear that the cascade is driven by two scenarios in balance, as the message feeds and Moments feeds account for similar proportion. Based on the statistics of final structural value, this cascade has medium mass value, low depth value, high relative breadth (the widest layer holding nearly half of all feeds), and low Wiener Index. The structural complexity is relatively low. From the dynamical evolution curve on timeline, the initial bursts in terms of all metrics are obvious. This observation indicates that, the major diffusion process has quickly come to the completed stage within only a few days after initial release. One explanation is that this topic of content could have drawn attention from all adults with different backgrounds. Therefore, the diffusion path is uniformly distributed on the underlying social network, presenting "star-like" pattern in final structure.

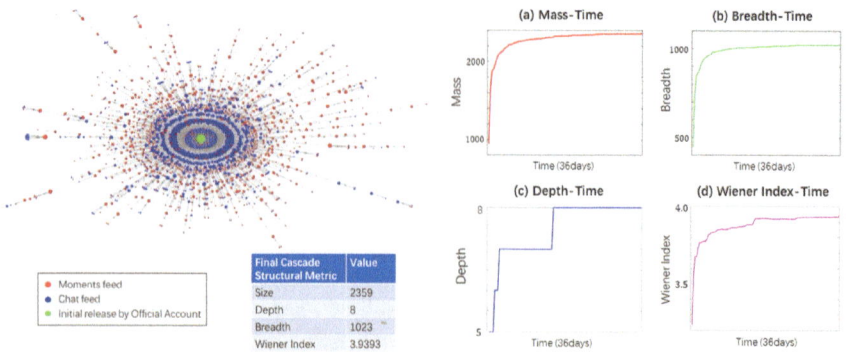

Fig. 7 "Star-like" case with visualization of structure and dynamics

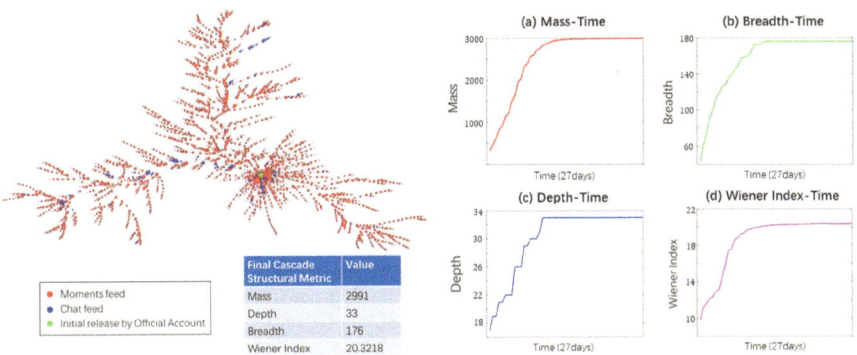

Fig. 8 "Chain-like" case with visualization of structure and dynamics

6.2 Case Two: "Chain-Like" Cascade

Case two examines the typical "chain-like" pattern. The article content is commercial advertisement driven by induced forwarding in title. The content topic could be categorized as "Commercial Marketing".

In this case, the cascade structure presents a "chain" like shape with several hierarchical branches as shown in Fig. 8. The cascade is dominated by the Moments scenario diffusion. Based on the statistics of final structural value, this cascade has medium mass, very high depth, quite low breadth value, and extremely large Wiener Index. The complexity of overall structure is rather high. The dynamical evolution curve presents a smooth growth process in terms of each metric value. The cascade consistently grew deeper and generated more branches along the timeline. This content is designed to attract specific people to buy certain goods, by using the approach of collecting "likes" in each user's Moments post. Therefore, it is not surprising that this cascade is Moments scenario-driven and diffused along the relatively determined path.

7 Conclusions

This paper built an analyzing system framework to discover patterns, test predictability and perform case study on WeChat information diffusion cascades. Several typical diffusion patterns emerged in terms of cascade structure, user behavior, and content topic. This finding indicated that the diffusion process on WeChat network has a huge complexity. The prediction performance obtained by advanced machine learning algorithms was satisfactory, which revealed the predictability of WeChat diffusion process. Cascade mass and breadth were more predictable than depth or Wiener Index. Prediction accuracy was stable at different observation sizes. By using case study with visualization, it was suggested that the content topic factor might intrinsically determine the pattern of structural and dynamical evolution. Our work enriched the knowledge of information diffusion mechanism on WeChat network, and facilitated the analysis and prediction practice. Future work includes obtaining more interesting diffusion patterns on comprehensive dataset, and improving prediction performance and by applying more effective features and more reliable learning algorithms.

Acknowledgements We gratefully acknowledge the support provided social diffusion research group in WeChat Team, Tencent Co., Ltd. This paper was also partially funded by Shenzhen Municipal Development and Reform Commission, Shenzhen Environmental Science and New Energy Technology Engineering Laboratory, Grant Number: SDRC [2016]172.

References

1. P. Maglio, C. Kieliszewski, J. Spohrer, *Handbook of Service Science* (Springer, New York, NY, 2010)
2. I. de Sola Pool, M. Kochen, Contacts and influence. Soc. Netw. **1**(1), 5–51 (1978)
3. D.J. Watts, S.H. Strogatz, Collective dynamics of small-world networks. Nature **393**(6684), 440–442 (1998)
4. A.L. Barabasi, R. Albert, Emergence of scaling in random networks. Science **286**(5439), 509–512 (1999)
5. S. Fortunato, Community detection in graphs. Phys. Rep. **486**(3–5), 75–174 (2010)
6. R. Pastor-Satorras, A. Vespignani, Epidemic spreading in scale-free networks. Phys. Rev. Lett. **86**(14), 3200 (2001)
7. C. Wang, X. Guan, T. Qin, Y. Zhou, Modelling on opinion leader's influence in microblog message propagation and its application. J. Softw **26**, 1473–1485 (2015)
8. J. Yang, J. McAuley, J. Leskovec, Community detection in networks with node attributes. in *IEEE 13th International Conference on Data Mining*, (Dallas, U.S.A., 2013), pp. 1151–1156
9. J.M. Hofman, A. Sharma, D.J. Watts, Prediction and explanation in social systems. Science **355**(6324), 486–488 (2017)
10. T. Martin, J.M. Hofman, A. Sharma, D.J. Watt, Exploring limits to prediction in complex social systems. in *Proceedings of the 25th International Conference on World Wide Web*, (Montréal, Canada, 2016), pp. 683–694
11. S. Vosoughi, D. Roy, S. Aral, The spread of true and false news online. Science **359**(6380), 1146–1151 (2018)
12. J. Cheng, L. Adamic, P.A. Dow, J.M. Kleinberg, J. Leskovec, Can cascades be predicted?. in *Proceedings of the 23rd International Conference on World Wide Web*, (Seoul, Republic of Korea, 2014), pp. 925–936
13. R. Guo, E. Shaabani, A. Bhatnagar, P. Shakarian, Toward order-of-magnitude cascade prediction. in *Proceedings of the 2015 IEEE/ACM International Conference on Advances in Social Networks Analysis and Mining*, (Paris, France, 2015), pp. 1610–1613
14. J. Ugander, L. Backstrom, C. Marlow, J. Kleinberg, Structural diversity in social contagion. Proc. Natl. Acad. Sci. **109**(16), 5962–5966 (2012)
15. C. Lagnier, L. Denoyer, E. Gaussier, P. Gallinari P, Predicting information diffusion in social networks using content and user's profiles. in *European Conference on Information Retrieval*, (Berlin, Germany, 2013), pp. 74–85
16. C.F. Camerer, *Behavioral Game Theory: Experiments in Strategic Interaction*. (Princeton University Press, 2011)

Study on the Relationship Between the Logistics Industry and Macroeconomic Factors in China Based on the Grey Incidence

Guangxing Chu

Abstract The logistics industry has entered industrial development stage but the statistical data of logistics industry is inaccurate and incomplete. As one of modern service industry, the development of logistics industry needs the support of scientific theory. There is a close relationship between economy and logistics, so the relationship between them needs to be studied to make the logistics industry scientifically develop. This paper analyzes the characteristics and functions of logistics industry development in China, and then uses the method of grey incidence to study the relationship between logistics scale and related economic factors reflecting the logistics scale with freight turnover, analyzes the reasons for the rapid development of China's logistics industry, and points out the future development direction of the logistics industry.

Keywords Logistics industry · Grey incidence · Macroeconomic factors

1 Introduction

With the deepening of global and regional economic integration and the extensive use of Internet information technology, the development of the global logistics industry has undergone profound changes and gained more and more attention. As a basic industry of the national economy, the logistics industry integrates many industries such as road transportation, warehousing, and information industry. It involves a wide range of fields and attracts a large number of employed people. It is known as the "accelerator" of economic development. The development of the logistics industry can promote the adjustment and upgrading of the industrial structure, and its development level has become one of the important indicators to measure the national economic strength. At the same time, economic development will bring about the demand for logistics, and there is a close relationship between economy and logistics.

G. Chu (✉)
Nanjing University of Aeronautics and Astronautics, Nanjing 210000, China
e-mail: yanfengc@nuaa.edu.cn

© Springer Nature Switzerland AG 2020
H. Yang et al. (eds.), *Smart Service Systems, Operations Management, and Analytics*, Springer Proceedings in Business and Economics,
https://doi.org/10.1007/978-3-030-30967-1_35

From the perspective of logistics and economic correlation, this paper uses the freight turnover to represent the scale of logistics, and studies the relationship between development scale of logistics industry and the total retail sales of total retail sales of consumer goods, GDP, first industrial output, second industrial output, tertiary industrial output, total import and export trades, output of the construction industry to analyze the important position and function of the logistics industry in the development of the national economy.

2 The Development of China's Logistics Industry

2.1 The Development of China's Logistics Industry During the "13th Five-Year Plan" Period

During the "13th Five-Year Plan" period, China's logistics industry continued developing steadily. According to the statistics of the National Bureau of Statistics and the China Federation of Logistics and Purchasing, China's total social logistics reached 252.8 trillion yuan in 2017. The annual growth rate was 6.7%, calculated at comparable prices, and the growth rate was 0.6 percentage points higher than the previous year. During the "13th Five-Year Plan" period, China's logistics industry has crossed the "starting period" and showed a steady and good development trend. During the period of 2010–2017, the total social logistics cost increased from 7.1 trillion yuan to 12.1 trillion yuan, with a compound annual growth rate of 7.91%, which indicates that China's logistics industry is in strong demand and the total logistics cost is expanding. During this period, the ratio of total logistics costs to GDP fell from 17.8 to 14.6%, which indicates that China's logistics efficiency has generally improved. In recent years, with the steady advancement of the "One Belt, One Road" policy, the state plans to build a number of export-oriented logistics hubs that are interconnected and smoothly connected with the "One Belt, One Road" countries, and improve the distribution capacity of import and export goods; "The implementation of the strategy will develop into the neighboring countries, form an internal and external infrastructure network and a large domestic and international logistics channel, and enhance the supporting role of logistics in the implementation of major strategies such as the "Belt and Road"; International production cooperation will also be carried out in various regions of the country. Synchronously promote the "One Belt, One Road" strategy, make good use of the existing cooperation mechanism and the investment and financing platform of the "Aesthetic Fund" of the AIIB, deeply participate in the construction of surrounding economic ports, promote mutual trust and cooperation of large projects, and achieve common development.

The logistics infrastructure conditions have improved significantly. By the end of 2016, the national railway operating mileage reached 124,000 km, including 22,000 km of high-speed railways, and the national highway mileage reached 4,696,300 km, an increase of 11.9 million kilometers compared with 2014. The total domestic river navigation mileage reached 127,100 km, the port has 2,317 berths of 10,000-ton and above, the number of scheduled flights to the airport reached 216, the national railway locomotive 21,000, and the highway operation vehicle was 14,435,700 [1].

China will continue to speed up the construction of integrated transportation systems, strengthen the coordinated operation of different modes of transportation of railways, highways, and waterways, and develop composite transportation, strengthen cooperation and cooperation with key logistics enterprises at the same time, and further improve the supervision and management of the logistics industry. Then China will actively participate in the research of transportation solutions for fresh products, and encourage more companies to participate in the "last mile" problem solving research.

2.2   The Promotion of the Logistics Industry to the Economy

China's logistics operation has grown steadily in 2017. The demand coefficient for GDP compared with the total logistics volume is 3.05, which means that China's one unit of GDP output needs 3.05 units of logistics quota. The increase of total logistics will promote China's GDP growth and promote the growth of China's economy simultaneously [2].

The rapid development of modern logistics is one of the important factors for China's economic circulation and efficiency growth. According to the statistics of the National Bureau of Statistics on industrial enterprises, the growth rate of finished goods inventory of industrial enterprises above designated size fell by 0.1 percentage points from the previous year; The turnover days of finished goods inventory was 13.8 days, a decrease of 0.4 days from the previous year; The growth rate of warehousing expenses in custody expenses decreased by 0.3 percentage points compared with the previous year, which can reduce the cost of warehousing logistics by about 40 billion yuan at the end of 2016. At the same time, the relevant capital occupation cost continued to decline. For example, the interest expense in the custody expenses decreased by 2.7% compared with the previous year, and the decline further expanded by 0.3 percentage points. The inventory turnover index in the national storage index averaged 53.8% for the whole year with growth of 6.5 percentage points, compared with the previous year.

Considering the development background of global economic development, the importance of logistics is constantly increasing. Combined with the status quo and trend of China's economic development; it is very urgent to develop modern logistics industry. In order to realize the optimal allocation of resources in the logistics industry at a deeper and broader level, and to promote the optimization and integration of various elements of logistics, logistics needs to be viewed as a whole industry. First, we must understand the development of China's logistics and analyze its influencing factors, then formulating a series of reasonable and reliable logistics industry policies, which is of great significance to the development of China's logistics industry.

3   The Reflection of Logistics Industrialization and Logistics Scale

3.1   Logistics Industrialization

The logistics industry refers to a collection of activities that combine the basic functions of transportation, storage, loading and unloading, transportation, packaging, distribution processing, distribution, and information processing. With the rapid development of logistics, many enterprises have established a professional logistics department. The government also regards logistics as an independent industry and formulates a unified development policy. China's logistics has entered the stage of industrialization development.

It is widely believed in developed countries that a very important factor in the macroeconomic quality is the degree of logistics industrialization. Logistics industrialization can improve the level of logistics system operation, effectively reduce logistics operation costs, and improve logistics operation efficiency. Macroscopically, the industrialization of logistics shows that the level of socialized intensive logistics activities is further improved, and the allocation of social logistics resources is more efficient. The industrialization of logistics can shorten the flow time of goods, thereby improving the efficiency of the use of social funds, accelerating the optimal allocation of social resources, and finally improving the quality of macroeconomic activities and realizing the growth and upgrading of the national economy.

Under the conditions of economic globalization, multinational corporations generally establish a global product production and sales system. Depending on their own logistics capabilities, they cannot achieve the establishment of a global commodity network but need to provide local logistics services. Therefore, the degree of industrialization of national logistics is one of the important bases for multinational companies to choose to invest and build factories. The degree of logistics industrialization is a key indicator to measure a country's investment environment.

The proportion of service industry in the modern economy is increasing. One common point of the economic structure of western developed countries is that the service industry accounts for a large proportion. China's economic development is not balanced, and the proportion of service industry is too small. Therefore, as an important part of the service industry, the logistics industry has received extensive attention from various regions.

Logistics industrialization has great benefits for the development of regional economy, which helps to enhance the overall attractiveness and influence of the region, improve the allocation efficiency of production factors, and thus drive the development of related industries such as the catering industry. This will not only increase the number of logistics services but also add high value of services to it, thus promoting the optimization and upgrading of the economic structure in a multifaceted manner.

3.2 Reflection of Logistics Scale

The scale of logistics is the reflection of the level of logistics development in a country, while the volume of goods is a numerical representation of the scale of logistics. The flow of material is the sum of the quantities (materials, parts, semi-finished products, finished products, etc.) produced in various operations during the flow of goods, and the total number of logistics activities. At present, there is no universal scientific measurement method to represent the flow of goods in China. The international standard industrial classification method regards transportation, storage, and postal electricity as a large category. Therefore, this paper considers the change of freight volume or turnover to express the flow of goods in China. For a period of time, the number of items displaced in space by various modes of transportation and transportation has certain representativeness of the logistics scale of the period. Freight volume refers to the quantity of goods actually transported by a transportation company within a certain period of time. The freight turnover refers to the product of the tonnage of the goods actually transported by the transport department and its transport distance within a certain period of time. The sum of the turnover of various modes of transportation can represent the scale of logistics in China to a certain extent.

4 Grey Incidence Analysis of Economic Factors and Logistics Scale

4.1 Selection of Relevant Factors

Grey system theory is a kind of method for studying the small data poor information uncertainty system with partial information and partial information unknown. Grey incidence analysis is a gray relational model with unclear operation mechanism or physical prototype. Grey incidence analysis model is established to show a method of gray relation. Grey incidence analysis has the functions of defining system boundaries, analyzing systems and behaviors in uncertain systems, analyzing major factors and secondary factors, and identifying patterns. There are defects in the relevant data statistics of China's logistics industry (such as inaccuracy, incompleteness, etc.), and the data has a large grayness. The social economy is an open system, which is affected by various complicated factors. Many factors still have the characteristics of interaction and effect, and the uncertainty is strong. There is currently no definitive mathematical model to characterize the relationship between various economic factors and the scale of logistics. Regression analysis and principal component analysis in traditional mathematical statistics can analyze the relationship between logistics scale and economy, but these methods either require a large amount of data or a sample satisfied a certain probability distribution. Because the data does not meet the conditions of the mathematical statistics method, the results of the analysis using

Table 1 Statistics of relevant economic factors unit: 100 million yuan

Item	Notation	2012	2013	2014	2015	2016
Freight turnover (billion tons)	X_0	173804.46	168013.80	181668.00	178356.00	186629.48
Total import and export trade	X_1	244160.20	258168.90	264241.77	245502.93	243386.46
Retail sales of consumer goods	X_2	214432.7	242842.8	271896.1	300930.8	332316.3
Industrial added value	X_3	208905.6	222337.6	233856.4	236506.3	247877.7
First industrial output	X_4	50902.3	55329.1	58343.5	60862.1	63672.8
Output of the construction industry	X_5	137217.86	160366.06	176713.40	180757.47	193566.78
Tertiary industrial output	X_6	244821.9	277959.3	308058.6	346149.7	383365.0
GDP	X_7	540367.4	595244.4	643974.0	689052.1	743585.5

Data in Table 1 is from the 2016 China Statistical Yearbook

the above methods may be distorted or reversed. Therefore, this paper uses the grey incidence analysis method to analyze the logistics scale. Considering the availability of data, this paper selects the freight turnover as the mapping value of logistics scale to study the relationship between the total import and export trade, retail sales of consumer goods, industrial added value, and first industrial output, second industrial output, tertiary industrial output, GDP and the development of logistics scale. The relevant data is shown in Table 1 [3].

The first, second and third industries have demand for logistics. The stronger the demand, the larger the scale of logistics. In turn, the development of the logistics industry will also affect various industries and promote economic development. Generally speaking, when a country's logistics industry develops at a high level, its high-quality logistics services will attract the attention of many industries. At this time, these industries will form more demand for logistics. When the demand for logistics in other industries in a country is small, it does not mean that the scale of logistics is low, because many industries will establish self-operated logistics.

4.2 Steps of Gray Incidence Analysis

System behavior sequence $X_0 = (x_0(1), x_0(2), \ldots, x_0(n))$, $n = 1, 2, 3, 4, 5$, correlation factor sequence $X_i = (x_i(1), x_i(2), \ldots, x_i(n))$, $i = 1, 2, 3, 4, 5, 6, 7$.

For $\xi \in (0, 1)$, then

$$\gamma(x_0(k), x_i(k)) = \frac{\min_i \min_k |x_0(k) - x_i(k)| + \xi \max_i \max_k |x_0(k) - x_i(k)|}{|x_0(k) - x_i(k)| + \xi \max_i \max_k |x_0(k) - x_i(k)|} \quad (1)$$

$$\gamma(X_0, X_i) = \frac{1}{n} \sum_{k=1}^{n} \gamma(x_0(k), x_i(k)) \quad (2)$$

ξ represents resolution coefficient, $\gamma(X_0, X_i)$ is the grey correlation degree of X_0 and X_i, gray absolute correlation degree, relative correlation degree, and comprehensive correlation degree can be further obtained [4].

Grey absolute correlation degree is only a measure of the degree of association between sequences, without considering other factors.

Gray relative correlation degree calculates the relationship of the sequence relative to the rate of change of the starting point. The closer the rate of change of the two sequences is, the greater the relative degree of gray is, and the smaller the vice versa.

Grey comprehensive correlation degree not only reflects the similarity degree of the two sequences but also reflects the closeness of the two sequences relative to the rate of change of the starting point. It is a quantitative index that comprehensively characterizes the close relationship between the sequences. The calculation results using the 7.0 version of the grey system theory modeling software of Nanjing University of Aeronautics and Astronautics are shown in Table 2 below.

Table 2 Gray correlation degree of logistics industry scale and related factors

Items	Grey correlation degree	Grey absolute correlation	Gray relative correlation degree	Grey comprehensive correlation degree
Total import and export trade	0.8568	0.6860	0.9468	0.8164
Retail sales of consumer goods	0.5716	0.5282	0.6821	0.6051
Industrial added value	0.7738	0.5763	0.8162	0.6963
First industrial output	0.7109	0.7310	0.7727	0.7519
Output of the construction industry	0.5866	0.5485	0.6944	0.6214
Tertiary industrial output	0.5702	0.5244	0.6808	0.6026
GDP	0.6437	0.5159	0.7288	0.6224

4.3   Analysis of Economic Factors Affecting the Development of Logistics Industry

From Table 2, we can find the gray correlation degree between China's freight turnover and X_i satisfies $X_1 > X_3 > X_4 > X_7 > X_5 > X_2 > X_6$, X_1 is optimal, which means that the total volume of imports and exports has the greatest impact on the turnover of goods. It also means that China's logistics industry has a large proportion of import and export logistics at present. Sort result in grey absolute correlation is $X_4 > X_1 > X_3 > X_5 > X_2 > X_6 > X_7$, X_4 is optimal, that is the relationship between first industrial output and the freight turnover is the highest, followed by the total value of imports and exports and industrial added value. Sort result in gray relative correlation degree is $X_1 > X_3 > X_4 > X_7 > X_5 > X_2 > X_6$, X_1 is optimal. Sort result in grey comprehensive correlation degree is $X_1 > X_4 > X_3 > X_7 > X_5 > X_2 > X_6$, X_1 is optimal.

We can analyze the current development status of China's logistics industry and further clarify the future development direction through the gray incidence analysis of the above data,

From the perspective of grey comprehensive correlation, the total volume of imports and exports has the greatest impact on the scale of logistics. The national economy is growing rapidly and the level of development of the logistics industry is constantly improving. On one hand, the improvement of the logistics level provides excellent transportation conditions for import and export trade. On the other hand, with the deepening of the global integration of the economy, the logistics demand for import and export trade is also increasing. For China's economy, the increase in total import and export is conducive to making full use of the products and resources of other countries, improving the efficiency of the market in resource allocation, and improving national welfare. Second, for the world economy, it is conducive to improving the contribution of the Chinese economy to global economic growth. Finally, the trade balance is conducive to the balance of international payments and is also conducive to the stability of the RMB exchange rate. Therefore, the demand for logistics services in China's import and export trade will be further strengthened.

The relationship between industrial added value and logistics scale is relatively large. Industrial logistics accounts for a large proportion of China's total logistics value. As China is still the dominant economic structure in China, industrial logistics is the main and key service target of China's logistics industry. At present, new industrialization construction of China will promote the further division of industrial division of labor, and on the other hand, it will put forward more professional requirements for the logistics industry. Therefore, the development of new industrialization will guide the gradual refinement of the division of labor in China's logistics industry.

References

1. China federation of logistics and purchasing, *China Logistics Yearbook*, vol. 12 (China Logistics and Purchasing, 2018), p. 82
2. J. Fangfang, An empirical study on the role of logistics industry development in economic growth. Explor. Econ. Issues **03**, 135–140 (2012)
3. China national Burean of statistics, *China Statistical Yearbook* (China Statistics Press, Beijing, 2016)
4. L. Sifeng et al., *Grey System Theory and Its Application* (Science Press, Beijing, 2017)

Index

© Springer Nature Switzerland AG 2020
H. Yang et al. (eds.), *Smart Service Systems, Operations Management, and Analytics*, Springer Proceedings in Business and Economics,
https://doi.org/10.1007/978-3-030-30967-1